Miss Margaret Paterson
31 Clarkston Dr.
Airdrie
ML6 7AH.

Optics

Optics

Ninth edition

W. H. A. Fincham,
MPhil, FInstP, FSMC
*Formerly Lecturer, Applied Optics
Department, Northampton Polytechnic
(now The City University), London*

M. H. Freeman,
BSc, MPhil, PhD, MInstP
*Company Scientist, Pilkington P–E Ltd,
Wales, UK.
Chairman, Applied Vision Association, 1976–1978*

Butterworths
London – Boston
Sydney – Durban – Wellington – Toronto

| United Kingdom | **Butterworth & Co (Publishers) Ltd** |
| London | 88 Kingsway, WC2B 6AB |

Australia	**Butterworth Pty Ltd**
Sydney	586 Pacific Highway, Chatswood, NSW 2067
	Also at Melbourne, Brisbane, Adelaide and Perth

| Canada | **Butterworth & Co (Canada) Ltd** |
| Toronto | 2265 Midland Avenue, Scarborough, Ontario, M1P 4S1 |

New Zealand	**Butterworths of New Zealand Ltd**
Wellington	T & W Young Building, 77-85 Customhouse Quay, 1,
	CPO Box 472

| South Africa | **Butterworth & Co (South Africa) Ltd** |
| Durban | 152-154 Gale Street |

| USA | **Butterworth (Publishers) Inc** |
| Boston | 10 Tower Office Park, Woburn, Mass. 01801 |

First published 1934
Second edition 1936
Third edition 1939
Fourth edition 1942
Fifth edition 1945
Reprinted 1947, 1949
Sixth edition 1951

Reprinted 1954, 1956, 1959
Seventh edition 1965
Reprinted 1969, 1972
Eighth edition 1974
Reprinted 1976, 1977, 1978, 1979 (twice)
Ninth edition 1980

ISBN 0 407 93422 7

British Library Cataloguing in Publication Data

Fincham, Walter Henry Angel
  Optics. —9th ed.
  1. Optics
  I. Title   II. *Freeman*, Michael Harold
  535   QC255.2   80-40274

  ISBN 0-407-93422-7

Typeset by Reproduction Drawings Ltd, Sutton, Surrey
Printed in Great Britain by The University Press, Cambridge

Preface to ninth edition

It is with deep regret that I record the death of W. H. A. Fincham in January 1977 while this ninth edition was being planned. Walter Fincham wrote the first edition of *Optics* in 1934 with the intention of meeting the need for a basic textbook for students with little or no previous experience of the subject. As a sometime lecturer using Fincham's *Optics,* I found it clear and unpretentious. To use a modern idiom, it is 'student orientated'.

In this ninth edition the opportunity has been taken to increase and improve the photographic plates. Some additions and deletions have been made to the earlier chapters. The chapters on polarization and aberrations have been re-written. In the latter case, the advent of programmable calculators at reasonable prices has prompted the inclusion of simple programs for ray tracing. It is left to the student to write programs for the numerical questions of the earlier chapters!

A further chapter has been added covering the optical properties of the (emmetropic) eye in terms of the concepts developed in the rest of the book. These changes are consistent with the policy set out in the original preface. It is recognised that the mathematical abilities of students using this book will be very varied. In geometrical optics only simple mathematics is required. However, I would encourage students with mathematical aptitude to pursue further the application of calculators and computers to the manipulation of aberration theory and Fourier methods at the eye/optics interface.

I am indebted to Mr. J. Archer Smith (University of Aston in Birmingham), Mr. Tudor Roberts (The City University London), Dr. W. N. Charman (University of Manchester Institute of Science and Technology) and Professor M. Millodot (University of Wales Institute of Science and Technology) for the extra questions which have been incorporated in this edition. Permission for their inclusion is gratefully acknowledged. I am also grateful to my colleagues Mr. E. Henderson, Mr. A. Scrivener and Dr. D. W. Swift for comment and discussion; and to Mrs. M. Thompson for typing from my manuscript. Other sources of specific assistance are noted in the text.

This ninth edition remains the essentially basic textbook that W. H. A. Fincham intended.

M.H.F.
Denbigh

Preface to first edition

During recent years considerable progress has been made in bringing the teaching of optics in line with practical requirements, and a number of text books dealing with Applied Optics have been published. None of these, however, caters for the elementary student, who has no previous knowledge of the subject and who, moreover, is frequently at the same time studying the mathematics required.

This book, which is based on lectures given in the Applied Optics Department of the Northampton Polytechnic Institute, London, is intended to cover the work required by a student up to the stage at which he commences to specialise in such subjects as ophthalmic optics, optical instruments and lens design. It includes also the work required by students of Light for the Intermediate examinations of the Universities.

The first eleven chapters deal with elementary geometrical optics, Chapters XII to XVI with physical optics, and the last three with geometrical optics of a rather more advanced character.

The system of nomenclature and sign convention adopted is that in use at the Imperial College of Science and the Northampton Polytechnic Institute, London. The sign convention is founded on the requirement that a converging lens shall have a positive focal length measured from the lens to the second principal focus. This is easily understood by the elementary student and is the convention commonly used throughout the optical industry. In ophthalmic optics—the most extensive branch of optical work at the present time—lenses are always expressed in terms of focal power; this idea has been introduced quite early and used throughout the work.

The solution of exercises plays an important part in the study of a subject such as Optics, and it is hoped that the extensive set of exercises with answers will be found useful. Typical examples from the examination papers of the London University, the Worshipful Company of Spectacle Makers and the British Optical Association are included by permission of these bodies.

My best thanks are due to my colleagues Messrs. H. T. Davey and E. F. Fincham for their valuable assistance in the preparation of the diagrams, which, together with the photographs, have been specially made for the book. I wish particularly to express my indebtedness to Mr. H. H. Emsley, Head of the Applied Optics Department, Northampton Polytechnic Institute, for reading the manuscript and for his very valuable help and suggestions given during the whole of the work.

January, 1934

W.H.A.F.
London

Contents

Chapter 1

The propagation of light

1.1 INTRODUCTORY

The branch of Science known as **optics** is concerned with the study of light and vision; it may also embrace the study of other radiations closely allied to light.

Light is a form of radiant energy, being similar in its essential nature to other forms, such as heat and radio waves. It is radiated out through space at an enormous speed by **luminous bodies**. Most of these are capable of emitting light by virtue of their exceedingly high temperature, as a result of which their constituent atoms are in a condition of considerable agitation, the effects of which are transmitted outwards from the body in all directions.

A poker, when cold, emits no radiation that we can detect. When gradually heated, it sets up a disturbance in the form of vibrations or waves in the surrounding medium, which radiate outwards at a speed of 186 000 miles per second; at some distance away we can 'feel' the effect as heat, i.e., we detect this form of radiation by our sense of touch. As the temperature rises and the vibrations become more rapid, the poker is seen to become red; the radiation is such that we can 'see' its effects; it is in the form of light. We detect this form of radiation by our sense of sight, the eye acting as detector. With further rise in temperature the poker passes to a 'yellow' and then to a 'white' heat.

The exact nature of light is not completely known; but a clear working idea of these 'wave motions' can be derived from the ripples set up on the surface of water into which a stone is dropped. The important characteristics of the disturbance are: the **velocity** with which it travels outwards, the **frequency** of the rise and fall of the undulations, and the distance between successive wave crests, called the **wave-length**. Clearly the velocity is equal to the product of the wave-length and the frequency or number of vibrations arriving at a given point in unit time.

In the case of light, heat and electrical radiations, the velocity has been found by experiment (Chapter 12) to be 186 000 miles or 300 million metres per second. As we should expect from the poker experiment, the frequency of the vibrations is greater for light than for heat, and greater for some colours of light than others. The wave-length is consequently longer for heat than light and longer for red light (which appeared first) than for other colours. The figures for a few selected radiations are given in the table.

In the table, only one typical value is given for each group of radiations. Actually, each group covers a wide range, and a more complete statement will be found in *Figure 12.8*.

Radiation		Velocity (metres per sec)	Frequency (per sec)†	Wave-length (nanometres*)
Radio		3×10^8	1×10^6	3×10^{11}
Heat	Thermal Infra-red	3×10^8	30×10^{12}	10000
	Near Infra-red	3×10^8	300×10^{12}	1000
Light	Red	3×10^8	395×10^{12}	759
	Yellow	3×10^8	509×10^{12}	589
	Violet	3×10^8	764×10^{12}	393
Ultra-violet		3×10^8	1000×10^{12}	300
X-rays		3×10^8	3×10^{18}	0.1

*1 nanometre equals 1×10^{-9} metre
†Frequency per sec is Hertz

1.2 RECTILINEAR PROPAGATION OF LIGHT

Any space through which light travels is an **optical medium**. Most optical media have the same properties in all directions and are said to be **isotropic**—although there are a few optical substances, notably certain crystals, where this is not the case and in which particular phenomena arise (Chapter 16). Most media, moreover, possess these same properties throughout their mass, so that they are **homogeneous**.

When light starts from a point source B (*Figure 1.1*) in an isotropic medium, it spreads out uniformly at the same speed in all directions; the position it occupies at a given moment will be a sphere having the source at its centre. Such imaginary spherical surfaces will be called **light fronts** or **wavefronts**. In the case of the water ripples, the disturbance is propagated in one plane only and the wavefronts are *circular*.

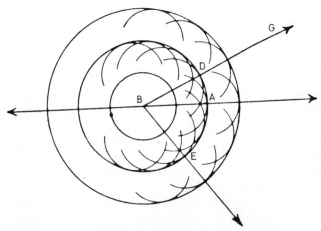

Figure 1.1 Wavefront. Huygens' Principle

Huygens (1629–1695), who is usually considered as the founder of the wave theory of light, assumed that any point on a wave surface behaved as a new source from which spherical 'secondary waves' or 'wavelets' spread out in a forward direction. The surface DAE, which at any instant touches or envelops all these wavelets, forms a new wavefront, and it is evident that while the light remains in the same isotropic medium, each new wavefront will be a sphere with its centre at B.

Huygens' Principle assumes that the wavelets travel out in a forward direction only, and that the effect of each wavelet is limited to that part which touches the enveloping new wave front. The first of these assumptions may be explained by the fact that the points from which the wavelets arise are not independent sources, but are set in motion as the result of a wavefront coming from the original source; while Fresnel (1788–1827) has shown (Chapter 15) by considering the 'interference' that takes place between the wavelets, that the second assumption is approximately correct.

From *Figure 1.1* it is evident that the path, such as BDG, travelled by any part of the disturbance, is a straight line perpendicular to the wavefront. We speak of such a line along which light travels as a **ray**. Huygens' construction for the new wavefront enables us to find the change in the form or direction of the wavefront on passing into a new medium, or on reflection at a surface, and will be used in this way in subsequent chapters.

A luminous source, such as B (*Figure 1.2*), will be seen only if light from it enters the eye. Thus an eye placed in the region between C and K, or between D and L, of the two screens CD and KL will see B, because the light reaches these regions without interruption. An eye placed between *c* and *d* will not see the object B, because the opaque obstacle *a b* shields the region *c a b d*, which is said to be in **shadow**. B will be visible, however, to an eye at *e*, just beyond the limits of the shadow region.

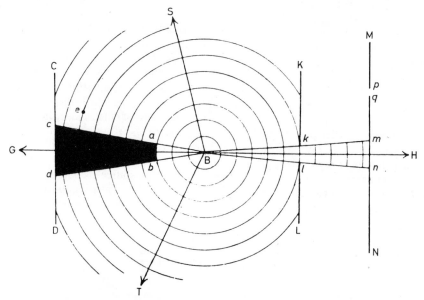

Figure 1.2 Rectilinear propagation of light. Aperture at kl; obstacle at ab

If we have two opaque screens at KL and MN, with circular apertures at $k\,l$ and $p\,q$ respectively, the screen MN will be illuminated over a circular patch $m\,n$, but will be dark elsewhere, and an eye placed at the aperture $p\,q$ will not see the source B.

The fact that B is invisible to an eye placed between $c\,d$ and visible to one placed within the region $k\,m\,n\,l$ shows, as was also seen from Huygens' principle, that the light effect at a given point is due sensibly to that portion of the light that has travelled along the straight line joining the point to the source.

If the conditions within the regions $c\,d$ and $m\,n$ were to be studied more closely, it would be found that the patch $m\,n$ is not uniformly illuminated, nor is the patch $c\,d$ completely dark and separated from the surrounding brightness on the screen CD by a sharp line of demarcation, especially if $a\,b$ and $k\,l$ are small. The light waves, in passing the edges of these apertures, bend round into the space behind them in the same way as water waves may be seen curving round the end of a breakwater. Light waves are so very small however, that this bending or **diffraction**, as it is called, takes place to an extent so minute, that special precautions have to be taken to observe it. (See Chapter 15).

1.3 PENCILS AND BEAMS

The light from a luminous point, or from any one point on a large source or illuminated object, after passing through a limiting aperture such as $k\,l$ (*Figure 1.2*) constitutes a **pencil** of light. The word **bundle** of rays is often used to mean the same thing. Ray bundles are more commonly associated with computer ray tracing and the calculation of geometrical aberrations (Chapter 18). We will use the word pencil for most of this book. The aperture (*Figure 1.2*) may be simply an opening in an opaque screen or the edges of a lens, mirror, etc. The pencil increases in width as the distance from the source increases, and the light is said to be **divergent**. Under certain circumstances, however, as for example, after the light has passed through a convex lens, the pencil may be modified in such a way that its width is gradually decreasing; the light is then said to be **convergent**, and it converges to a point or **focus**. This focus will be the **image** of the object point from which the light started. Beyond the focus the pencil again diverges, its width still being limited by the original aperture. When the object point or the focus is at a great distance as compared with the width of the aperture the edges of the pencil will be practically parallel. Divergent, convergent and parallel pencils, with the corresponding wavefronts, are illustrated in *Figure 1.3*. The ray passing through the centre of the aperture, will be the **principal** or **chief ray** of the pencil.

The collection of pencils arising from an extended source or object (*Figure 1.4a*) constitutes a **beam**. The edges of the beam may be divergent or convergent, irrespective of the divergence or convergence of the pencils constituting the

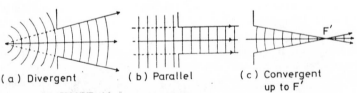

(a) Divergent   (b) Parallel   (c) Convergent up to F′

Figure 1.3 Pencils showing light fronts

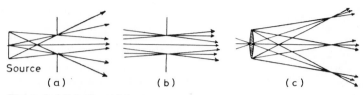

Figure 1.4 Pencils and beams

beam. Thus, if light from the sun passes through an aperture (*Figure 1.4b*) the individual pencils from each point on the sun will be parallel, but the pencils are not parallel to each other, and the edges of the beam are divergent. Similarly, the beam from a lens (*Figure 1.4c*) may be divergent, while its constituent pencils are convergent. *The terms divergent, convergent, or parallel light refer to the form of the pencils and not to that of the beam.*

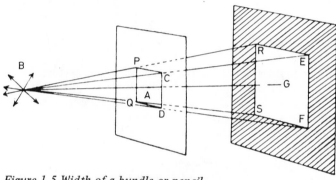

Figure 1.5 Width of a bundle or pencil

In *Figure 1.5* BEF is the section of a pencil diverging from a luminous point B, and limited by an aperture CD, EF being the width of the cross section of the pencil in a plane parallel to the plane of the aperture. In the similar triangles CDB and EFB

$$\frac{EF}{CD} = \frac{BG}{BA} \qquad (1.01)$$

As the light spreads out uniformly in all directions,

$$\frac{\text{area EFSR}}{\text{area CDQP}} = \left(\frac{BG}{BA}\right)^2 \qquad (1.02)$$

that is, *the area of the cross section of a pencil varies as the square of the distance from the source.*

Since the amount of light in a pencil depends only upon the amount given out by the source, the amount falling upon a unit area of an illuminated surface—the **illumination**—varies inversely as the area of the cross section of the pencil; hence, *the illumination of a surface placed perpendicularly to the direction in which the light is travelling varies inversely as the square of the distance of the surface from the source.* This is the well-known **Law of Inverse Squares**, which is of great importance in questions of light measurement, illumination, etc.

1.4 VERGENCE

For a given aperture, the divergence or convergence of a pencil and the curvature of the wavefront at the aperture decrease as the distance of the luminous point or of the focus from the aperture is increased. As the effect of a lens or curved mirror is to change the divergence or convergence of light, it is necessary to have some means of expressing the amount of divergence or convergence of a pencil in any given position. This is done by defining the divergence or convergence of a pencil at any particular position as the reciprocal of the distance *from* that position *to* the luminous point or the focus. It will be seen in Chapter 5 that this quantity also represents the curvature of the wavefront. The general term **vergence** has been suggested to denote either divergence or convergence, the difference being shown by a difference of sign. The unit of vergence is the **dioptre**, the vergence of a pencil one metre from the luminous point or focus (see section 5.1).

1.5 THE PINHOLE CAMERA

We have an interesting verification of the law of rectilinear propagation in the pinhole camera. The light from each point on an illuminated object, on passing through a small aperture in an opaque screen, gives rise to a narrow pencil, and if the light is received on a second screen at some distance from and parallel to the plane of the aperture, each pencil produces a patch of light of the same shape as the aperture. As the light travels in straight lines, the patches of light on the screen occupy similar relative positions to those of the corresponding points on the object, and the illuminated area of the screen is similar in shape to the original object, but inverted (*Figure 1.6a*). If the aperture is made quite

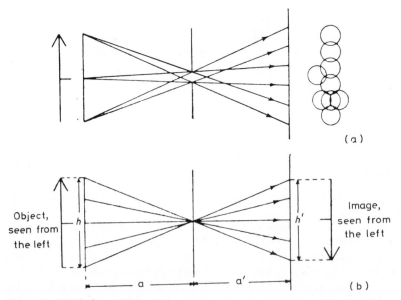

Figure 1.6 Pinhole camera

small, the individual patches of light will overlap to only a small extent, and a fairly well defined picture or image of the object is formed (*Figure 1.6b*). As will be seen from *Figure 1.6b*, the size of the inverted image of any object will depend on the distances of object and image from the aperture, thus

$$\frac{h'}{h} = \frac{a'}{a}$$

The degree of sharpness of the pinhole image can never be of a very high order, because if the diameter of the aperture is reduced beyond a certain amount, additional blurring becomes evident, due to diffraction effects. Also, the illumination of the image is very low as compared with that formed by a lens, owing to the very small aperture used. The image is free from distortion and the depth of focus is very great, i.e., images of objects at greatly varying distances are reasonably sharp in one plane.

1.6 SHADOWS AND ECLIPSES

The properties of shadows may easily be deduced from the law of rectilinear propagation. If the source is a point, the boundary of the shadow is sharply defined (neglecting diffraction) and its section, as received on a screen perpendicular to the typical direction of the light, will be the same shape as the opaque obstacle would appear when viewed from the position of the source. Thus the shadow of a sphere will have a circular section, while that of a flat circular body will be circular or elliptical according to how it is tilted. The size of the shadow in any position may be determined in the same way as the cross section of a pencil (section 1.3).

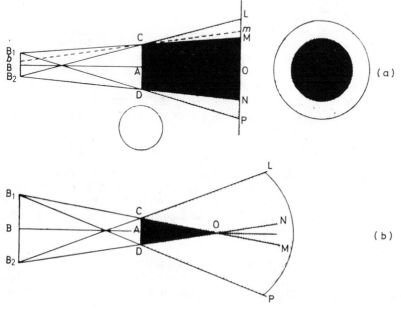

Figure 1.7 Shadows. Umbra and penumbra

Usually, the light will be coming from an extended source and we must consider the shadow as formed by light from different points on the source. In *Figure 1.7a* $B_1 B_2$ is a circular source, CD a circular opaque object and LP a screen. No light from the source can enter the space CDNM, and on the screen there is a circular area of diameter MN in total shadow, known as the **umbra**. Surrounding the umbra is a space, CLM, DNP, in partial shadow, known as the **penumbra**. In the penumbra the illumination gradually increases from total darkness at the edges of the umbra to the full illumination at the outer edges of the shadow; for example, the point *m* is receiving light from the portion $B_1 b$ of the source. The shadow received on the screen consists, then, of a central dark portion surrounded by an ill-defined shaded edge. The diameters of the umbral and penumbral portions of the shadow received on a screen, may be found from similar triangles.

When the source is larger than the opaque object (*Figure 1.7b*), the umbra is in the form of a cone having the object as its base and at a certain distance from the object the umbra disappears, and the shadow is wholly penumbral. This may readily be seen in the case of the shadows of objects in sunlight. As the sun subtends an angle at the earth of 32 minutes, the length of the umbra in sunlight is 105 times the diameter of the opaque object.

In *Figure 1.7b*,

$$\frac{AO}{BO} = \frac{CD}{B_1 B_2} \tag{1.03}$$

$$AO = \frac{CD \times BA}{B_1 B_2 - CD} \tag{1.04}$$

Eclipse of the sun

At certain times the moon comes between the sun and the earth and as the distance between the moon and the earth is less than the length of the umbra of the moon's shadow, the shadow on the earth's surface will consist of an umbral portion surrounded by a penumbra. An observer in the area covered by the umbra will see the moon completely covering the sun and the eclipse will be total; in the penumbra the eclipse will be partial, only a portion of the sun being covered.

Eclipse of the moon

When the earth is directly between the sun and the moon, the shadow of the earth is seen projected upon the moon's surface, and the eclipse will be total when the moon lies wholly within the umbra. When the eclipse is partial the penumbra of the shadow may be seen as a partial darkening of the moon's surface outside the complete shadow.

Shadows play an important part in our visual interpretation of the forms of solid objects; for example, a distant white sphere, evenly illuminated, is indistinguishable from a circular flat surface, while an illumination giving a shadow on some part of it at once gives it its solid appearance. Much can be done in disguising the form of objects by darkening various parts to create a false impression of shadows.

1.7 THE NATURE OF WHITE LIGHT

In the poker experiment, it was seen that the light emitted changed from red, which appeared first, to yellow and finally to white when the temperature was sufficiently high. The nature of white light was investigated by Sir Isaac Newton (1642–1727) in his classical experiments in 1666. On passing a narrow beam of sunlight through a glass prism, Newton found that the patch of light received on a screen was broadened out into a band of colours, in which he recognised seven distinct colours in the following order: red, orange, yellow, green, blue, indigo and violet, each colour gradually shading into the next. This coloured band is called a **spectrum** and the white light is said to be **dispersed**. Recombining the various coloured beams by passing them through a similar prism with its base in the opposite direction to the first, or through a convex lens, Newton found that the light was again white: he therefore concluded that white light is composed of a mixture of light of the seven spectrum colours. There is nothing magical about the number seven. In fact, Newton was colour blind! His assistant named the colours. Recently the spectrum has been divided into 15 colours. These are listed in Chapter 13.

It is now known that each of these colours consists of vibrations of a given range of frequency, red having the lowest frequency and therefore the longest wave-length, and violet the highest frequency and the shortest wave-length, the other colours occupying intermediate positions in the scale of frequencies according to their position in the spectrum. Any solid body, when raised to a 'white' heat, is emitting vibrations of all these frequencies and, in addition, a certain amount which lie outside the range of frequencies to which the eye is sensitive. Of these, those of shorter wave-length than the violet are known as the **ultra-violet** radiations and those of longer wave-length than the red, as the **infra-red** (see Chapter 13).

CHAPTER 1   EXERCISES

1. The velocity of light *in vacuo* is 3×10^{10} cm per second. What will be the wave-lengths corresponding to the following frequencies?

Orange 4.5×10^{14} Hertz
Green 5.69×10^{14} Hertz
Blue 6.96×10^{14} Hertz

2. Describe an experiment to show that light travels in approximately straight lines.

3. A person holding a tube 6 inches long and 1 inch in diameter in front of his eye just sees the whole of a tree through the tube. What is the apparent (angular) height of the tree, and what is its distance if its actual height is 40 feet?

4. Explain the terms, convergent, divergent and parallel light. Illustrate your answer with diagrams showing the form of the wavefronts in each case.

5. A pencil of light diverges from a point source through a rectangular aperture 3 cm × 4 cm at 30 cm from the source. Find the area of the patch of light on a screen parallel to, and 120 cm from the aperture.

6. A pencil of light on passing through a lens of 40 mm diameter converges to a point 60 cm from the lens. Find the areas of the cross sections of the pencil at 20, 50 and 75 cm from the lens.

7. What are meant by pencils and beams of light? Illustrate your answer with diagrams.

A spherical source of 5 cm diameter is placed 30 cm from a circular aperture of 8 cm diameter in an opaque screen; find the size and nature of the patch of light on a white screen 50 cm from and parallel to the plane of the aperture.

8. A small source of light is 30 cm from a rectangular aperture 18 cm × 8 cm. A screen is placed at 105 cm from and parallel to the plane of the aperture; what will be the area of the illuminated patch on the screen?

9. What is the height of a tower that casts a shadow 65 feet 3 inches in length on the ground, the shadow of the observer who is 6 feet high being at the same time 7 feet 6 inches in length?

10. A man 5 feet 11 inches in height is standing at a distance of 4 feet 6 inches from a street lamp. What will be the length of the man's shadow, if the lamp is 9 feet above the roadway?

11. A circular opaque disc of 2 feet diameter is 6 feet from an arc lamp; find the size of the shadow on a screen 20 feet from and parallel to the disc. What will be the form and size of the shadow if the arc is surrounded by a diffusing globe of 1 foot 6 inches diameter?

12. Explain carefully the formation of the image in a pinhole camera. How does the character and size of the image depend on:

    (*a*) the size of the aperture.
    (*b*) the shape of the aperture,
    (*c*) the distance of the aperture from the screen,
    (*d*) the distance from the aperture to the object.

13. If the distance between an object and its image formed by a pinhole is 5 feet, what will be the position of the pinhole for the image to be one-tenth the size of the object?

14. A pinhole camera produces an image 2.25 inches diameter of a circular object, and when the screen is withdrawn 3 inches further from the pinhole, the diameter of the image increases to 2.75 inches. What was the original distance between the pinhole and the screen?

15. If the sun subtends an angle of 32 minutes at the earth, what must be the distance between the aperture and the screen of a pinhole camera in order that the image of the sun shall have a diameter of $^1/_2$ inch?

16. Explain the formation of an image in a pinhole camera. Sunlight is reflected from a small plane mirror about 2 mm square and the reflected light falls normally on a white screen. Describe and explain the patch of light formed on the screen when this is (*a*) a few inches from the mirror, (*b*) about 10 feet from the mirror.

17. A circular opaque object 3 inches diameter is placed 12 inches from and parallel to a circular source of 5 inches diameter. Find the nature and size of the shadow on a screen perpendicular to the line passing through the centres of the source and the object and 3 feet from the object.

18. Explain with diagrams the way in which total and partial eclipses of (*a*) the sun and (*b*) the moon are produced.

19. Find the diameter of the umbra and penumbra of the earth at the distance of the moon (240 000 miles) taking the earth's diameter as 8000 miles, and the visual angle subtended by the sun from the earth as 32 minutes. (Assume the light from the sun to be parallel light).

20. Why are shadows much sharper in the case of an arc lamp without a surrounding diffusing globe than with one? Explain with a diagram.

21. A dark room 10 feet square, with white walls, has a small hole in one wall. The image of a man 6 feet high outside the room is formed on the wall and is 5 inches high. How far away is the man? What will be the size of the image of a tree 50 feet high and 300 feet away?

22. An opaque circular disc is interposed between a luminous disc of larger diameter and a screen so that the shadow consists of a central umbra and surrounding penumbra. The discs and screen are parallel and the line joining the centres of the discs is perpendicular to them. Show that the width of the penumbra ring is independent of the size of the opaque disc.

Chapter 2

The behaviour of light on reaching a new medium

2.1 INTRODUCTION

The light from a source continues to travel out until it meets the surface bounding some new medium. Here a number of effects are produced, which depend upon the nature of the two media and of the surface. In all cases some of the light is sent back or *reflected*, while the remainder passes into the new medium, where some of it will be transformed into some other form of energy, or *absorbed*, and some will be *transmitted*. These three modifications of the light always take place, but one often predominates over the others. It will generally be convenient to consider separately the light which is reflected at the surface and that which passes into the new medium, but it must be remembered that both effects are always taking place at a surface.

2.2 REFLECTION—SPECULAR AND DIFFUSE

When the surface is a polished one, practically the whole of the reflected light travels back in definite directions, as though coming from the source placed in some new position, the surface acting as an aperture in limiting the reflected beam (*Figure 2.1*). An observer in this reflected beam sees the image of the original source, but is not conscious of the presence of the reflecting surface if this is perfectly polished. Such reflection is said to be **regular** or **specular**. When the surface is not polished, every irregularity of the surface will reflect the light in a different direction, and the light will not return as a definite beam. The surface itself will be visible from all directions, and the light may be considered as spreading out from every point on the surface. This reflection is said to be **diffuse** or **irregular**, although the actual reflection at each minute portion of the surface will be regular. All illuminated objects are visible by reason of the light irregularly reflected at their surfaces.

The amount of light reflected at a surface will vary greatly with different media, as may be seen in the amount of light diffusely reflected by, say, white, grey and black paper, while the quantity reflected by a polished silver surface is much greater than that reflected from polished steel. The angle at which the incident light meets the reflecting surface also has a considerable effect on the quantity of light reflected. This is very apparent when light is reflected from a polished glass surface at different angles.

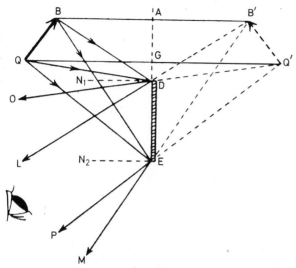

Figure 2.1 Specular reflection. Position of image formed in plane mirror (see also section 3.2)

In the case of transparent substances, such as glass, water, etc., the percentage of light reflected for any angle of incidence may be found from Fresnel's law (section 16.4). It will depend not only upon the angle of incidence, but also upon the difference in the refractive indices (section 2.5) of the two media. For example, if some pieces of glass are placed in glycerine, or, better still in canada balsam, media which have almost the same refractive index as glass, practically no light is reflected at the surface of the glass, and the pieces are almost invisible. A similar effect is seen in the case of a grease spot on paper. The grease filling the pores of the paper has a refractive index more nearly that of the paper fibres than the air has, and so less light is reflected, and the spot appears darker than the surrounding paper.

The light reflected by some substances is coloured. This is due to the fact that, of the various colours contained in white light, some only are reflected, and the others are absorbed by the substance. Such reflection is said to be *selective*. A red object, for example, when illuminated with white light, reflects mainly the red, the other colours passing into the substance and being absorbed. Such an object, illuminated with green light, would appear practically black, as the green light contains no red.

2.3 THE LAW OF REFLECTION

In *Figure 2.2* BA is a ray of light incident on a surface DE. The perpendicular to the surface at the point A is known as the **normal to the surface at the point of incidence**, and the plane containing the incident ray BA and the normal AN is the **plane of incidence**. The angle BAN, which the incident light makes with the normal is the **angle of incidence**, i. The light after reflection will travel along AF, and the angle FAN is the **angle of reflection**, r.

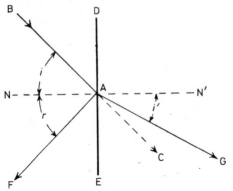

Figure 2.2 Reflection at a plane surface (see also section 3.4)

The law of reflection, which may easily be found experimentally, has been known from very early times. It may be stated as follows:

The incident and reflected rays and the normal to the surface lie in the one plane and the incident and reflected rays make equal angles with the normal on opposite sides of it.

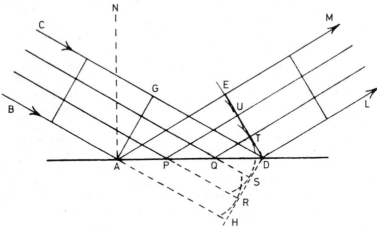

Figure 2.3 Reflection of plane wave at plane surface. Huygens' construction

The way in which the light is reflected at a surface and thence the law of reflection, may be found by applying Huygens' principle. In *Figure 2.3* BCGA represents a parallel pencil of light incident on a plane reflecting surface AD, AG being the position of a plane wavefront at a certain instant. Had there been no change in medium, the new wavefront would, after a time, have reached the position HD. Assuming the speed of the light to be the same before and after reflection, in a given time each part of the reflected light will have travelled back from the surface a distance equal to that which it would have travelled forward, had there been no reflection. Each point on the surface between A and D becomes, in succession, the centre of a reflected wavelet as the incident wavefront meets it. The reflected wavelets arising, for example, at the points A, P and Q will have travelled out distances AE, PU and QT, equal to AH, PR and QS respectively, while the edge of the wavefront at G is reaching the surface at D. From Huygens' principle, the envelope or common tangent, ED, to the wavelets will be the reflected wavefront. It may easily be proved that the angles GAD and EDA are equal, and therefore the angle of incidence equals the angle of reflection.

2.4 ABSORPTION AND TRANSMISSION

The subsequent behaviour of the light that enters the new medium will depend upon the nature of the substance. In most cases the whole of the light, after travelling a short distance, is changed into some other form of energy, such as heat or chemical action, and the light is said to be absorbed. Substances through which light cannot pass are said to be **opaque**. The degree of absorption will depend upon both the nature of the substance and the thickness through which the light travels. Gold, in a thin film, will transmit a considerable amount of light.

When all colours are equally absorbed, as will usually be the case when the substance is of any considerable thickness, the absorption is said to be *neutral*; but with some substances, coloured glasses, dyes, etc., the absorption is *selective*; some colours are absorbed and others are transmitted. We shall deal with this again when considering transmission. Certain substances possess the property of absorbing energy of short wave-length, the ultra-violet and the violet and blue light, and emitting it as radiations of longer wave-length. This is known as fluorescence and is considered more fully in section 11.11.

Some substances absorb very little light and practically all of it is transmitted; such substances are **transparent** and are of great importance in optical work. If the surfaces of such a transparent substance are polished, the light passes through as a definite beam. No substance is perfectly transparent and some part of the light is always absorbed. Good quality glass absorbs about 0.5% of the incident light in passing through a thickness of 10 mm, although glass developed for optical fibres may have a transmission better than 50% per kilometre.

Other substances, which contain fine particles of a different optical nature from the surrounding mass, or which are composed of minute particles of transparent material with polished faces, such as minute crystals, transmit the light but diffuse or scatter it on the way, and objects cannot be seen clearly through such substances. These media, such as opal glass, paraffin wax, tracing paper and cloth, smoke and fog, are said to be **translucent**. The same effect is obtained with a transparent substance having an unpolished surface, as in ground or 'frosted' glass. Screens made of translucent substances are frequently used to diffuse the light from a source, or as screens on which to receive images.

As was previously mentioned, the absorption and therefore the transmission of certain media are *selective* with respect to the colour or wavelength of the light absorbed or transmitted. Thus, when white light is incident on a piece of, say, green glass, some of the light is reflected at the polished surface and this will still be white; of that which passes into the glass, only green light is transmitted, the other colours composing the white being absorbed. All coloured glasses, dyes, etc., behave in this way. A good white glass, whilst being transparent to all the colours of the visible spectrum, is quite opaque to radiations below a certain wave-length in the ultra-violet. Some of the glasses used for protective purposes in ophthalmic work are designed to be opaque to ultra-violet or infra-red, or both, while transparent to ordinary light.*

2.5 REFRACTIVE INDEX

The changes in the direction of the light that occur when light passes from one medium to another can be explained on the assumption that the velocity of light is different in different media. Foucault, in 1850, determined experimentally the velocity of light, both in air, and in water, and found that the velocity was less in water than in air. This important result was in accordance with the wave theory explanation of the behaviour of the light on passing between these two media. The action of lenses and prisms is entirely due to this change of velocity in different media, and it is necessary to have some means of expressing the relative velocity of the light in any medium.

If we take the velocity of light *in vacuo* as a standard, then its relative velocity in any other medium may be expressed in terms of the ratio velocity *in vacuo*/velocity in medium. This ratio is known as the **absolute refractive index** and will be denoted by n. As the velocity in any transparent medium is less than that *in vacuo*, the value of the absolute refractive index of any medium is greater than unity and, with the exception of gases, will be between 1.33, the refractive index of water and 2.42, that of diamond.

When the light passes from one medium to another, such as from air to glass, glass to water, etc., it is sometimes convenient to express the change in velocity as a ratio of the velocities in the two media; thus

$$\frac{\text{velocity in air}}{\text{velocity in glass}} = {_a}n_g$$

this is called the **relative refractive index** from air to glass. If n_a is the absolute refractive index of air and n_g the absolute refractive index of glass then the relative refractive index from air to glass

$${_a}n_g = \frac{\text{velocity in air}}{\text{velocity in glass}} = \frac{n_g}{n_a}$$

As the velocity of light in air is very nearly that *in vacuo*, the average absolute refractive index of air being 1.00029, the relative refractive index from air to any substance is usually given as the refractive index of that substance.

The terms 'rare' and 'dense' are frequently used in a comparative sense in referring to media of low and high refractive index respectively.

It will be seen, when we consider the action of a prism (section 4.7), that the dispersion of white light is due to the refractive index being different for each of the colours composing the white light, the refractive index being highest for the violet and lowest for the red in all substances with a few exceptions. In stating any refractive index, therefore, it is necessary that the colour of the light shall be

*The action of filters is dealt with in Chapter 11.

specified; this is considered more fully in Chapter 13. Usually, a mean value of refractive index is taken as that for a particular yellow. This used to be the light emitted by incandescent sodium vapour. Then light from helium vapour was used as this allowed a more accurate measurement. Lately, yellow-green light from mercury vapour has been used although values for sodium light are still commonly quoted. The following are the mean refractive indices of some optical substances. More values are given in Appendix 5.

Water	1.3336
Optical Plastics	1.46–1.6
Optical Glass	1.5–1.85
Spectacle Crown Glass	1.5230
Quartz	$\begin{cases} 1.5442 \\ 1.5534 \end{cases}$
Diamond	2.4173

2.6 REFRACTION

When light is incident obliquely on the surface between two media of different refractive index, its direction is changed on passing into the new medium, and it is said to be **refracted**. The new direction of the light may be found by applying

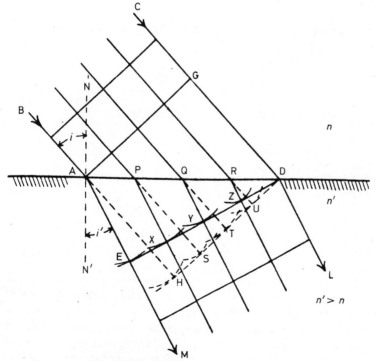

Figure 2.4 Refraction of plane wavefront at plane surface. Huygens' construction

Huygens' principle as follows: *Figure 2.4* represents a parallel pencil of light BADC incident on the surface AD between two media of refractive index n and n', n' being greater than n. AG is a plane wavefront one edge of which has just reached the surface at A. Had there been no change in medium, the incident light would have travelled on with velocity V without change of direction, and the wavefront after a certain time would have been in the position HD. As the incident wavefront meets the surface, each point on the surface, such as A, P, Q, R and D, becomes in succession the centre of a system of wavelets travelling in the new medium. These wavelets travel in the second medium at a velocity V', where $V'/V = n/n'$, and by the time the light at G has travelled to D in the first medium, the radii of the wavelets in the second medium will be

$$AE = AH \times \frac{n}{n'}, \quad PX = PS \times \frac{n}{n'}, \quad QY = QT \times \frac{n}{n'}, \text{ etc.}$$

The envelope of all these wavelets will be the refracted wavefront ED. It will be found that the refracted wavefront is still a plane surface and that when, as in *Figure 2.4*, the light is passing from a rarer to a denser medium, it is refracted *towards the normal*. A similar construction for light passing from a denser to a rarer medium will show that in this case it is refracted *away from the normal*. The angle BAN is the angle of incidence i and the angle MAN', the angle of refraction i'.

Law of Refraction

From the definition of refractive index (section 2.5).

$$n\text{GD} = n'\,\text{AE} \quad (Figure\ 2.4)$$

$$n\frac{\text{GD}}{\text{AD}} = n'\frac{\text{AE}}{\text{AD}}$$

$$n \sin \text{DAG} = n'\, \sin \text{ADE}$$

but

$$\angle\text{DAG} = i, \text{ and } \angle\text{ADE} = i'$$

$$\therefore n \sin i = n' \sin i' \tag{2.01}$$

and

$$\frac{\sin i}{\sin i'} = \frac{n'}{n} \tag{2.01a}$$

n'/n is the relative refractive index from the first to the second medium, and is a constant for these two media and for light of any one colour. The law of refraction may be stated as follows:

The incident and refracted rays and the normal to the surface at the point of incidence lie in the one plane on opposite sides of the normal, and the ratio of the sine of the angle of incidence to the sine of the angle of refraction is a constant for any two media for light of any one colour.

This law, which was first stated, but in a somewhat different form from the above, by Willebrord Snell (1591-1626), is often known as *Snell's Law.* *

The refraction at a plane surface and the law of refraction may be demonstrated by the following simple experiment. A block of glass having two parallel polished surfaces is placed on a sheet of paper with its polished surfaces vertical. Two pins A and B (*Figure 2.5*) are placed a few inches apart to represent the path of a narrow beam of light incident obliquely on one surface; on looking through the glass, the pins will be seen to occupy the apparent positions A' B'. Two further pins C and D are then placed in apparent line with A and B, as seen through the glass. The line AEFD then represents the path of the light between the points A and D. By repeating the experiment for a number of different angles of incidence and measuring these angles and the corresponding angles of refraction, the law of refraction may be experimentally verified and the refractive index of the glass obtained.

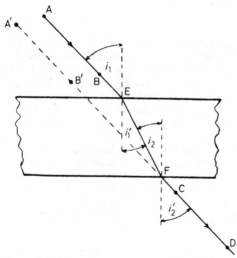

Figure 2.5 Experimental verification of the law of refraction

2.7 THE PRINCIPLE OF LEAST TIME

The laws of reflection and refraction may be summed up in a general law, which was first stated by Fermat (1601-1665), and is known as the **principle of least time**; this may be stated as follows:

The actual path travelled by light in going from one point to another is that which, under the given conditions, requires the least time.

In certain cases, however, when the bounding surface is curved, the time may be a maximum instead of a minimum, but must always be one or the other.

*Snell stated the law of refraction as a constant ratio of the cosecants of the angles of incidence and refraction. The statement of the law as a constant ratio of the sines of the angles was first given by Descartes (1596-1650) in 1637.

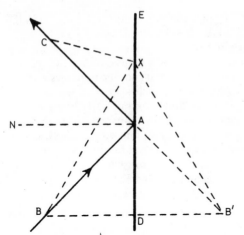

Figure 2.6 Path of least time on reflection

In *Figure 2.6* DE is a reflecting surface and BAC the path of the light reflected at the surface in travelling from B to C. It can easily be proved that the path BAC, which is in accordance with the law of reflection, is shorter than any other path such as BXC. Let BB′ be a perpendicular to DE and DB′ = BD, then B′A = BA and B′X = BX; therefore the path BAC = B′C, and the path BXC = B′X + XC. But in the triangle B′CX the sides B′X and XC are together greater than the side B′C; therefore BAC is the shortest path and hence the path of least time. The principle of least time, in the case of reflection at a plane surface, was discovered by Hero of Alexandria (150 BC).

In the case of refraction the path that the light travels in the least time will not be the shortest distance between the two points. In *Figure 2.7* DE represents the surface between two media of refractive index n and n', n' being greater than n, and B and C two points between which the light is travelling. Since the light is travelling more slowly in the second medium than in the first, the straight line BXC, although the shortest path between B and C, is not the path of least time, for a greater part of the path lies in the second medium than

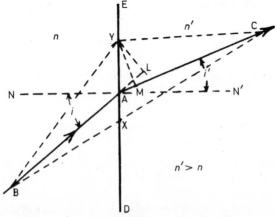

Figure 2.7 Principle of least time. Refraction at plane surface

is the case with the path, say BAC. For some other path BYC, the time gained by the further shortening of the distance in the second medium may be more than offset by the increase in the distance travelling in the first medium, and it can be shown that the path BAC that is in accordance with the law of refraction is the path of least time between B and C.

In *Figure 2.7*

$$\frac{\sin NAB}{\sin N'AC} = \frac{n'}{n} = \frac{V}{V'}$$

where V and V' are the velocities of the light in the first and second media respectively. The time taken by the light in travelling over the path BAC is

$$t = \frac{BA}{V} + \frac{AC}{V'}$$

and it is required to show that this is less than the time $BY/V + YC/V'$ taken over any other paths such as BYC. Drawing YL perpendicular to BA and YM perpendicular to AC, then $\angle AYL = \angle NAB$ and $\angle AYM = \angle N'AC$.

therefore,

$$\frac{\sin AYL}{\sin AYM} = \frac{n'}{n}$$

and

$$\frac{AL}{AM} = \frac{n'}{n} \text{ or } \frac{AL}{V} = \frac{AM}{V'}$$

$$\frac{BA}{V} + \frac{AC}{V'} = \frac{BA}{V} + \frac{AM + MC}{V'} = \frac{BL}{V} + \frac{MC}{V'}$$

But

$$\frac{BL}{V} < \frac{BY}{V} \text{ and } \frac{MC}{V'} < \frac{YC}{V'}$$

therefore

$$\frac{BA}{V} + \frac{AC}{V'} < \frac{BY}{V} + \frac{YC}{V'}$$

2.8 OPTICAL LENGTH

From the definition of the refractive index of a medium, $n = V_0/V$ where V_0 is the velocity of light *in vacuo* and V the velocity of light in the medium concerned of refractive index n.

Therefore,

$$V = \frac{V_0}{n}$$

The time taken by light in travelling a distance l *in vacuo* = l/V_0, while the time taken in travelling a distance l in medium of refractive index $n = l/V = nl/V_0$.

When the light is travelling through a number of different media, if l_1, l_2, l_3, etc. are the distances in the separate media of refractive index n_1, n_2, n_3, etc. respectively, the total time taken to travel this distance will be

$$t = \frac{1}{V_0}(n_1 l_1 + n_2 l_2 + n_3 l_3 + \ldots) = \frac{1}{V_0} \Sigma nl.$$

nl is termed the **optical length** of a path l in medium of refractive index n. Paths of equal optical length are paths of equal time of travel.

The optical length of path is of considerable importance in dealing with the interference of light. (Chapter 14).

2.9 THE FUNDAMENTAL LAWS OF GEOMETRICAL OPTICS

The study of optics may conveniently be divided into two parts (1) that dealing with the physical nature of light, usually called **physical optics** and (2) that concerned chiefly with the way in which the directions in which the light travels are modified by mirrors, prisms and lenses. This is known as **geometrical optics**.

Geometrical optics, ignoring diffraction effects due to the wave nature of light, assumes:

(1) Neighbouring rays of light are independent of one another.
(2) The propagation of light is rectilinear, i.e., light travels in straight lines.
(3) Law of reflection.
(4) Law of refraction.

Physical optics is considered in Chapters 12 to 16 while the remainder of the book is concerned chiefly with geometrical optics but giving due regard to physical effects.

CHAPTER 2   *EXERCISES*

1. State the law of reflection and show how it may be verified experimentally.

2. Describe Huygens' principle, and show how it may be used to explain the laws of reflection and refraction, on the wave theory.

3. Using Huygens' principle construct the reflected wavefront when a parallel pencil of light is incident on a plane surface (a) normally, (b) at $30°$, (c) at $70°$.

4. Describe and give reasons for the appearances of the following substances:

    (a) Grey paper;
    (b) Finely crushed glass;
    (c) Paraffin wax, solid and molten.

5. Why is a lump of sugar practically opaque, while the small separate crystals of which the lump is composed are themselves transparent?

6. Explain what is meant by the refractive index of a substance. Why is it necessary to specify the colour of the light referred to in accurately stating a refractive index?

7. The velocity of light *in vacuo* is 300 million metres per second. Find the velocity (a) in water of refractive index 1.33; (b) in oil of refractive index 1.45; and (c) in dense flint glass of refractive index 1.65.

8. The velocity of yellow light in carbon disulphide is 183.8×10^6 metres per second, in ether it is 221.5×10^8 metres per second. Find the relative refractive index (a) for refraction from ether into carbon disulphide, (b) from carbon disulphide into ether.

9. The refractive indices of a few substances are given.

Water 1.33. Dense Flint Glass 1.65.
Crown Glass 1.50. Diamond 2.42.

Find the relative refractive index for the case of light being refracted from:

(a) Water into crown glass.
(b) Water into diamond.
(c) Diamond into water.
(d) Dense flint glass into air.
(e) Crown glass into dense flint glass.

10. Given that the speed of light in air is 3×10^{10} cm per second and in water 2.25×10^{10} cm per second, construct the emergent wavefront from a small source situated in the water 10 cm below the surface, and from this determine the position of the image.

11. State the law of refraction, and describe a method of measuring the refractive index of a block of glass having plane parallel faces.

12. Show that if a ray of light passes from a medium in which its velocity is V_1 to another in which it is V_2, $V_2 \sin i = V_1 \sin i'$ where i and i' are the angles of incidence and refraction respectively. What is the ratio V_1/V_2 called?

13. A poster has red letters printed on white paper; describe and explain its appearance when illuminated (a) with red light and (b) with green light.

14. Give diagrams showing the change in direction of the wavefronts of a parallel pencil on passing from air to water when the pencil is incident at (a) $30°$, (b) $60°$ to the normal (n of water = 4/3).

15. Explain why objects cannot be seen through a fog, although the water particles of which fog is composed are transparent.

16. A parallel pencil of light is incident at $45°$ to the upper surface of oil ($n = 1.47$) floating on water ($n = 1.33$). Trace the path of the light in the oil and the water.

17. A parallel pencil is incident at an angle of $35°$ on the plane surface of a block of glass of refractive index 1.523. Find the angle between the light reflected from the surface and that refracted into the glass.

18. A white stone lies on the bottom of a pond. Its edges are generally observed to be fringed with colour, blue and orange. Explain this and state which is the blue edge.

19. State the fundamental laws of geometrical optics and describe an experimental verification of each.

20. What happens when some irregular fragments of glass are immersed in a liquid of the same refractive index? How may this device be employed to measure the refractive index of such fragments?

21. A ray incident at any point at an angle of incidence of $60°$ enters a glass sphere of refractive index $\sqrt{3}$ and is reflected and refracted at the further surface of the sphere. Prove that the angle between the reflected and refracted rays at this surface is a right angle. Trace the subsequent paths of the reflected ray.

Chapter 3

Reflection at a plane surface

3.1 IMAGES–VIRTUAL AND REAL

On looking into a plane reflecting surface one sees, apparently behind the surface, images of any objects that are in front of the surface. Some of the light from each object point is reflected at the surface, and enters the eye as though it were coming from points behind the surface. Apparent images of this kind from which the light is diverging are termed **virtual images**. When, by reflection or refraction at a curved surface, the light from object points is made to converge again through points, an image is formed that actually exists at the position to which the light converges. Such images, which can be received on a screen, as with the image formed by the lens of the camera or projection lantern, are termed **real images**.

3.2 POSITION OF IMAGE FORMED BY PLANE MIRROR

The position of the image of any object before a plane reflecting surface can be determined from the law of reflection, or it may be found by using Huygens' construction to determine the form of the reflected wavefront. Thus, in *Figure 3.1*, B is an object point in front of a plane mirror DE. If the mirror had not been present, DLE would be the form of a wavefront at a particular instant. As, however, the incident wavefront meets the mirror, each point on the mirror becomes the centre of a series of wavelets travelling back at the same speed as the incident light was travelling forward. Thus the portion of the incident wavefront meeting the mirror at A gives rise to wavelets that travel back to M in the same time as the incident light would have travelled to L. Similarly, the wavelets originating at G travel back to P in the same time as the incident light would have travelled to N, and so on over the whole mirror surface. The reflected wavefront DME will be the envelope of all the reflected wavelets corresponding to points on the incident wavefront. It can easily be proved from the figure that DME is an exactly similar curve to DLE, but in the opposite direction, and the reflected light is, therefore, diverging as though from a virtual image point B', where AB' = BA.

The same result may be obtained from the law of reflection. An object BQ (*Figure 2.1*) is placed in front of a plane mirror DE. Rays BD and BE are reflected along DL and EM as though from an image at B', while rays QD and QE are reflected along DO and EP as though from an image at Q', B'Q' is, therefore, the virtual image of BQ, formed by the mirror. From the law of reflection and the

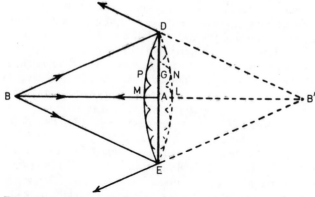

Figure 3.1 Image formed by reflection at a plane surface. Huygens' construction

geometry of the figure, it may easily be proved that BB′ and QQ′ are perpendicular to the mirror surface and that AB′ = BA, GQ′ = QG.

Hence *the image formed by a plane mirror is situated as far behind the mirror as the object is in front, and the line joining the object and image is perpendicular to the plane of the mirror surface.*

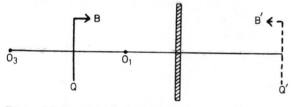

Figure 3.2 Reversion of image formed by plane mirror

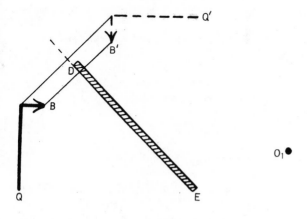

Figure 3.3 Reverted image. Light turned through 90 degrees

It follows that in shape and size, the image will be an exact reproduction of the object. The point of the object nearest the mirror is represented by the point of the image nearest the mirror, and the top and bottom of the image corres- pond with the top and bottom of the object; that is, the image is erect. The question of whether the image is or is not reversed left for right, or *reverted*, as it is sometimes called, depends on the position and posture of the observer in viewing both object and image. Thus, an observer at O_1 (*Figure 3.2*), in viewing the object BQ directly, sees B to his right; on turning round to view the image B′ Q′, B′ the image of B is now to his left. If the observer is at O_3 he will see both B and B′ to his left, but in this case he will see opposite sides of object and image; that is, if the side of the object nearer to him is called the front, he will see the back in the image. Similarly, if the light is turned through 90° by reflec- tion, as is frequently done in optical instruments, an observer at O_2 (*Figure 3.3*) will see the image reverted as compared with the object seen directly from O_1.

3.3 THE FIELD OF VIEW OF A PLANE MIRROR

The extent of image in a plane mirror that can be seen by an eye in any position will depend on the size of the mirror and the position of the eye. This extent of image seen is termed the **field of view** of the mirror.

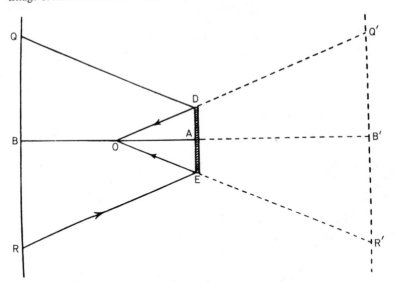

Figure 3.4 Field of view of a plane mirror

If an eye at O (*Figure 3.4*) is looking into a plane mirror DE, DO and EO will be the limiting directions along which the eye can see images reflected in the mirror. The reflected rays DO and EO correspond to incident rays from the object points Q and R respectively, and QR is, therefore, the extent of object in this plane of which the image, Q′R′, can be seen from the position O. It will be seen that the field of view is exactly the same as would be obtained if an eye situated at O was looking at an object occupying the position of the image, through an aperture in the position of, and the same size and shape as the mirror. From *Figure 3.4* we have

$$Q'R' = \frac{DE \times OB'}{OA} = QR$$

It will often be required in practice to find the least size and the position of a mirror in which the whole of a given object may be seen reflected from a given position. Such problems can be solved by treating the mirror as an aperture, and considering the image of the object formed behind the mirror.

3.4 DEVIATION PRODUCED BY A MIRROR

A plane reflecting surface is frequently used to produce a change in direction of a beam of light. The angle through which the beam is turned—**the angle of deviation**—will clearly depend on the angle of incidence at the mirror. In *Figure 3.5* a beam incident in the direction BA on a mirror DE is reflected along AF. Then $\angle CAF$ is the angle of deviation and $\angle CAF = 180° - 2i$.

3.5 REFLECTION FROM A ROTATING MIRROR

If a mirror is rotated through a given angle, a beam reflected from the mirror will be rotated through twice this angle, that is, the beam rotates twice as rapidly as the mirror. A beam incident in the direction BA (*Figure 3.5*) on the mirror D_1E_1, will be reflected in the direction AF_1. On the mirror being turned through an

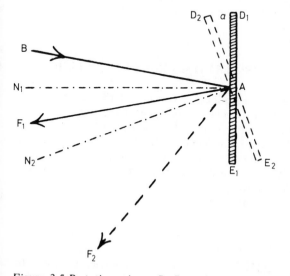

Figure 3.5 Rotating mirror. Reflected beam turned through twice the angle the mirror turns through

angle a to D_2E_2 the beam is reflected in the direction AF_2. Since the mirror and the normal turn through the angle a, the angle of incidence and, therefore, the angle of reflection are each increased by a, and the angle between the directions of the incident and reflected beams by $2a$. Hence $\angle E_1 AF_2 = 2a$.

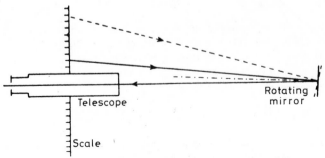

Figure 3.6 Rotating mirror galvanometer

Use of this fact is made in the measurement of small angular deflections. A beam of light, reflected from a mirror attached to the moving part of the apparatus, takes the place of the index or pointer, and this optical pointer turns through twice the angle the moving part turns through; also, the beam may be made of any length without adding weight to the apparatus, and thus quite small angular movements can be read with considerable accuracy. This method is frequently used in reading the small deflections of a galvanometer. A small light mirror is attached to the moving magnet, and a narrow beam of light is focused, after reflection at the mirror, in a small patch on a scale placed at some distance from the galvanometer. Another arrangement shown in *Figure 3.6* is also often used; a telescope is focused on the image of a scale formed by reflection at the mirror attached to the moving part.

3.6 DEVIATION ON REFLECTION AT TWO MIRRORS IN SUCCESSION

When a ray incident in a plane containing the normals of the two mirrors is reflected from two mirrors in succession, it undergoes a total deviation that depends only on the angle between the mirrors. The total deviation produced is, therefore, independent of the angle of incidence at the first mirror, and a rotation of the mirrors about an axis perpendicular to the plane containing their normals, and keeping the angle between them constant, produces no change in the final direction of the reflected light.

In *Figure 3.7* the total deviation of the ray

$$= \angle C_1 A_1 A_2 + \angle C_2 A_2 F$$

$$= 2(\angle EA_1 A_2 + \angle EA_2 A_1) = 2\angle A_1 EH.$$

Therefore, total deviation $= 2(180° - a)$ where a is the angle between the mirrors.

Successive reflection from two mirrors is used in all possible cases where considerable accuracy is required in the direction of a reflected beam. Providing the

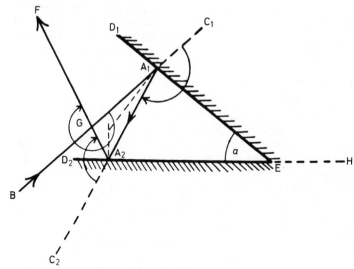

Figure 3.7 Deviation at two mirrors

angle between the reflectors is accurately fixed, and this is frequently done by making them the surfaces of a prism (Chapter 4), the correct placing of the mirrors with respect to the incident light need not be made with any great accuracy.

3.7 MULTIPLE IMAGES FORMED ON REFLECTION AT TWO MIRRORS

If an object is situated between two plane mirrors, more than the two direct images will be formed, because some of the light reflected at one mirror may be again reflected at the other mirror. The number of such images that will be seen

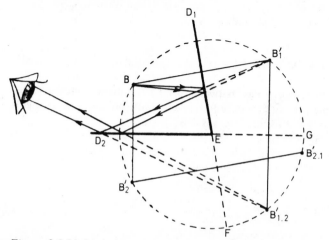

Figure 3.8 Multiple images formed by reflection at two mirrors

will depend on the angle between the mirrors, the position of the object, and, to a certain extent, on the position of the eye viewing the images. The positions of the images may be found by considering any image so found as acting as an object for the mirror of which it is in front.

Thus light from a luminous point B (*Figure 3.8*) is reflected at the mirror D_1E giving rise to the image B'_1. Some of this reflected light reaches the mirror D_2E where it is again reflected, and as the incident light is travelling as though from B'_1, the reflected light is coming apparently from the image $B'_{1.2}$. In the same way, the light that is first reflected at D_2E travels as though from an image at B'_2, and on being reflected again at D_1E, it is coming apparently from the image $B'_{2.1}$. As both the images $B'_{1.2}$ and $B'_{2.1}$ lie behind the mirrors, the light coming as though from them does not meet the mirrors, and no further images are formed in this case. The object and each of the images will lie on a circle having its centre at E, the intersection of the mirrors. The path of the light by which an eye at O, looking into the mirror D_2E sees the image $B'_{1.2}$ is shown in the figure.

The following facts concerning the number of images formed should be verified by constructing the images for a number of cases with different angles between the mirrors and different positions of the object. The results may also easily be verified by experiment.

Calling the angle D_1ED_2 between the mirrors a, the angle D_1EB, x and the angle BED_2, y, the total number of images in the series that starts with light reflected at D_1E, that is, the images B'_1, $B'_{1.2}$, $B'_{1.2.1}$, etc., is given by the integer next higher than $(180° - x)/a$. The total number of images in the other series, that is, the images B'_2, $B'_{2.1}$, $B'_{2.1.2}$, etc., is given by the integer next higher than $(180° - y)/a$.

The exception to this rule is when the angle a is contained in $(180° - x)$ or $(180° - y)$ an exact whole number of times. The number of images in the series, for which this is so, is given by the actual integer obtained by the division, the last image of the series in this case falling on the projection of the mirror, EG or EF. When the angle a is divisible an integral number of times into $180°$, the last images of each series coincide, and therefore the total number of images is $2(180°/a) - 1$.

A well-known application of the multiple images formed with inclined mirrors is the instrument known as the **Kaleidoscope**.

3.8 TWO PARALLEL PLANE MIRRORS

If an object B (*Figure 3.9*) is placed between two plane mirrors, that are facing each other and parallel, the light will be reflected backwards and forwards between the mirrors giving rise to the images B'_1, $B'_{1.2}$, etc. behind the mirror D_1E_1 and the images B'_2, $B'_{2.1}$, etc. behind the mirror D_2E_2, each image acting as an object for the mirror it is facing. It is clear that as the mirrors are parallel, the images and the object will be on a straight line and

$$A_1B'_1 = BA_1 \qquad A_2B'_2 = BA_2$$

$$A_1B'_{1.2} = B'_2A_1 \qquad A_2B'_{2.1} = B'_1A_2$$

$$A_1B'_{1.2.1} = B'_{2.1}A_1 \qquad A_2B'_{2.1.2} = B'_{1.2}A_2$$

Theoretically, the number of images will be infinite, but, as only a certain

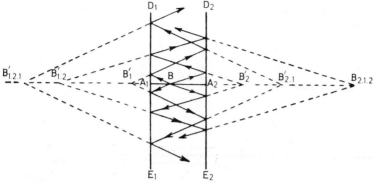

Figure 3.9 Multiple images in parallel mirrors

fraction of the incident light is reflected each time, each successive image is fainter than the preceding one, and the number of images seen will depend on the brightness of the original object.

3.9 REFLECTION AS A MEANS OF ERECTING AN INVERTED IMAGE

Reflection from a number of plane surfaces is often used as a means of erecting the inverted real image formed by a lens; the commonest application of this is in the prism binocular. Each half of the binocular consists essentially of a telescope of the astronomical type, which as explained in Chapter 10 produces an image that is inverted in all directions. The light is reflected at two pairs of plane reflectors, the surfaces in each pair being at right-angles, and the common edge of the

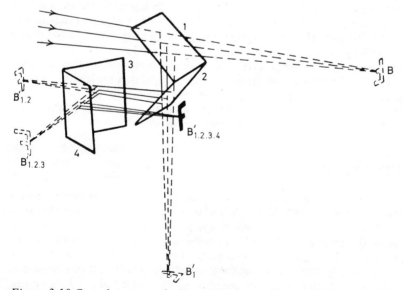

Figure 3.10 Complete reversal of image by four reflections, light travelling in original direction. Prism binocular principle

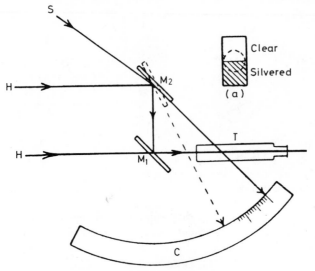

Figure 3.11 Principle of the sextant

surfaces in the one pair being perpendicular to the common edge of the other pair. After these four reflections, the light is travelling parallel to its original direction and the image has been completely reversed as shown in *Figure 3.10.*

B is the real inverted image formed by convergent light from the first lens— the objective—of the telescope. After reflection at surface 1, the light is converging towards the image B'_1, and after a further reflection at surface 2, towards the image $B'_{1.2}$. It will be seen that the image has now been reversed up and down as compared with the original image B. The light is then intercepted by a second pair of reflecting surfaces, 3 and 4, having their common edge perpendicular to that of the first pair. These two further reflections give rise in turn to the images $B'_{1.2.3}$ and $B'_{1.2.3.4}$, and a left for right reversal is produced. The result of the four reflections is, therefore, a reversal of the image in two perpendicular planes, and in this way the inversion of the original image formed by the lens is corrected.

An additional advantage is obtained by the 'folding-up' of the light path causing a considerable reduction in the length of the instrument. In the actual instrument, the reflectors consist of two 90° prisms.

3.10 THE SEXTANT

Another instrument that depends on the successive reflection at two mirrors is the **Sextant** (Hadley 1682–1744). When it is required to measure the angle between two distant objects, as in surveying, the usual method is to employ some form of sighting device, such as a telescope, attached to a divided circle, and to measure the angle turned through by the telescope when directed to the two objects in turn. Such an instrument is called a **theodolite** but this method is, of course, impracticable when the measurement is to be made from a moving object, as, for example, in the measurement of the angle of elevation of the sun or a star from a moving ship, and it is for measurements of this kind that the sextant is used.

The principle of the sextant is shown in *Figure 3.11*, T is a telescope in front of which is fixed a mirror M_1 known as the *horizon glass*, half of which is silvered and the other half clear glass, as shown in *Figure 3.11a*. The position of the telescope is such that it receives light from each half of this horizon glass. A second mirror M_2 known as the *index glass*, can be rotated and its position read by an index arm on a circular scale C. When the two mirrors are parallel, the telescope is receiving light from a distant object directly through the clear half of the horizon glass and also by reflection at the two mirrors, and a single image is seen in the telescope. The index attached to the index glass should then be reading zero. To measure the angle between two objects, say the sun and the horizon, the telescope is directed to the horizon, and the index mirror turned until the light from the sun, after reflection from the two mirrors, also travels along the telescope axis. An image of the sun and of the horizon are then seen in coincidence in the telescope.

As the light is turned through twice the angle turned through by the mirror M_2, the angle SM_2H between the two objects, the angle of elevation of the sun, will be twice the angle turned through by the index glass from its zero position. To facilitate reading, the divisions on the scale are numbered with double their actual value; that is, half degrees are marked as degrees, and so the value of the angle SM_2H can be read directly. The scale is usually divided from $0°$ to $140°$ in $10'$ divisions and is read by vernier to $10''$.

CHAPTER 3   EXERCISES

1. State the law of reflection. Show that the line joining an object and its image formed by a plane mirror is bisected at right angles by the mirror surface.

2. Describe, with a diagram, an experiment to prove that the image formed by a plane mirror is as far behind the mirror as the object is in front.

3. At 50 cm from a plane mirror, 20 cm by 15 cm in size, and on a normal to its middle point, is situated a small source of light. What will be the size of the patch of reflected light received on a screen 2 metres from and parallel to the mirror?

4. A plane mirror 2 feet high is fixed on a wall of a room with its lower edge 3 ft above the floor. If a man, whose eyes are 6 feet above the floor, stands directly in line with the centre of the mirror and 3 feet from it, what will be the length of floor that he can see by reflection?

5. A sight-testing chart, measuring 1.2 m by 0.8 m, the longer dimension being vertical, is to be viewed by reflection in a plane mirror. Find the smallest size of mirror that can be used if the chart is 3.25 m and the observer 2.75 m from the mirror. If the observer's eye is 1.2 m and the lower edge of the chart 1.5 m above the ground, what must be the height of the bottom edge of the mirror?

6. Show that the length of the smallest mirror in which a person may see the whole of himself reflected is half the person's height.

7. On one wall of a room is a picture, 6 feet by 4 feet, and on the opposite wall 15 feet away is a plane mirror. What is the least size of the mirror in which a person 8 feet away may see the whole of the picture reflected?

8. Two plane mirrors are placed together with their edges touching so as to be approximately in the same plane, and an observer standing 8 feet away sees two images of his face which are just touching one another. If the width of his face is 6 inches, what is the angle between the mirrors?

9. A plane mirror is set up vertically facing a house 35 feet high and 25 feet wide and 100 feet away from it. What size and how high must the mirror be in order that a man standing 20 feet in front of it and whose eyes are 5 feet 3 inches above the ground may just be able to see the whole of the house reflected in the mirror?

10. What must be the minimum size of a rectangular plane mirror on one wall of a room which will enable a person standing in the middle of the room to see the whole of another person 5 feet 6 inches high and 2 feet wide entering a doorway behind him?

11. On one wall of a room 6 m long hangs a picture of horizontal width AB = 24 cm. On the opposite wall hangs a plane mirror. Seated at a distance of 3 m from the mirror, a man views a reflection of the picture; the side A of the picture is to his right and B to his left. If the separation of his eyes (P.D.) is 6 cm, find the least horizontal width of mirror in order that he may just see the total width of the picture with *each* eye.

If C is the middle point of AB, what portion or portions of the mirror width are required in order that the right eye can see only AC and the left eye only BC?

12. What will be the rotation of a mirror reflecting a spot of light on to a straight scale 1 metre from the mirror if the spot of light moves through 5 cm?

13. Show that the images of an object placed between two inclined mirrors all lie on a circle.

14. Show how you would arrange an experiment to prove that when parallel light falls upon a plane mirror which is turned through an angle, the reflected beam turns through twice that angle. What conclusion can be drawn from this experiment?

15. Explain how an image is formed by reflection at a plane surface. In what respect do the object and image differ one from the other? How would you find by experiment the position of the image?

16. Show how by means of two plane mirrors a person standing in front of one of them can see the image of the back of his head. Trace the path of the light from the back of his head to his eye, and show clearly where the image is formed.

17. Account for the formation of multiple images by two plane mirrors hanging on adjacent walls of a room, and show in plan the object and images.

18. A kaleidoscope is made with three plane mirrors each 2½ inches wide, and a small square object is viewed by it. Draw a diagram showing the appearance that would be seen.

19. Two vertical plane mirrors are fixed at right angles to one another with their edges touching and a person walks past the combination on a straight path at 45° to each mirror. Show how his image moves.

20. It is required to turn a beam of light through 80° by means of mirrors; explain, giving diagrams, three ways in which this may be done. State the advantages and disadvantages of each method.

21. Explain the formation of an image by reflection in a plane mirror.

Two plane mirrors are placed so that each makes an angle of 45° with the other, and a luminous object is placed between them. Determine the position of all the reflected images.

22. Two plane mirrors are fixed together with their common edge vertical and their normals pointing North and North-East respectively; what deviation will a horizontal beam undergo after being reflected at both mirrors? Find the final direction of a beam incident on the first mirror in a direction 15° North of East.

23. Two plane mirrors are placed 1 metre apart with faces parallel. An object placed between them is situated 40 cm from one mirror; find the positions of the first six images formed by the mirrors.

24. Two mirrors are inclined at 35°. At what angle must a ray be incident on the first mirror so that after reflection at the two mirrors it returns along its own path?

25. Two plane mirrors are inclined at an angle of 60° to one another, and a small object is placed between them and 1 inch from one mirror. Show on a diagram the position of each of the images and the path of the light by which an eye sees the last image formed. What practical application is made of this arrangement?

26. Prove that if two plane mirrors are inclined to one another at an angle x then any ray, after reflection once at each mirror, is turned through an angle of $(360 - 2x)$.

By treating the case of two plane mirrors inclined at 30°, prove the following statement.

If two plane mirrors are inclined to one another at an acute angle x degrees, an incident ray parallel to one of them will retrace its original path after $(180/x)-1$ reflections at the two mirrors when this quantity is a whole number: and the incidence is normal at the $(90/x)$th reflection.

27. Explain with a diagram the principle of the sextant. How may the altitude of the sun above the horizon be found when the horizon is not visible?

28. Two objects 60 yards apart are observed by means of a sextant from a position on a line perpendicular to that joining the two objects. In order that the two objects may be seen in coincidence the rotating mirror of the sextant has to be rotated through 5°; what is the distance away of the objects?

29. How would you arrange two mirrors so as to be able to see the side of your head when looking straight forward? Give a diagram showing the path of the light.

30. A small source of light S lies midway between the observer's eye E (considered as a point) and a plane mirror, the line joining the eye and source being perpendicular to the mirror and intersecting it at C. The mirror is tilted about the point C through a small angle, say 2°. Find approximately the inclination to the line ESC at which the light now enters the eye, after reflection at the mirror. Draw a scale diagram showing the path of the light.

Chapter 4

Refraction and internal reflection at a plane surface — prisms and optical fibres

4.1 REFRACTION BY A PLANE PARALLEL PLATE

It follows from the law of refraction, that if a parallel pencil of light is incident normally on a plane surface separating two different media, the light will continue to travel in the new medium without change of direction. If a second surface is parallel to the first, the light reaches this normally, and again suffers no change in direction. Thus, light will be unchanged in direction on passing normally through a plane parallel plate of different refractive index from the surrounding medium, such, for example, as a parallel plate of glass in air.

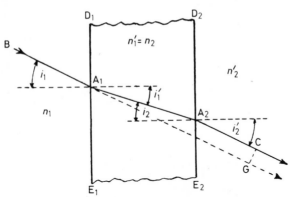

Figure 4.1 Refraction at plane parallel plate

When, however, the light is incident obliquely on such a plate, its direction will be changed both on entering and leaving and the directions can be found from the law of refraction. In *Figure 4.1* $D_1E_1D_2E_2$ is a plate having a refractive index $n'_1 = n_2$; the medium in front of the plate has a refractive index n_1, and that behind the plate a refractive index n'_2. Then, if BA_1 is the direction of the incident light, i_1 is the angle of incidence and i'_1 the angle of refraction at the first surface, i_2 and i'_2 are the angles of incidence and refraction at the second surface and

$$n'_1 \sin i'_1 = n_1 \sin i_1$$

$$n'_2 \sin i'_2 = n_2 \sin i_2$$

As the surfaces are parallel, $i_2 = i'_1$.

It follows that, when the media on the two sides of the plate are equal, that is, $n_1 = n'_2$, as in the case of a plate of glass in air, $i_1 = i'_2$, and the light emerges parallel to its original direction. It will however have suffered a lateral displacement GC, the extent of which will depend on the thickness and refractive index of the plate and on the angle of incidence at the first surface. As the light emerges from the plate parallel to its original direction, there will be no change in the apparent position of a *distant* object seen through the plate.

The apparent displacement of a *near* object or image, when it is viewed through a thick parallel plate of glass and the plate is rotated about an axis perpendicular to the line of sight, has been made use of in measuring the size of small objects—*parallel plate micrometer* (Clausen 1841)—and two such plates at an angle to one another were used by Helmholtz as the doubling device in his *ophthalmometer* for measuring the curvature of the cornea (1854).

4.2 REFRACTION BY A SERIES OF PARALLEL PLATES

When light traverses a number of media in succession, as in *Figure 4.2*,

$$n_1 \sin i_1 = n'_1 \sin i'_1$$

$$n_2 \sin i_2 = n'_2 \sin i'_2$$

$$n_3 \sin i_3 = n'_3 \sin i'_3$$

$$n_4 \sin i_4 = n'_4 \sin i'_4$$

and

$$n'_1 = n_2, n'_2 = n_3, \text{etc.}$$

If the surfaces are parallel, $i'_1 = i_2, i'_2 = i'_3$, etc., and substituting these values in the above expressions, we have

$$n_1 \sin i_1 = n_2 \sin i_2 = n_3 \sin i_3 = n_4 \sin i_4 = n'_4 \sin i'_4$$

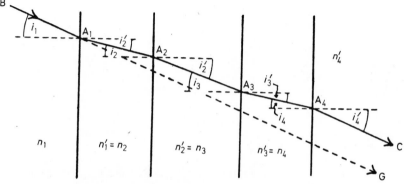

Figure 4.2 Refraction by a series of parallel plates

Thus it will be seen that the final direction of the light is the same as if it had undergone a single refraction at a surface separating the first and last media, and, if these are of equal refractive index, the final direction of the light is parallel to its original direction. There is, of course, a lateral displacement.

4.3 REVERSIBILITY OF OPTICAL PATH

An important general law in optics is the **principle of the reversibility of the optical path**. It may easily be shown that if the light in *Figure 4.2* is reversed in direction it will follow exactly the same path in a reversed direction. This is true of the light passing through almost any system, and problems may often conveniently be solved by considering the light as travelling in the opposite direction to which it is in the actual case.

4.4 CRITICAL ANGLE AND TOTAL REFLECTION

Light on passing from a denser to a rarer medium is refracted away from the normal to the surface, and the angle of refraction in the rarer medium increases more rapidly than the angle of incidence. For a certain angle of incidence the angle of refraction will be $90°$, that is, the refracted light will be just grazing the surface (*Figure 4.3*). This angle of incidence is termed the **critical angle**, i_c, between the two media. Light meeting the surface at a greater angle of incidence than the critical angle cannot therefore enter the second medium, and the whole of the light will then be reflected at the surface. This is known as **total reflection** or sometimes **total internal reflection (TIR)**. The direction of this totally reflected light will be in accordance with the ordinary law of reflection. It must be remembered that some light is always reflected at the bounding surface between two media, and the amount reflected gradually increases as the angle of incidence increases, so that there is not a complete change in condition at the critical angle (see *Figure 16.13*). The increase in the amount of the reflected light when the critical angle is exceeded is nevertheless considerable.

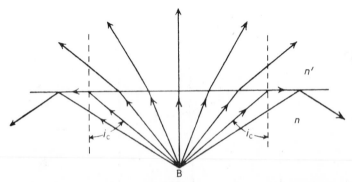

Figure 4.3 Refraction at a plane surface, $n' < n$. Critical angle and total reflection

In *Figure 4.3* light is travelling from a medium of refractive index n to one of refractive index n', where n is greater than n', and

$$n \sin i = n' \sin i'$$

When i is the critical angle, i' is $90°$ and

$$n \sin i_c = n' \sin 90° = n'.$$

Hence

$$\sin i_c = \frac{n'}{n} \tag{4.01}$$

The refractive indices and therefore the critical angle between two media will, of course, depend on the colour of the light. The critical angle will be smaller for blue light than for red, and in the case of incident white light, the red light may be passing into the second medium, while the other colours are totally reflected.

The following table gives the critical angle for yellow light of a few transparent substances when in contact with (*a*) air and (*b*) water.

	Refractive Index n_D	Critical Angle (in contact with Air)	Critical Angle (in contact with Water)
Water	1.3336	48° 34′	—
Crown Glass	1.5175	41° 14′	61° 32′
Light Flint Glass	1.5760	39° 23′	57° 46′
Dense Flint Glass	1.6214	38° 5′	55° 21′
Diamond	2.4173	24° 26′	33° 30′

The characteristic brilliancy of the diamond is due to its very high refractive index and therefore small critical angle. The stone is cut in such a way that a very large part of the light entering the stone is totally reflected and passed out through the *table*, the large plane surface of the stone.

Another interesting example of total reflection is seen in the case of the **mirage**, of which there are two distinct forms. One of these, which is much the commoner, is seen over land, and the other is occasionally seen at sea. The first form occurs in hot calm weather, especially over large expanses of sand, when inverted images of distant objects are seen reflected as though in a pool of water. This phenomenon is produced by the air close to the ground being hotter and therefore of lower refractive index than that a little higher up. Light travelling downwards from any point on an object is gradually refracted away from the normal, as it passes through air of decreasing refractive index, until near the ground it is totally reflected at a layer of low refractive index and enters the eye as though from an image below the surface of the ground. The reflection of the sky, trees, etc., gives the impression of the presence of a lake of water. This form of mirage may often be seen when motoring on an asphalt coated road, the surface of the road a hundred yards or so ahead appearing to be covered with water although actually dry.

The second type of mirage is seen at sea and most commonly in Arctic regions. In this case the air close to the sea is cold and therefore of comparatively high refractive index, becoming warmer and of lower refractive index higher up. Light travelling upwards from an object, such as a ship, is gradually refracted away from the normal until it is totally reflected downwards, and an eye receiving this reflected light sees an inverted image of the ship apparently floating in the sky.

Total reflection has many practical applications, both in replacing the reflection from a silvered surface over which, in many cases, it has definite advantages, and as the basis of methods of measuring refractive index. As these applications of total reflections in most cases involve the use of prisms they will be considered later.

4.5 GRAPHICAL CONSTRUCTION FOR REFRACTION

Problems involving the law of refraction are often conveniently solved by graphical construction. With the point of incidence A (*Figure 4.9*) as centre, draw two circles having radii in the ratio of n to n'. The incident ray BA cuts the circle corresponding to medium n at B; from B draw a line parallel to the normal at the point of incidence, meeting the other circle at G. GA produced is the direction of the refracted ray.

$$\frac{BM}{AB} = \sin i, \quad \frac{GL}{AG} = \sin i'$$

$$GL = BM \text{ (from construction)}$$

Therefore

$$\frac{\sin i}{\sin i'} = \frac{BM}{AB} \bigg/ \frac{GL}{AG} = \frac{AG}{AB} = \frac{n'}{n}$$

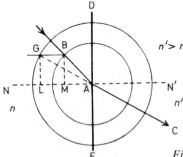

Figure 4.4 *Graphical construction for refraction*

The same construction may be used to determine the critical angle between two media. In this case, as the refracted light in the medium of lower refractive index n' is grazing the surface, the construction line BG must be tangential to the smaller circle and i is then the critical angle. Sometimes, when using the construction for light travelling from a denser to a rarer medium, it may be found that the construction line does not meet the smaller circle. This is a case of total reflection, and on producing the construction line to meet the larger circle again at G, GA produced is, of course, the direction of the reflected light. A more convenient construction for curved surfaces is described in Chapter 5.

4.6 IMAGE FORMED BY PLANE REFRACTING SURFACE

When all the rays of a pencil are diverging from or converging to one point, the pencil is said to be **homocentric**. In order that a perfect image may be formed, it is necessary that the homocentric pencils from points on the object shall still be homocentric after reflection or refraction. Except in reflection at a plane mirror, this ideal condition will never be realised and homocentric pencils in the object space are no longer homocentric in the image space. When however the pencils are very narrow, the rays in the image space may be considered, in many cases, to be diverging from or converging to a point with sufficient accuracy for a reasonably good image to be formed.

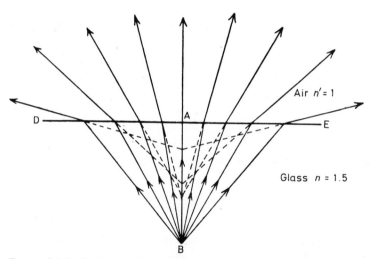

Figure 4.5 Light from object point in glass refracted on passage into air

In *Figure 4.5*, B represents a luminous point in a medium of refractive index n, which is separated from a second medium of index n', smaller than n, by the plane surface DE. Light diverges from B, and is refracted on passing into the second medium; the direction of the refracted rays may be calculated, or found by the graphical methods described above. It will be found that the refracted light is no longer travelling as though from a point. While all rays meeting the surface at equal angles travel, after refraction, as though coming from a point on the normal to the surface through the object point, the greater the obliquity of the rays, the nearer to the surface will be the point from which the refracted rays are apparently coming.

As the pupil of the eye is small, the narrow pencils entering an eye, looking normally through the surface, will be reasonably homocentric and a fairly good

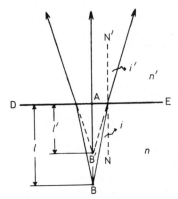

Figure 4.6 Narrow pencil refracted from dense to rarer medium

image will be seen. A narrow pencil from B (*Figure 4.6*) will, after refraction, appear to be coming from B', and as the light travelling along BA normally to the surface is unchanged in direction, B' must lie upon BA. As the angles are small,

$$\frac{\sin i'}{\sin i} = \frac{\tan i'}{\tan i} = \frac{AB}{AB'} = \frac{n}{n'}$$

Calling AB and AB', l and l' respectively, we have

$$\frac{l'}{n'} = \frac{l}{n} \tag{4.02}$$

and when the second medium is air, $n' = 1$,

$$l' = \frac{l}{n}$$

Thus objects in a denser medium, when viewed from a rarer medium, appear closer to the surface than they actually are, a well-known appearance in the case of objects in water or bubbles in a block of glass.

The principle explained above may be applied to the measurement of the refractive index of a transparent substance. A microscope, fitted with a scale and vernier for reading the position of the microscope body, is focused on an object in the usual way and the position of the microscope body is read on the scale. The glass or other substance to be measured, in the form of a block with plane parallel polished surfaces, is now placed on the stage in contact with the object, and the microscope is racked up until the object seen through the block is focused; the position is read on the scale. Finally the microscope is focused on a mark on the upper surface of the block, and a further reading taken. The difference between the first and third readings then gives the actual thickness of the block, l, and the difference between the second and third readings gives its apparent or *reduced* thickness, l'. Then the refractive index of the block is $n = l/l'$.

4.7 PRISMS

By a prism in optical work is meant a transparent substance bounded by plane polished surfaces inclined to one another. In the simplest form of prism, as used for refraction, only two surfaces, the **refracting surfaces**, need be considered, the light entering the prism at the first and leaving at the second. In the case however of the various reflecting prisms, there is a variety of forms having three or more polished surfaces at which the light is refracted or reflected.

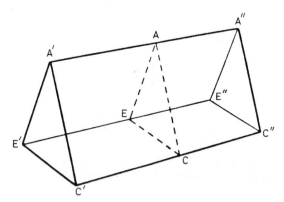

Figure 4.7 Refracting prism

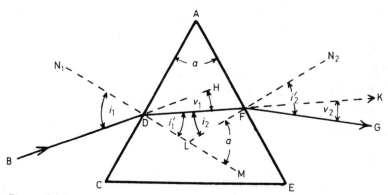

Figure 4.8 Path of light through a prism

Figure 4.7 shows a simple refracting prism with refracting surfaces A'C'C"A" and A'E'E"A". The line of intersection of these two surfaces is the **refracting edge** or **apex** of the prism. Any section through the prism perpendicular to the edge, such as ACE, is a **principal section**, and the angle CAE of such a section is the **refracting** or **apical angle**.

A ray BD (*Figure 4.8*) incident in a principal section of a prism ACE will remain in the same plane after refraction, and its path may be traced by considering the refraction at each surface in turn. We shall consider only the case of a prism of refractive index n surrounded by air. Then we have the following:

at first surface, $\sin i'_1 = \dfrac{\sin i_1}{n}$ (4.03)

at second surface, $\sin i'_2 = n \sin i_2$ (4.03a)

and from the geometry of the triangle ADF

$$i'_1 + i_2 = a$$ (4.04)

where a is the refracting angle of the prism.

The light is deviated through an angle v_1 at the first surface and v_2 at the second surface and from the figure

$$v_1 = i_1 - i'_1 \,; v_2 = i'_2 - i_2$$

then the total deviation

$$v = v_1 + v_2 = i_1 - i'_1 + i'_2 - i_2$$

$$= i_1 + i'_2 - a$$ (4.04a)

It will be seen that when the prism is of higher refractive index than the surrounding medium, as with a prism in air, the light is always deviated away from the refracting edge.

Example: Refracting angle of prism $60°$, $n = 1.53$, angle of incidence at first surface, $i_1 = 45°$.

$$\sin i'_1 = \frac{\sin i_1}{n} = \frac{0.7071}{1.53} = 0.4620$$

$$i'_1 = 27° \; 31'$$

$$i_2 = a - i'_1 = 32° \; 29'$$

$$\sin i'_2 = n \sin i_2 = 1.53 \times 0.5370 = 0.8216$$

$$i'_2 = 55° \; 15'$$

Therefore

$$v = i_1 + i'_2 - a = 40° \; 15'$$

4.8 TOTAL INTERNAL REFLECTION IN A PRISM

In certain cases, the angle of incidence at the first surface of a prism will be such that the light meets the second surface at an angle greater than the critical angle, and is therefore totally reflected. This reflected light will meet a third surface of the prism and, if this is a plane polished surface, the light will emerge or again be totally reflected. Its direction after leaving the third surface may easily be found. It should be noted that total reflection will occur when the value obtained from expression (4.03a) is greater than unity.

For any given prism there will be a limiting angle of incidence at the first surface at which light can pass through the prism, and for a given glass there will be a maximum limiting value of the angle between the faces of the prism through which any light can pass. These limiting values may be found as follows: As the greatest possible angle of emergence i'_2 is $90°$ (*Figure 4.9*) i_2 is the critical angle i_c between the glass and air, then

$$i'_1 = a - i_c$$

and $\sin i_1 = n \sin (a - i_c)$

this value of i_1 gives the smallest angle of incidence for a ray that will pass through the prism; any ray closer to the normal will be totally reflected at the second surface. When the angle a is smaller than i_c the limiting direction of the

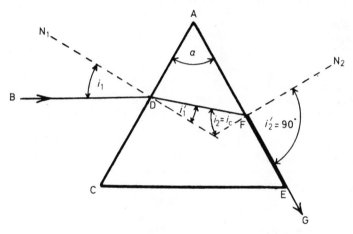

Figure 4.9 Refraction by a prism. Grazing emergence

incident ray lies on the other side of the normal to that shown in *Figure 4.9*. The greatest angle that a prism may have for light to pass through it will be such that the incident and emergent light is grazing the surface as in *Figure 4.10*.

Then

$$i_1 = i'_2 = 90°$$

and

$$i'_1 = i_2 = i_c$$

Therefore

$$a = 2i_c$$

Hence *no light can pass through a prism, the refracting angle of which is greater than twice the critical angle.* For a prism of crown glass in air, the refracting angle must not exceed about $82°$.

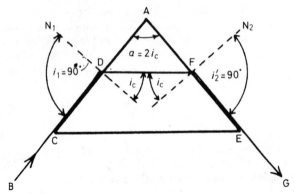

Figure 4.10 Maximum angle of refracting prism, $a = 2i_c$

4.9 MINIMUM DEVIATION

A special and important case of refraction by a prism occurs when the prism is in such a position that the light passes through it symmetrically, that is, when the angle of incidence at the first surface equals the angle of emergence at the second. It has been shown in section 4.7 that the deviation produced by a given prism depends on the incident angle of the light, and it may be shown experimentally or by tracing a number of rays through a given prism that, when the incident and emergent angles are equal, the deviation is a minimum. The prism is then said to be in its position of **minimum deviation**. An exact theoretical proof is beyond the scope of this book, but the student should calculate the deviation produced by a given prism for various angles of incidence and plot the deviations against the angles of incidence.

It will be found experimentally that there is only one position of a prism in which the deviation is a minimum, and this fact provides an experimental proof that minimum deviation only occurs when the angles of incidence and emergence are equal. For if these angles were not equal, from the principle of reversibility of path, there would be two positions of minimum deviation, the angle of incidence in the first position being the angle of emergence in the second and vice-versa.

For a more formal reasoning we divide 4.03 by 4.03a to obtain

$$\frac{\sin i_1}{\sin i'_2} = \frac{\sin i'_1}{\sin i_2} \qquad (4.05)$$

This equation is valid for all angles of incidence and deviation.

If we differentiate 4.04a we have

$$\frac{dv}{di_1} = 1 + \frac{di'_2}{di_1}$$

The minimum value of v occurs when $\dfrac{dv}{di_1} = 0$ and so

this is when $1 + \dfrac{di'_2}{di_1} = 0$ or $di_1 = -di'_2$ $\qquad (4.05a)$

If also we differentiate 4.03, 4.03a and 4.04 we obtain

$$\cos i_1 \; di_1 \;\; = \;\; n\cos i'_1 \; di'_1$$

$$n \cos i_2 \; di_2 = \; \cos i'_2 \; di'_2$$

$$di'_1 \qquad\quad = \; -di_2$$

Putting these together with 4.05a we have

$$\frac{\cos i_1}{\cos i'_2} \;=\; \frac{\cos i'_1}{\cos i_2} \tag{4.05b}$$

This equation is only valid for minimum deviation.

By multiplying 4.05b by 4.05 we obtain

$$\frac{\sin i_1 \; \cos i_1}{\sin i'_2 \; \cos i'_2} = \frac{\sin i'_1 \; \cos i'_1}{\sin i_2 \; \cos i_2}$$

which may be rendered as:

$$\frac{\sin 2i_1}{\sin 2i'_2} = \frac{\sin 2i'_1}{\sin 2i_2} \tag{4.05c}$$

Because sine is as a non-linear function of angle, 4.05 and 4.05c can only be both valid when $i_1 = i'_2$ and $i_1' = i_2$.

In the position of minimum deviation, as the path of the light is symmetrical, we have

$$i_1 = i'_2$$

$$i'_1 = i_2 = a/2$$

$$v_1 = v_2 = v/2 = i'_2 - i_2 = i'_2 - a/2$$

Therefore

$$i'_2 = \frac{a}{2} + \frac{v}{2} = \frac{a+v}{2}$$

$$\sin i'_2 = n \sin i_2$$

Hence

$$\sin \left(\frac{a+v}{2}\right) = n \sin \left(\frac{a}{2}\right) \tag{4.06}$$

or $\qquad n = \dfrac{\sin\left(\dfrac{a+v}{2}\right)}{\sin\left(\dfrac{a}{2}\right)} \tag{4.06a}$

Expression (4.06a) is the basis of a standard method of determining the refractive index of glass. The sample of glass to be measured is made into a prism with refracting angle of about 60°, and the refracting angle and the angle of minimum deviation for light of a given colour are carefully measured. These measurements are made on the instrument known as the **spectrometer**, which is described in Chapter 13.

Example: The same prism as in section 4.7, $a = 60°$, $n = 1.53$.
When the prism is in the position of minimum deviation $i'_1 = i_2$

$i'_1 + i_2 = a$ therefore $i'_1 = i_2 = 30°$

$\sin i_1 = n \sin i'_1 = 1.53 \times 0.5 = 0.765$

$i_1 = i'_2 = 49° 54'$

Hence minimum deviation v

$= i_1 + i'_2 - a = 39° 48'$

4.10 NORMAL INCIDENCE AND EMERGENCE

When light enters the prism normally, $i_1 = 0°$, no deviation takes place at the first surface, and therefore $i_2 = a$. The total deviation produced

$v = i'_2 - i_2 = i'_2 - a$

or

$i'_2 = a + v$

Then as

$\sin i'_2 = n \sin i_2$

$\sin (a + v) = n \sin a$ (4.07)

As the path of the light is reversible, this expression applies equally to the case of light emerging normally at the second surface.

4.11 OPHTHALMIC PRISMS

Prisms are frequently used in ophthalmic work for the measurement, relief and correction of muscular defects of the eyes. These prisms are of comparatively small refracting angle, rarely exceeding 15°, and are generally placed normally to either the incident or emergent light. As for small angles the sines are very nearly proportional to the angles, expression (4.05) may, for *thin* prisms, be written

$v = ni'_1 + ni_2 - a$

$= na - a$

$v = (n - 1)a$ (4.08)

Ophthalmic prisms* are usually numbered in terms of the deviation produced; this may be expressed in degrees of deviation or, more conveniently in this class of work, in units which give the lateral displacement of a beam at a given distance. The unit generally adopted is the angle whose tangent is 1/100, and the deviation is measured on a scale, the divisions of which are one hundredth of the

*For further particulars of ophthalmic prisms see Emsley and Swaine *Ophthalmic Lenses*.

distance between the scale and the prism. This angle is called the **prism dioptre** and is denoted by Δ. Thus 1Δ gives 1 cm deviation at 1 metre. Using e to denote lateral displacement at 1 metre, equation 4.08 becomes

$$e = 100(n-1)\tan a \qquad\qquad (4.08a)$$

or

$$e = 100(n-1)a \quad \text{if } a \text{ is in radians}$$

4.12 IMAGES SEEN THROUGH PRISMS

Different portions of the light diverging from a near object point B (*Figure 4.11*) will meet the prism at different angles and, as the deviation produced by a given prism is dependent upon the angle of incidence, the various rays of the pencil will be deviated different amounts. The light leaving the prism is therefore no longer proceeding as though from a point, and a perfect image of the object point will not be formed. As, however, the pencils entering an eye are narrow, there will be little difference in the deviations of the rays in these narrow pencils, and the light entering the eye will be coming approximately from a point. This is particularly the case when the prism is in the position of minimum deviation, as changes of incident direction near the minimum deviation position produce small changes in deviation. As the light is deviated away from the apex of the prism, the eye sees the image apparently displaced towards the apex as in *Figure*

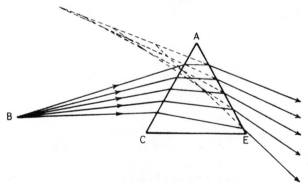

Figure 4.11 Light from a near object point refracted by prism

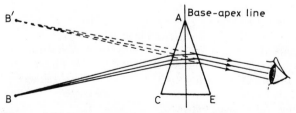

Figure 4.12 Image seen through prism

4.12 and along a line perpendicular to the refracting edge. This line, in the case of ophthalmic prisms, is known as the **base-apex line** of the prism.

On looking through a prism two other effects will usually be noticed. Unless the light from the object is monochromatic, that is, of one colour only, it will be dispersed by the prism into its component colours, the violet light being the most deviated and the red the least. Thus a small *white* object appears, when seen through the prism, as a spectrum with the violet end the most deviated. A larger white object appears fringed with violet or blue on one edge and with red on the other, the colours from intermediate points on the object overlapping and giving white.

The other effect is seen on looking at an object such as a line parallel to the edge of the prism. Light from only one point on the line will be passing to the eye through a principal section of the prism and, as the effective refracting angle in an oblique section of the prism is greater than the angle in a principal section, the light in any oblique direction is deviated more than that in a principal section. Therefore if a horizontal line is viewed through a prism held with its principal section vertical and refracting edge upwards, the image will be displaced upwards, and distorted to a curve as shown in *Figure 4.13*, the light from the ends of the line being deviated more than that from the centre.

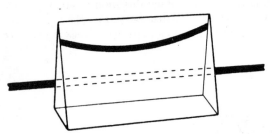

Figure 4.13 Appearance of image of horizontal bar seen through prism

4.13 REFLECTING PRISMS

The use of prisms for purposes of reflection often has definite advantages over the use of the silvered reflecting surface. The ordinary mirror silvered on the back surface cannot generally be used in optical instruments owing to the multiple reflections that take place between the two surfaces. Front surface mirrors are now usually produced by aluminizing a glass surface (section 14.17). When conditions allow the use of total reflection a better reflection is obtained than with a metal surface. The making of the reflecting surfaces in the form of a prism ensures that in those cases where two or more successive reflections are required the angle between the reflecting surfaces remains fixed.

Numerous forms of reflecting prisms are used in various optical instruments, and a few of the more important types will be considered here. In *Figure 4.14a* ACE is the section of a prism having angles of 45° at C and E. Light falling normally on the surface AC is incident on the surface CE at an angle of 45° and, this being greater than the critical angle between glass and air, the light is totally reflected and emerges normally from the surface AE, having been deviated

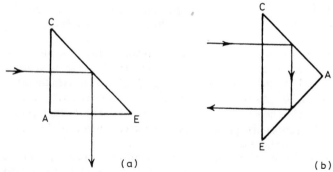

(a)   (b)

Figure 4.14 Totally reflecting prisms

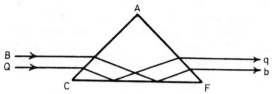

Figure 4.15 Dove prism rotator

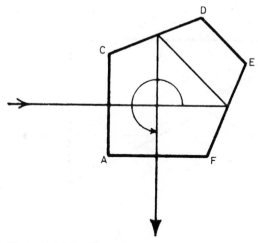

Figure 4.16 Pentagonal prism

through 90°. The same form of prism is also used to deviate a beam through 180° by means of two reflections (*Figure 4.14b*), the light being totally reflected at the faces AC and AE. Such prisms are used in the prism binocular, section 3.9.

A 90° reflecting prism may be employed as an erecting prism, as in *Figure 4.15*. A prism of this form usually known as a **Dove prism** is sometimes used for erecting the inverted image of an object but because only one reflection is used the image remains reverted. If this prism is rotated about an axis passing through the faces AC and AE and parallel to its hypotenuse face, the image will be rotated through twice the rotation of the prism. Prism systems which do this are called **rotators**.

When a constant deviation of 90° is required a prism of the form shown in *Figure 4.16* is often used. The reflecting surfaces CD and FE are at an angle of 45° to one another and the light is deviated through 270°, which gives the same result as a deviation of 90°. As the angle of incidence at the reflecting faces is less than the critical angle, these surfaces have to be silvered. This form of prism, which is known as the **pentagonal prism**, is used as the end prisms of the range-finder and as one form of optical square used in surveying.

By a suitable arrangement of the surfaces, the complete reversal of an image, both up and down and left for right can be effected with a single prism. One such form is shown in *Figure 4.17*. Reflection at the surfaces ABCD and CDEF gives a reversal of the image up and down, while reflection at the 'roof' surfaces HGLM and HKLN produces a left for right reversal. Such prisms, of which there are many forms, are known as **roof prisms**, the two surfaces at the top of the prism illustrated in *Figure 4.17* forming the roof. The roof surfaces must always be perpendicular to one another. If the roof surfaces are replaced by a single surface the system becomes a rotator with a light path similar to that of a dove prism.

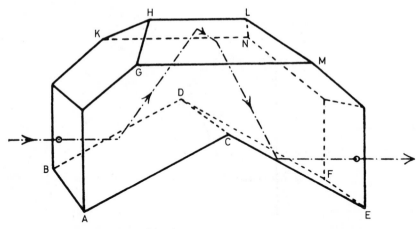

Figure 4.17 Roof prism, Abbe form

4.14 REFRACTOMETRY BY TOTAL INTERNAL REFLECTION (SEE ALSO SECTION 13.5)

The refractive index of a substance may be determined by measuring the critical angle between the substance and a glass of higher and known refractive index. This method is especially useful when only small samples of the substance to be measured are available, and with suitable instruments a high degree of accuracy can be obtained. Instruments for this purpose, of which there are a number of different forms, are known as **refractometers**.

In the **Pulfrich Refractometer** (*Figure 4.18a*) ACE is a prism of very dense flint glass, the refractive index n of which is accurately known. The specimen to be measured, S, is placed on the upper surface of the standard prism with a thin film of liquid, of a higher refractive index than the specimen, between the surfaces. (As was shown in section 4.2 this liquid has no effect on the final direction of the light.) A convergent beam of monochromatic light is incident horizontally on the end face of the test specimen close to the contact surface, and some of this light will enter the standard prism and emerge from the surface CE. The only light entering the standard prism will be that contained within the critical angle i_c between the substance being measured and the standard glass and if this angle can be measured the refractive index n' of the test specimen can be found, as $n' = n \sin i_c$.

The light emerging from the surface CE is focused by a telescope attached to a vertical divided circle, which turns about its centre at which the prism is mounted, the position of the telescope being read by a fixed vernier. The setting consists in bringing the cross-line or pointer of the telescope on to the edge of the 'shadow', that is, the limiting direction of the light leaving the prism, and the refractive index corresponding to the particular angular reading of the telescope is found from prepared tables. The refractive index of a liquid may be determined by placing the liquid in a short length of glass tube cemented to the upper surface of the standard prism.

The **Abbe Refractometer** (*Figure 4.18b*) is based on the same principle. As used for the determination of the refractive index of liquids, two equal prisms of dense flint glass, having angles of $30°$, $60°$ and $90°$, are placed with their hypotenuse faces together and a thin film of the liquid to be measured between

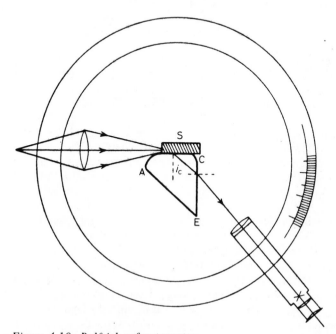

Figure 4.18a Pulfrich refractometer

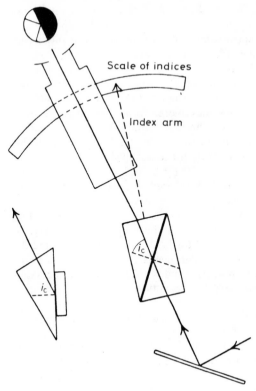

Scale of indices

Index arm

Figure 4.18b Abbe refractometer

them. Light is reflected from a mirror on to the short face of the lower prism, and some of this light will emerge from the short face of the upper prism. The limiting direction of the light that can emerge will be that of the light which is at grazing incidence at the surface between the liquid and the upper prism, that is, the light that enters the upper prism at the critical angle between the glass and the liquid.

The light emerging from the prisms is received by a telescope, and the prisms are turned about a horizontal axis to bring the limiting direction—the 'shadow' edge—on to the cross-line of the telescope. Attached to the prisms is an arm carrying an index that reads on a scale divided to give the refractive index of the substance tested. As the refractive index and therefore the critical angle varies with the colour of the light, the shadow edge would be coloured and badly defined when white light is used. This colour effect is neutralised by an equal but opposite dispersion given by two Amici direct-vision prisms (see Chapter 13), which rotate in opposite directions about the telescope axis, the achromatised 'shadow' edge corresponding to that for the yellow light.

The Abbe refractometer may be used to determine the refractive index of a solid by removing the lower prism and placing the substance to be measured on the hypotenuse face of the upper prism with a thin film of a high refractive index liquid such as monobromonaphthaline between the surfaces. The prism is then turned as before until the edge of the 'shadow' comes on the cross-line in the telescope, when the scale reading gives the refractive index.

4.15 OPTICAL FIBRES

A further use of Total Internal Reflection is in optical fibres. If a rod of glass is illuminated at one end it is found that most of the light is trapped inside the rod and emerges at the other end. Even if the diameter of the rod is reduced to that of a fine fibre only a few microns across the trapped light is still transmitted through the glass. For maximum transmission the glass must be very transparent and its surface must be very smooth. *Figure 4.19a* shows how the light must not

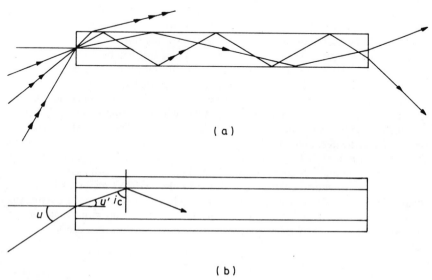

(a)

(b)

Figure 4.19 Transmission of light by fibre. (a) Light transmission by a single fibre (b) Light transmission by a clad fibre

be greater than a certain angle to the axis of the fibre. Inside the glass this is equal to i_c the critical angle and therefore before entry it is given by Snell's law at the input face, so that

$$\sin u = n \sin (90 - i_c) = n \cos i_c = (n^2 - 1)^{1/2} \qquad (4.09)$$

Although fibres can be used to transmit light (nowadays as information over many kilometers using special glass) the main reason for inclusion in this book is their use in transmitting images.

Images are transmitted by using many fibres, each one carrying a particular point of the image. Problems arise when neighbouring fibres touch. Even if they are not in complete optical contact light is able to transfer from one fibre to another, and therefore degrade the image. In order to avoid this fibres are made with an internal core glass of high refractive index surrounded by a cladding glass of lower refractive index. The acceptance angle of such **clad fibres** is given by

$$\sin u = \left(n_f^2 - n_c^2\right)^{1/2} \qquad (4.10)$$

where n_f and n_c are the refractive indices of the fibre and cladding respectively and the reasoning follows the same lines as (4.09) but applied to the fibre to cladding interface, *Figure 4.19b*. Even if the input face is not square the overall acceptance angle remains the same.

For fibres of circular cross section the acceptance angle is a cone of half-angle *u* and this hardly changes with other shapes. No significant change in this angle occurs when the fibre follows a curved path provided the minimum bend radius remains about 20 times the diameter of the fibre.

The manufacture of clad fibres proceeds by melting the two glass types in separate crucibles and then pouring the core glass followed by the cladding glass into a drawing crucible so that the latter floats on top of the former. If, by dipping through the upper glass, a ribbon of core glass is pulled out of the crucible a covering of the cladding glass adheres to it. This ribbon can be given a circular cross-section about 2 or 3 mm diameter. Subsequently this so-called **clad rod** can be heated at one end and drawn into fine fibre.

As an alternative to the two-glass method a cladding can be formed in a single homogeneous glass rod drawn in a similar way but with the core glass only in the drawing crucible. The drawn rod is heat treated so that its components separate within the rod and one of these can be etched away by chemical action. The porous rod resulting may have its spaces filled with other materials so that there is a difference between the mix at the centre of the rod to that at the surface. When this is heated and drawn it is found to have a different refractive index between centre and surface as in the usual clad-fibre case. The index change may be abrupt or gradual but containment of the light occurs in both cases.

These gradual index change or **Gradient Index (GRIN)** materials have been suggested for use on a larger scale as lenses which would not need curved surfaces! To date, however, the index changes obtained do no more than modify the power of the lens by a few percent although this could prove useful for aberration control.

4.16 IMAGE CONDUITS AND FACE PLATES

The transmission of optical images by fibres needs an individual fibre for each point on the image rather like the dots of a newspaper picture. The arrangement of the fibres in the cross-section of the fibres bundle must be the same at one end as at the other if the various parts of the image are to remain in the same relationship to each other. This important requirement leads to the name **coherent fibre bundle** or **image conduit** for these systems. It may be thought that, for a given image size, a better transmission of information is achieved by having as many fibres as possible which means the smallest possible diameter for each fibre. However, more and smaller fibres not only entails greater cost. It is also found that the action of imperfections such as inclusions in the core or irregularities in the core-cladding surface have a greater effect on the finer fibres. Thus for a given length the transmission of the smaller fibres is less, the chance of breakage of individual fibres (giving black dots on the image) is higher.

A common use of flexible image conduits is for inspection of otherwise inaccessible regions such as internal organs in the body or internal parts of complex mechanisms. The latter are called **borescopes** while the medical systems tend to be described by the part of the body on which it is used, e.g. **gastroscopes**. The word **fiberscope** is a useful general term.

For lengths up to a metre the fibre diameter is commonly about 10 microns. The manufacture of flexible systems is usually done by winding a single fibre many times round a drum, the circumference of which is equal to the required length of fibre. At a particular part of the circumference the fibres are packed together and fused into a solid unit in which each glass core retains its cladding glass round it. When this fused area is cut across the two surfaces exposed have identical patterns of fibres and the bundle may be unwound from the drum without changing this. In a suitable plastic protective tube the system has the flexibility of the unfused length and the coherence necessary at the two ends.

In some applications, flexibility is not required, and rigid image conduits may be made by laying thick glass fibres carefully together, fusing them and drawing the bundle in the same way as a single fibre. These bundles may be fused together and the process repeated. Such rigid systems generally have a fibre length much shorter than flexible fiberscopes and so their individual fibre diameter may be as low as 3 microns, when the length of the system is only one or two centimetres. As the image size may be 50 or 100 mm or more the term **fibre optic face plates** has come into use for these.

This is particularly applied to electronic devices where an image, carried by electron beam pattern as in a cathode ray tube or image intensifier, is formed on a phosphor screen. Considerable advantage is gained when this phosphor screen is allowed to be concave to the electron beam, while being flat to the following optical system. This can be done by laying down the phosphor on a curved face plate. Such a curved face plate is shown in Plate 5 but without any phosphor, allowing the drawing on the paper to reappear on the upper curved surface of the face plate. The same effect would occur if the incident image were projected onto the first face, although in this case the image on the second face is no longer an aerial image but must be regarded as a real self-luminous object.

CHAPTER 4   *EXERCISES*

1. Describe and prove a graphical construction for refraction. Use the construction to find the direction of the refracted ray in glass ($n = 1.53$) when the incident ray in air meets the surface at an angle of $45°$.

2. Explain what is meant by critical angle and internal total reflection, and obtain the relation between the critical angle and the refractive indices of two media. Find by a graphical construction the critical angle between glass $n = 1.67$ and water $n = 1.33$.

3. A tank having a plane glass bottom 3 cm thick is filled with water to a depth of 5 cm. Trace the path of a narrow parallel beam of light, showing the angles at each surface, when the beam is (*a*) normal (*b*) at an angle of incidence of $45°$ at the upper surface of the water. (Refractive index of the glass 1.62).

4. A narrow beam of light is incident on one surface of a plane parallel plate of glass ($n = 1.62$) 5 cm thick at an angle of $45°$ to the normal. Trace the path of the light through the plate and determine the lateral displacement of the beam on emergence.
What will be the direction of the emergent light if the second surface of the glass plate be immersed in oil of refractive index 1.49?

5. Light is diverging from a small object inside a block of glass towards a polished plane surface of the glass. Describe and show by means of a diagram what will happen to the light on reaching the surface.

6. Describe fully a method of determining the refractive index of a liquid contained in a tank with a plane glass bottom. Explain the principles upon which the method depends.

7. Explain how a microscope may be used to determine the refractive index of a sample of glass in the form of a parallel-sided slab. In a certain case the slab was 1¾ inches thick and made of glass of refractive index 1.6. Give figures for the settings of the microscope.

8. An object held against a plate glass mirror of refractive index 1.54 appears to be 0.375 inch from its image. What is the thickness of the mirror?

9. Explain the conditions under which total reflection occurs at the boundary between two transparent media. How would you make use of this phenomenon to measure refractive index?

10. The apparent thickness of a slab of glass of actual thickness t and refractive index

n is $\dfrac{t}{n}$

Prove this statement (*a*) by considering the direction of a ray close to the normal and (*b*) by considering the change in curvature of the wavefront. Describe a practical application of this fact.

11. Find the critical angle for light passing from glass of refractive index 1.523 to (*a*) air, (*b*) water, (*c*) oil of refractive index 1.47.

12. A 4 inch cube of glass of refractive index 1.66 has a small object at its centre; show that if a disc of opaque paper exceeding a certain diameter be pasted on each face, it will not be possible to see the object from anywhere outside the cube. Find the minimum diameter of disc required.

13. Rays of light are emitted upwards in all directions from a luminous point at the bottom of a trough containing liquid to a depth of 3 inches; the refractive index of the liquid is 1.25. Find the radius of the circle lying in the surface of the liquid with its centre vertically above the luminous point, outside which all rays will be totally reflected.

14. Explain with a diagram why an oar partly immersed in water appears to be bent.

15. An object placed 1 foot from a tank of water 1 foot wide and having thin plane glass sides is viewed through the tank, the line of sight being normal to the sides. Where will the object appear to be?

16. Show that the film of liquid placed between the substance to be measured and the prism of a refractometer has no effect on the reading of the instrument.

17. Given that the refractive indices of a liquid for red, yellow and blue light are 1.512, 1.532 and 1.543 respectively, find what will happen to a beam of white light travelling in the liquid and meeting the surface of the liquid separating it from air at the critical angle for yellow light.

18. A plate glass mirror 2 cm thick is fixed in a vertical position. An object is placed 2 cm in front of the glass surface and 4 cm above the top edge of the mirror. Show by a full-size diagram how multiple images of the object are formed by the mirror. (Take the refractive index of the glass to be $^3/_2$.)

19. The refractive index of the air gradually increases towards the earth's surface. Show by means of a diagram how this will affect the apparent time of rising and setting of the sun or moon.

20. A rectangular trough has vertical sides of glass 1 inch thick and of refractive index 1.54. The bottom of the trough is covered by a layer of mercury above which is a layer of water 2 inches deep and of refractive index 1.33. A narrow beam of light strikes the surface of the water at an inclination of 45° and at a point 1½ inches from the one side of the trough, the beam being in a plane perpendicular to that side and turned towards it. Trace the course of the light after striking the surface of the water.

21. A strip of white paper is viewed through a glass prism. Explain carefully the character of the image seen, describing (*a*) its position and (*b*) the colours observed. Give a diagram.

22. A prism of 30° refracting angle is made of glass having a refractive index of 1.6. Find the angles of emergence and deviation for each of the following cases:

(*a*) Angle of incidence at first face = 24° 28′

(b) Angle of incidence at first face = 53° 8′
(c) Incident light normal to first face.

23. A parallel beam of yellow light is incident at an angle of 40° to the normal to the first surface of a 60° prism of refractive index 1.53. Trace the path of the light through the prism and find the deviation produced on the beam. What will be the minimum deviation produced by this prism?

24. In measuring the refractive index of a prism for yellow light on a spectrometer, the refracting angle is found to be 60° 14′ and the angle of minimum deviation 42° 26′. Find the refractive index.

25. A prism has a refracting angle of 60° and a refractive index of 1.5. Plot a graph showing the relation of the total deviation to the angle of incidence of the light on the first surface and find the angle of incidence for which the deviation is a minimum. Show that this angle of incidence is equal to the angle of emergence.

26. A parallel pencil of light is incident upon one face of a 30° prism at an angle of 54° with the normal and after refraction at this face the light reaches the second face, which is silvered, in such a way that it is reflected along its original path. Find the refractive index of the glass.

27. Light from a distant object is incident normally at the front surface of a prism of 3° angle, $n = 1.53$. Part of the light is reflected at the first surface and part at the second; find the angle between the two reflected beams leaving the prism.

28. A prism of refractive index 1.65 has one angle of 90° and another of 75° and a ray of light strikes the face between these two angles at an inclination of 65° to the normal on the side of the 90° angle. Trace its course through the prism and find the total deviation produced.

29. A parallel beam of light is incident at 30° to the normal to one face of an equilateral prism ($n = 1.53$), the three faces of which are polished. Trace the path of the light through the prism and find the total deviation produced. For what angle of incidence will the deviation be a minimum?

30. A narrow parallel beam of white light is incident at 40° to the normal to one face of a prism of 60° refracting angle made of glass having refractive indices 1.605 and 1.647 for red and blue light respectively. What will be the length of the spectrum on a screen 2 m from the prism?

31. A narrow parallel beam of white light is incident at 50° on a prism of 30° refracting angle, $n = 1.53$. Some of the light is reflected at the first surface, and the remainder passes into the prism and is reflected at the second surface, which is silvered. Find the angle between the two beams on emergence from the prism and explain carefully any difference in their nature.

32. Show that if a prism of refractive index n is to give a minimum deviation v, its refracting angle a is given by

$$\tan \frac{a}{2} = \frac{\sin \dfrac{v}{2}}{n - \cos \dfrac{v}{2}}$$

Hence find the refracting angle of a prism of refractive index 1.65 to give a minimum deviation of 60°.

33. A candle is mounted 1 foot away from and on the normal to one face of a thin prism of 1° refracting angle and refractive index 1.53. Find the position of the successive reflected images from its front and back surfaces.

34. A plane mirror 2 feet 6 inches wide has its front edges bevelled, and an observer standing at 8 feet from the mirror sees three images of his eye, one in the centre of the mirror and the two others in the bevelled portions. If the refractive index of the glass of the mirror is 1.53, what is the angle at which the edges are bevelled?

35. The angle of a prism of nominally 90° is tested by directing a parallel beam of light on to the hypotenuse face of the prism and measuring the angle between the two emergent beams which have travelled through the prism in opposite directions. What will be the error in the angle of the prism if the angle between the emergent beams be 14 minutes? (n of prism 1.516).

36. Two isosceles right angle prisms ($n = 1.53$) are placed with their hypotenuse faces together. Find the limiting angle of incidence at the first face of one prism in order that light may pass through the two prisms when there is (a) air, (b) water between them.

37. Explain briefly the principle and optical construction of the Abbe refractometer. Give a diagram.

The right angle prisms of an Abbe refractometer are made of dense flint glass of $n = 1.75$, and the angle between the hypotenuse and the end face is 60°. Find the angle of emergence of the limiting ray from the last face of the upper prism block when testing substances with refractive indices (a) 1.35 (b) 1.50 (c) 1.65.

38. A thin prism is tested by observing through it a scale 80 cm away, the scale divisions being 1 cm apart. The deviation is found to be 1.75 scale divisions. Express the deviation in prism dioptres.

A ray of light falls normally on one face of this prism, is reflected by the second face, and emerges from the first face making an acute angle of 7.5 degrees with its original direction. Find the apical angle and refractive index of the prism.

39. A convergent pencil of light is intercepted before reaching its focus by a right-angled isosceles prism. The angle between the edges of the pencil is 18° and the central ray meets one of the short faces of the prism normally. What must be the minimum refractive index of the prism in order that the whole of the pencil may be totally reflected at the hypotenuse face?

40. A pencil of parallel light is incident on the face PQ of a right-angled isosceles prism (right angle at P) in a direction parallel to the hypotenuse face QR. The width of the pencil is equal to one quarter of the length of face QR.

The prism is such as *just* to permit the whole width of the pencil to be reflected at face QR and then to emerge into air from face PR. Find the refractive index of the prism.

41. A ray of light is incident at grazing incidence (angle of incidence 90°) on the face AB of a right angle prism ABC (right angle at B) of refractive index 1.72; the ray travels in the direction from A to B. What happens to this ray at the face BC?

When a block of glass of refractive index n is placed in optical contact with the face AB, the same incident ray then emerges from face BC into air at an angle of 45° to the normal. Find the value of n.

Describe very briefly, with a sketch, an instrument in which this principle is employed.

42. A glass prism of refracting angle 72° and refractive index 1.66 is immersed in a liquid of refractive index 1.33. What is the angle of minimum deviation for a ray of light passing through the prism?

Chapter 5
Curvature: refraction at
a curved surface

5.1 CURVATURE

Most of the curved surfaces used in optical work are of spherical form, although other forms of curved surfaces are sometimes used. Thus in ophthalmic work, cylindrical and toroidal curves are quite common, while the reflectors of head-lamps and searchlights, and the mirror of the reflecting telescope, are of para-boloidal form. For the present, however, the consideration of curved surfaces will be confined to spherical surfaces.

The **curvature** of a surface may be defined as *the angle through which the surface turns in a unit length of arc.*

Let DAE (*Figure 5.1*) be a portion of a circular track having its centre at C. An object travelling round the track in a clockwise direction will be travelling, when at D, in the direction DF, and when it reaches E it will be travelling in the

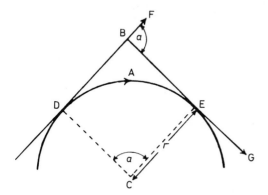

Figure 5.1 Curvature of a surface

direction EG, having turned through the angle FBE in a length of arc DE. As DF and EG are tangents to the curve, the angle FBE = the angle DCE = a; then from the definitions

$$\text{Curvature} = \frac{a}{\text{arc DE}}; \quad a \text{ in radians} = \frac{\text{arc DE}}{r}$$

Hence curvature $= \dfrac{1}{r} = R.$

The curvature of a surface may therefore be expressed as the *reciprocal of its radius of curvature*.

The common unit of curvature is reciprocal metres (m^{-1}) although reciprocal millimetres (mm^{-1}) or reciprocal inches (in^{-1}) can be used.

Another unit of curvature is the **dioptre (D)** which is again the curvature of a surface having a radius of one metre.

$$\text{Curvature in dioptres} = \frac{1}{r \text{ in metres}} = \frac{100}{r \text{ in cm}} = \frac{39.37}{r \text{ in inches}}$$

This second unit is used principally for the curvature of wavefronts which is described in section 5.3 and hence related to vergence in section 5.6. It is good practice to restrict the use of dioptres to vergences and to the extent to which optical elements affect the vergence of light, the focal power (section 5.11). For actual optical surfaces the use of reciprocal metres is recommended.

5.2 MEASUREMENT OF CURVATURE

As the curves to be considered are usually very small portions of the complete sphere, a direct measurement of curvature will not be possible. The curvature of a spherical surface can, however, be determined from two measurements that can be readily made. In *Figure 5.2* DAE represents a small portion of a spherical curve, having its centre of curvature at C. AN is known as the **sagitta** or **sag**, s, of the curve for a given chord 2y, being so called from its resemblance to an arrow on a bow.

In the triangle ECN

$$r^2 = y^2 + (r-s)^2$$

$$r = \frac{y^2}{2s} + \frac{s}{2} \tag{5.01}$$

When s is small compared with r, as it frequently is,

$$r = \frac{y^2}{2s} \text{ (approx.)}$$

and, Curvature, $R = \dfrac{2s}{y^2}$ (approx.)

Curvature is in m^{-1} when y and s are in metres.

For small portions of curves the curvature is approximately proportional to the sag, s.

The mechanical determination of the curvature of a surface by means of various forms of **spherometers** consists in measuring the sag s for a given chord 2y of the curve, when the radius or curvature may be found from expression (5.01). One of the simplest forms of spherometer is the **optician's lens measure**

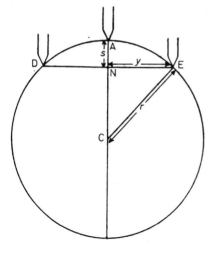

Figure 5.2 Measurement of curvature

Front

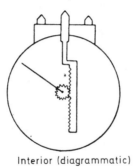

Interior (diagrammatic)

Figure 5.3 Optician's lens measure or 'Lens Clock'

illustrated in *Figure 5.3*. This instrument has two fixed points D and E, midway between which is the moving point A, the position of which is read by means of a series of levers actuating a pointer over a circular scale. The movement of the centre point from its zero position, when the three points are resting on a plane surface, to its position when the points rest on the curved surface being measured, gives the distance *s*, the distance between the two fixed points being 2*y*. The scale of the lens measure is divided to give the focal power (*see* section 5.1) of the surface for the refractive index 1.523 of the glass of which spectacle lenses are usually made. Thus the device measures in dioptres and not reciprocal metres.

As the three points of the lens measure are in a straight line the instrument may be used to measure the curvature of any meridian of cylindrical and toroidal surfaces. This arrangement, however, introduces errors in that it is impossible to ensure that the instrument is standing perpendicularly on the surface. To overcome this source of error the more accurate spherometers usually have the moving point at the centre of a ring, or equidistant from three fixed points that lie on a circle.

The classic form of simple spherometer is shown in *Figure 5.4*. It consists of a metal stand with three pointed legs arranged at the corners of an equilateral

Figure 5.4 Spherometer (old form)

triangle, the fourth leg is formed by a micrometer screw having a pointed end and passing through the stand equidistant from the three fixed legs. A divided head is attached to the screw to read fractions of a complete turn; thus if the pitch of the screw is $^{1}/_{2}$ mm and the head is divided into one hundred equal divisions, a rotation of one division represents a movement, up or down, of $^{1}/_{200}$ mm. Complete rotations of the screw are read on a short length of scale divided in $^{1}/_{2}$ and 1 mm divisions attached to the stand against the divided head. To measure a surface, the instrument is first placed on a plane surface, and the screw adjusted until all four points touch the surface; this position is best found by holding one of the fixed legs lightly between finger and thumb, and noting the position of the screw where the slightest further movement of the screw downwards allows the instrument to rotate freely about the centre point. The mean of a number of settings gives the reading on the plane or the zero setting of the instrument. The instrument is then placed on the surface to be measured, and reset in the same manner until the four points touch the curved surface. The difference between the readings on the plane and curved surfaces gives the value s. The value y, the distance between either of the fixed points and the centre point is best found by measuring the distance between each of the three fixed points, and dividing the mean by $\sqrt{3}$.

A more accurate instrument is the **Abbe spherometer**. In this the surface to be measured rests on an accurately turned metal ring, a number of such rings being provided to accommodate surfaces of various diameters. Passing through the centre of the ring is a plunger, which is kept in contact with the surface by means of a counterbalance weight. A finely divided silver scale, attached to the plunger and observed with a microscope fitted with a micrometer eyepiece, enables one to read movements of the plunger to $^1/_{1000}$ mm. The displacement of the plunger from its position when in contact with a plane to that when in contact with the curve gives the value s, and the diameter of the ring is $2y$. As the edge of the ring has a certain thickness, its inner diameter must be taken when measuring convex curves and its outer diameter when measuring concave curves.

The spherometer with sharply pointed feet or a sharp-edged ring is very liable to error due to the points or edge wearing and thus altering the value of y; also there is always the danger of damage to a polished surface. To overcome these difficulties, Aldis has suggested the modification of fitting to the points or at three equally spaced positions around the ring, small polished steel spheres. With the **Aldis spherometer** readings are taken in the usual way, and a correction for the small spheres is applied as in the following.

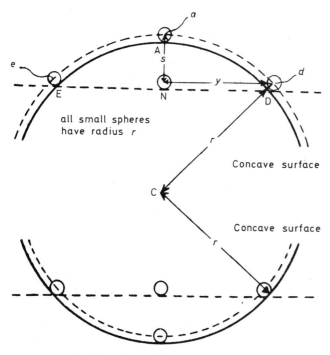

Figure 5.5 Principle of Aldis spherometer

In *Figure 5.5* the positions of two of the fixed feet and the movable centre foot are shown as resting on a plane surface END and on a curved surface EAD. If y is taken as the distance between the centres of the outer and the centre spheres and s, as before is the displacement of the centre leg, then, as may be seen from the figure, expression (5.01) gives the radius of the curve *ead* passing through the centres of the small spheres. The radius of the surface is therefore found by subtracting the radius r' of the small spheres from the calculated value in the case of a convex curve and adding r' to the calculated value for a concave curve.

5.3 CURVATURE OF WAVEFRONTS

The effects of mirrors, prisms and lenses on the light may be studied by considering the changes produced in the curvature of the wavefronts (section 1.2). For example, *Figure 5.6* shows the action of a convex lens in changing the light diverging from an object point B into light converging to a focus B'. The effect of the lens has been to change the curvature of the wavefront, and if the amount of this change for a given lens and the curvature of the incident wavefront at the lens is known, the new curvature and therefore the position of the focus of the emergent light is easily found.

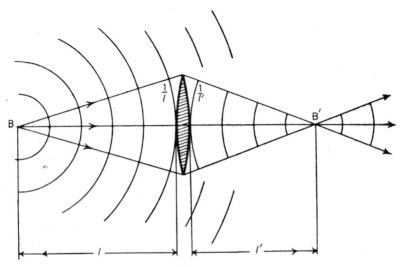

Figure 5.6 Curvature of wavefronts

The curvature of a wavefront is expressed in the same way as that of a surface. Thus if l is the distance of the object point B from the lens, then $1/l$ will be the curvature of the incident wavefront at the lens, and similarly if l' is the distance of the focus B', $1/l'$ is the curvature of the emergent wavefront at the lens.

In the divergent pencil the curvatures of the wavefronts are continuously decreasing as they recede from the object point and in the convergent pencil continuously increasing as they approach the focus. The curvature of the wavefront may be used to express the amount of divergence or convergence of the light at any point. In *Figure 5.6* the light has a divergence $1/l$ on reaching the lens and a convergence $1/l'$ immediately on emergence.

5.4 SIGN CONVENTION

To denote the direction in which various distances, lengths and angles are to be measured, positive and negative signs are given to the values. Thus in *Figure 5.6* the distances l and l' are each measured from the lens, and as these are distances in opposite directions the values will be given opposite signs. The sign convention to be adopted throughout the work is as follows (*Figure 5.7*):

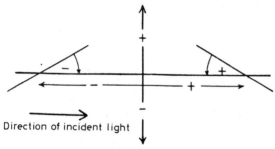

Figure 5.7 Sign convention

(a) Distances along the lens or mirror axis.
All distances are measured *from* the lens or mirror. Those in the same direction as the *incident* light is travelling are positive, and in the opposite direction negative.

In diagrams the incident light will oridinarily be represented as travelling from left to right.

(b) Distances perpendicular to the axis.
All distances are measured from the axis. Those above are positive and below, negative.

(c) Angles.
Anti-clockwise rotation is positive.

Slope of a ray.   This is defined as the acute angle described by a ruler in being moved from the direction of the ray to the direction of the axis.

Angles of incidence, reflection and refraction.   Angles are to be measured *from* the normal *to* the ray.

In *Figure 5.14* all the distances and angles are positive; it may be referred to as the 'all positive diagram.'

Applying the sign convention to *Figure 5.6*, it will be seen that l and therefore $1/l$ are negative, and l' and $1/l'$ positive. The radius and curvature of the first surface of the lens are positive and of the second surface, negative.

An arrow head shown on a line fixing the limits of a given distance is to indicate the direction in which that distance is measured.

The question of signs is of fundamental importance in optical work, and the student should make himself thoroughly conversant with the above convention. There is a strong temptation in such cases as above to regard l as 'being negative' and to insert $-l$ into the formulae of this chapter and those following. *This is wrong and leads to incorrect answers.* In *Figure 5.6*, the distance l equals a negative value, say -50 mm. The minus sign is part of the value and not part of l. Thus -50 is inserted into the relevant formula without changing the signs in the formula. If $l = -50$ mm and $f' = +30$ mm then the formula

$$\frac{1}{l'} = \frac{1}{l} + \frac{1}{f'} \tag{6.03a}$$

for instance, is used by writing:

$$\frac{1}{l'} = \frac{1}{(-50)} + \frac{1}{(+30)} \quad \text{and calculating from there on.}$$

5.5 REFRACTION AT A PLANE SURFACE (CHANGE IN CURVATURE OF THE WAVEFRONT)

In section 4.6 it was shown that when a narrow pencil of light, diverging from an object point, meets a plane surface bounding two different media it will, after refraction, be diverging from an image point in some other position depending on the refractive indices of the media. The same result may be obtained by considering the change that takes place in the curvature of the wavefront on passing from one medium to another.

In *Figure 5.8* DAE is a plane surface separating two media having refractive indices n and n' respectively, n being the greater. Light is diverging from the object point B, and DME represents the form of a wavefront if the light had continued to travel in the first medium. As the light on passing into the second medium has its velocity increased, the light, which would have been at M, now reaches M' in the same time and light, that would have reached any other point on the front DME, is advanced a corresponding amount. The curve of the incident wavefront DME is therefore changed on refraction to a curve DM'E.

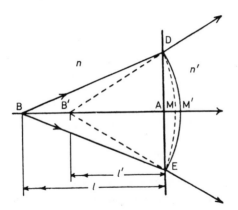

Figure 5.8 Refraction at a plane surface, $n > n'$

As the light travels a distance AM' in the medium n' in the same time as it would have travelled AM in the medium n, from the definition of refractive index (section 2.5)

$$n' \, AM' = n \, AM$$

As the sags AM and AM' are small when the pencil DBE is a narrow one, they are proportional to the curvatures $1/l$ and $1/l'$ of the wavefronts DME and DM'E.

Hence
$$\frac{n'}{l'} = \frac{n}{l}$$

or
$$l' = \frac{n'}{n} l$$

5.6 REDUCED DISTANCE AND VERGENCE

When the second medium is air, $n' = 1$ and $l' = l/n$. This is called the *equivalent air distance* or the **reduced distance** corresponding to the real or *absolute* distance l existing in the medium of index n. Similarly $1/l' = n/l$ is the *equivalent air curvature or vergence* or the **reduced vergence**. The idea of reduced distances and vergences is very useful in calculations on lenses and lens systems.

A reduced vergence will be represented by the capital letter corresponding to the small letter used for the corresponding absolute distance, thus

$$L = \frac{n}{l} \qquad L' = \frac{n'}{l'}$$

It should be specially noticed that when an image is virtual, its distance l' is measured back from the surface, but the emergent light is actually travelling in the second medium and n' is the refractive index involved in the reduced distance. The unit of vergence is the dioptre (see section 5.1).

5.7 REFRACTION AT A CURVED SURFACE

The normal to the surface at the point of incidence will be a radius of the curve passing through that point. The line joining the **centre of curvature**, C, to the **vertex** or **pole**, A, of the surface, that is, the centre of the aperture, is known as the **optical axis**. In most optical systems the various surfaces have a common axis, and such systems are said to be *coaxial* or *centred*. If the curves are spherical, the centred system is symmetrical about its optical axis. Almost all optical systems are symmetrical.

When light falls on a curved surface between two different media, some of it is reflected (see Chapter 7), and the remainder is refracted on passing into the second medium. The direction of any part of the refracted light may be calculated from the law of refraction or may be found by a graphical construction, such as that given in section 4.5.

5.8 YOUNG'S CONSTRUCTION

A simple and more convenient construction for the refraction at a curved surface is that due to Thomas Young (1773--1829).*

In *Figure 5.9* QD is the direction of a ray incident on a spherical refracting surface DAE, having its centre at C. With C as centre and with radii equal to

$$\frac{n'}{n} \times r \quad \text{and} \quad \frac{n}{n'} \times r$$

draw the arcs a and a'. The incident ray QD intersects the arc a at G, and the straight line joining G and C intersects the arc a' at H; then DH is the direction of the refracted ray. The construction for a ray incident on a concave surface is shown in *Figure 5.10*.

*Thomas Young, *Lectures on Natural Philosophy*, Vol. 2 Art. 425, 1807.

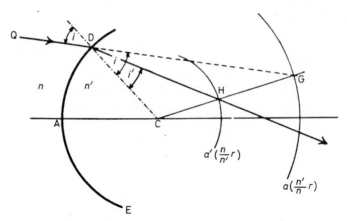

Figure 5.9 Young's construction. Positive surface, $n'/n = 1.5$

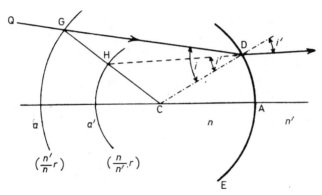

Figure 5.10 Young's construction. Negative surface, $n'/n = 1.5$

Proof

$$\text{As } CG = \frac{n'}{n} \times r, \quad CH = \frac{n}{n'} \times r, \quad \text{and } CD = r$$

$$CG:CD = CD:CH = n':n$$

and therefore the triangles CDG and CHD are similar.

$$\text{Hence } \angle CDG = \angle CHD$$

$$\text{and} \quad \angle CDH = \angle CGD$$

In the triangle CDG

$$\sin CDG : \sin CGD = CG:CD = n':n$$

As $\angle CDG = i$ and $\sin i : \sin i' = n' : n$ then $\angle CGD = \angle CDH = i'$, and the line DH is the path of the refracted ray.

The ray plotter

A simple and convenient method of applying Young's construction is by means of the **ray plotter**. From the above proof it is clear that so long as CG = n'/n CD and CH = n/n' CD, C may be any point on the normal to the surface and not necessarily the centre of curvature. In this way it is possible to construct a simple instrument to give the direction of the refracted ray at any surface between two given media.

One form of ray plotter devised by F. G. Smith* consists of a piece of celluloid in the form shown in *Figure 5.11*, with the points D, C, H and G marked on a straight line, small holes being drilled through these points. DC is taken of any convenient length, say 5 cm, and CH is made equal to n/n' DC and CG equal to n'/n DC, where n and n' are the two refractive indices for which the instrument is intended to be used.

The instrument is placed on the diagram with D over the point of incidence and the line DG over the normal, and a needle is put through the hole at C to act as a pivot. The instrument is turned until the point G lies on the incident ray, produced if necessary, when the refracted ray will pass through the point H, the position of which can be pricked through to the paper.

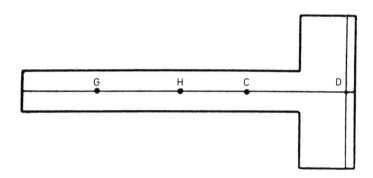

Figure 5.11 The ray plotter

The line through D perpendicular to DG allows of the instrument being used by placing this line tangentially to the surface, D being over the point of incidence. The construction is then carried out as before, pivoting the instrument at C.

5.9 SPHERICAL ABERRATION. CAUSTIC CURVE

In *Figure 5.12* light is shown diverging from an object point B on the axis of a spherical refracting surface. The direction of any part of the refracted light may be found as described in the previous paragraph, and a number of refracted rays are shown in the figure. It will be seen that any pair of rays at an equal distance from the axis intersect at some point on the axis. As the surface is symmetrical about the axis, all light from an axial point meeting the surface at a given distance from the axis, that is, in the same zone of the surface, passes through the axis at the same point. The greater the distance of a particular zone of the surface from the axis, the closer to the surface will the light from that zone meet the axis, so that light which started from an object point will not be brought to a

*F. G. Smith, *A Ray Plotter*. Trans. Optical Society. Vol. xxi, No. 3 (1919-20).

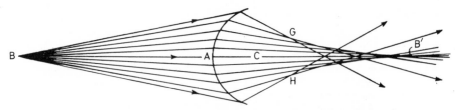

Figure 5.12 Refraction at spherical surface. Spherical aberration, caustic curve

point again after refraction. This effect is known as **spherical aberration**. A similar effect is produced when the light is divergent after refraction, the light from each zone of the surface diverging from a different point on the axis.

It will be noticed in the figure that consecutive pairs of rays intersect on a double curve GB'H, which is known as a **caustic curve**. There will consequently be a concentration of light over a surface of the form generated by the revolution of the caustic curve about the optical axis. This surface is a **caustic surface**. The point or *cusp* of the curve will lie on the axis at the point at which rays near the axis intersect. Aberrations are considered more extensively in Chapter 18.

5.10 PARAXIAL RAYS

Rays which lie very close to the axis of a spherical surface and which therefore meet the surface near the vertex will, after refraction, pass approximately through a point focus, as will be seen in *Figure 5.12*, or will diverge as though from a point. Such rays are known as **paraxial rays**, and the region very close to the axis as the **paraxial region**.

For light in the paraxial region the angles of incidence and refraction will be very small, and may be considered as proportional to their sines in applying the law of refraction. Also the apertures of the wavefronts being very small, their curvatures are proportional to the sags. Whenever therefore the aperture of an *axial* pencil is quite small we may assume that all the refracted light will be converging to or diverging from a *point* on the axis, the refracted or reflected pencil thus being homocentric.

Much of the future work will apply strictly only to the paraxial light with surfaces of quite small aperture and object points very close to the axis. It is possible, however, by a suitable choice of curvatures, separations of surfaces, and refractive indices to produce lens systems that will give approximately a point image of a point object for quite large apertures and for object points at considerable distances from the axis. Many of the results obtained for the paraxial region may be applied to such *corrected systems*. The mathematics giving these results was extensively developed by the German mathematician **Karl Friedrich Gauss** (1777-1855). For this reason paraxial optics is sometimes called **Gaussian Optics**. For reasons given in Chapter 18 it is also called **First-order Optics**.

5.11 REFRACTION AT A SPHERICAL SURFACE (CHANGE IN VERGENCE)

In *Figure 5.13* light converging towards a focus at B is intercepted by a spherical surface DAE, separating the media of refractive indices n and n'. The incident

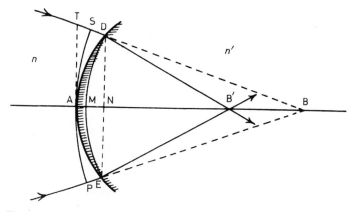

Figure 5.13 Refraction at a spherical surface (n'>n). Paraxial region. Change in vergence. All positive diagram

wavefront SAP meets the vertex of the surface at A and, if n' is greater than n, the centre of the wavefront is retarded with respect to the *rim* portion on passing into the second medium. DME represents the form of the wavefront immediately after refraction. If the light is confined to the paraxial region (the aperture is shown greatly exaggerated in the figure), the refracted wavefront is spherical and the light is now converging to a focus at B'. B' is therefore the image of the *virtual object* B. As the light which travels along the axis meets the surface normally, it is unchanged in direction and therefore, if the object point is on the axis, the image must also be formed on the axis.

The centre of the refracted wavefront travels from A to M in medium n' in the same time as the rim portion travels from S to D in medium n.

n SD $= n'$ AM

n (TD $-$ TS) $= n'$ (AN $-$ MN)

n' MN $- n$ TS $= n$ AN $- n$ TD

For small portions of the curves, that is, in the paraxial region, TD = AN, and the sags are proportional to the curvatures hence:

$$\frac{n'}{l'} - \frac{n}{l} = \frac{n' - n}{r} \tag{5.02}$$

$$\text{or better } \frac{n'}{l'} = \frac{n}{l} + \frac{n' - n}{r} \tag{5.02a}$$

where l and l', the distances of object and image from the refracting surface, are the radii of curvature of the incident and refracted wavefronts respectively, and r is the radius of curvature of the surface. As n/l and n'/l' are the reduced vergences of the incident and refracted light respectively, equation (5.02) may be written

$$L' - L = \frac{n' - n}{r} = (n' - n)R \tag{5.02b}$$

$L' - L$ is the change produced by the surface in the reduced vergence of the light, and is known as the **focal power** of the surface. It will be denoted by F.

$$\text{Hence} \quad \left. \begin{array}{l} L' - L = F \\ L' = L + F \end{array} \right\}$$

(5.03)

or

and

$$F = (n' - n)R$$

(5.04)

The focal power is thus a constant for a particular surface, that is, a given surface will always produce the same total change in the vergence of the light. Focal power is usually expressed in dioptres and will be positive for a surface that tends to produce convergence (*Figure 5.13*), and negative for one that tends to produce divergence.

The equations (5.02a) or (5.02b) will be called the **fundamental paraxial equation** for a surface.

When the first medium is air ($n = 1$) and the second, glass of refractive index n, equation (5.04) becomes

$$F = (n - 1)R$$

This is the quantity given by the opticians lens measure (section 5.2).

5.12 REFRACTION AT A SPHERICAL SURFACE (CHANGE IN RAY PATH)

The expression (5.02) may also be derived by considering the change produced by the surface in the path of a single ray, as follows:

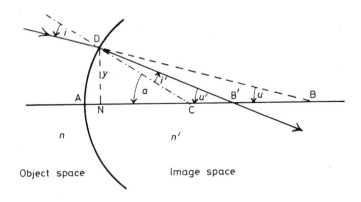

Figure 5.14 Refraction at a spherical surface. Change in ray path. All positive diagram

Figure 5.14. From the law of refraction

$$n \sin i = n' \sin i'$$

and when i and i' are small

$$n i = n' i'$$

When a, u and u' are small, the arc AD may be considered as a short line y perpendicular to the axis; then

$$u = \frac{y}{AB}; \quad u' = \frac{y}{AB'}; \quad a = \frac{y}{AC}$$

From the figure, in Δs DNC and DNB

$$i = (90 - u) - (90 - a) = a - u$$

and in Δs DNC and DNB'

$$i' = (90 - u') - (90 - a) = a - u'$$

therefore

$$n(a - u) = n'(a - u')$$

$$n'u' - nu = (n' - n)a$$

Substituting the above values for u, u' and a, and dividing through by y

$$\frac{n'}{AB'} - \frac{n}{AB} = \frac{n' - n}{AC}$$

$$AB' = l'; \quad AB = l; \quad AC = r$$

Hence

$$\frac{n'}{l'} - \frac{n}{l} = \frac{n' - n}{r} \tag{5.02}$$

5.13 CONJUGATE FOCI. PRINCIPAL FOCI. FOCAL LENGTHS

It follows from the above that for any object point B there will be a corresponding image point B', the position of which may be found from expressions (5.02) or (5.02a), giving the appropriate signs to the various distances. As the path of the light is reversible, the positions of object and image are interchangeable. These pairs of object and image points are therefore known as **conjugate points** or **foci** and the distances of object and image from the surface as the **conjugate distances.**

Two of these pairs of conjugate points are of special importance. When the object is at an infinite distance and the incident light therefore parallel (*Figures 5.15a* and *b*), the refracted light will be converging to a real image in the case of a surface of positive power, or diverging as though from a virtual image in the case of a surface of negative power. This image, the conjugate of the infinitely distant object, is termed the **second principal focus**, F', of the surface, and its distance *from* the surface is the **second focal length**, f'. There must also be an object position, such that the image is at an infinite distance and the refracted light therefore parallel. This will be a real object position from which the incident light is diverging in the case of a positive surface, and a virtual object position towards which the incident light is converging in the case of a negative surface. This object position is termed the **first principal focus**, F, and its distance *from* the surface is the **first focal length**, f.

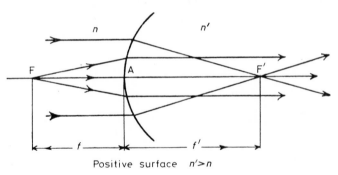

Positive surface   *n′>n*

Figure 5.15a Refracting surface. Principal foci and focal lengths

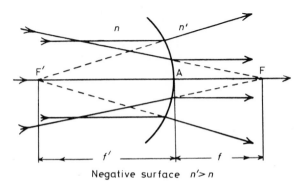

Negative surface   *n′> n*

Figure 5.15b Refracting surface. Principal foci and focal lengths

It will be seen that both principal foci are real with a positive surface and virtual with a negative surface. *The two focal lengths are of opposite sign; the sign of the second is that of the focal power.*

The planes perpendicular to the axis passing through the principal foci are known as the **focal planes**.

Any incident ray parallel to the axis will, after refraction at a spherical surface, pass through or diverge from the second principal focus, and any incident ray passing through or travelling towards the first principal focus will, after refraction, travel parallel to the axis. These facts are useful in graphical construction (section 6.14).

Focal Lengths

When l' is infinity, $L' = 0$ and

$$-L = -\frac{n}{l} = F$$

Similarly when l is infinity,

$$L' = \frac{n'}{l'} = F$$

These special values of l and l' are the first and second focal lengths respectively, hence

$$-\frac{n}{f} = \frac{n'}{f'} = F = (n' - n)\,R \qquad\qquad (5.05)$$

and

$$\frac{f'}{f} = -\frac{n'}{n}$$

The power of a surface is equal to the reciprocal of the 'reduced' second focal length.

Also we have $f' + f = r$

and

$$CF' = -f; \quad CF = -f'$$

Example: An *aphakic* eye, that is, one from which the crystalline lens has been removed, may be regarded as a single spherical surface, the cornea, of + 7.8 mm radius of curvature between air, $n = 1$, and a medium of refractive index, $n' = 1.336$. Find the focal power and the positions of the principal foci of such an eye. What will be the position of the object of which an image is formed on the retina 24 mm from the cornea?

$$r = + 7.8 \text{ mm} = 0.0078 \text{ m}$$

$$F = \frac{0.336}{0.0078} = + 43.1 \text{ D}$$

$$f = -\frac{1000}{43.1} = -23.2 \text{ mm}$$

$$f' = \frac{1336}{43.1} = +31.0 \text{ mm}$$

Therefore the first principal focus is 23.2 mm in front of the cornea, and the second principal focus 31.0 mm behind the cornea.

$$l' = +24.0 \text{ mm} = +0.024 \text{ m}$$

$$L' = \frac{n'}{l'} = \frac{1.336}{0.024} = +55.66 \text{ D}$$

$$L = L' - F = 55.66 - 43.1 = +12.56 \text{ D}$$

$$l = \frac{1000}{12.56} = +79.6 \text{ mm}$$

The incident light must therefore be converging towards a point 79.6 mm behind the cornea in order that, after refraction, it may focus on the retina.

5.14 IMAGES OF OBJECT POINTS NOT ON THE AXIS

As, in practice, any object will be of finite size, only one point of the object can lie on the optical axis. It will be necessary therefore to consider the light coming from points off the axis—**extra-axial points**. A good image of such a point will only be formed by a single surface when the light is in the paraxial region, that is, when the light makes a small angle with the axis, and the aperture is small. For the present we shall limit the work to such cases.

Of the light from any object point Q (*Figure 5.16*), one ray will be incident on the surface normally, and will therefore be unchanged in direction on passing into the second medium. This ray is called the chief ray or **principal ray** from the particular point and will, of course, be the ray that passes through the centre of curvature C. C is frequently called, on this account, the **nodal point** of the surface. Since, in the paraxial region, all rays from an object point meet at the corresponding image, this image must be on the principal ray from the object point. If the direction of a second ray is known, such as an incident ray parallel to the axis, which after refraction must pass through the second principal focus, the position of the image is fixed by the intersection of these two rays.

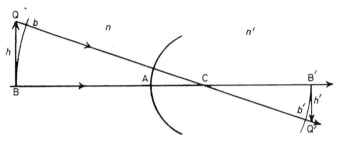

Figure 5.16 Spherical refracting surface. Image of extra-axial point

If Bb is an arc drawn with C as centre, the image of b is b′ on an arc B′b′. Clearly then, the image of the perpendicular BQ is a curved line more curved than B′b′, as Q is further from the surface than b, and its image therefore closer to the surface than b′. This image curvature is discussed in Chapter 18. When BQ is very small, however, we can take B′Q′ as the image, and in the paraxial region we shall assume that for a plane object perpendicular to the axis the image is also in a plane perpendicular to the axis. Object and image are said to be in **conjugate planes.**

In *Figure 5.17* BQ is an object in front of a spherical refracting surface DAE. The position of the axial image point B′ can be found from the fundamental paraxial equation. The position of any extra-axial point such as Q can be found graphically by considering any two rays, the refracted directions of which are known. Thus, the chief ray QC will be undeviated, a ray QD parallel to the axis will be refracted to pass through the second principal focus F′, and a ray QE passing through the first principal focus F travels parallel to the axis after refraction. The intersection of any two of these rays gives the position of the image Q′. As the chief ray QC is undeviated,

$$\frac{h'}{h} = \frac{CB'}{CB}$$

where h and h' are the sizes of object and image respectively. The ratio of the size of the image to the size of the object is termed the **lateral magnification** of the image, and will be denoted by m.

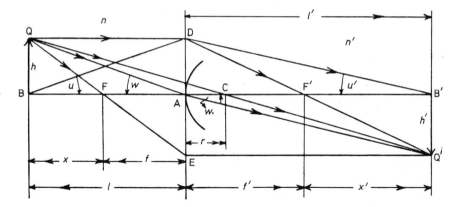

Figure 5.17 Spherical refracting surface. Formation of image. The construction applies only to the paraxial region, and the position of the surface is therefore represented by the straight line DE. (See section 6.15)

Then $m = \dfrac{h'}{h} = \dfrac{l'-r}{l-r}$

As the angles w and w' are very small

$$\frac{\tan w'}{\tan w} = \frac{\sin w'}{\sin w} = \frac{n}{n'}$$

then

$$\frac{h'}{h} = \frac{l'\tan w'}{l\tan w} = \frac{nl'}{n'l} = \frac{L}{L'} \tag{5.06}$$

The magnification will be positive when the image is erect, and negative when the image is inverted. Images will be erect when they lie on the same side of the surface as the object, and inverted when on the opposite side.

Example:
Seen through the cornea the pupil of an eye appears to be situated 3 mm from the cornea and of 4.5 mm diameter. Treating the cornea as a single spherical surface of 7.8 mm radius of curvature between air and a medium of refractive index $^4/_3$, find the actual position and size of the pupil.

In solving problems the student will always be helped by making a diagram showing all the given data.

In this example the actual pupil is the object and light from it is refracted at the cornea on leaving the eye and diverges as though from the virtual image, the apparent pupil, situated −3 mm from the cornea and of 4.5 mm diameter. Thus $n = ^4/_3$, $n' = 1$, $l' = -3$ mm and $r = -7.8$ mm. (Surface is concave towards the incident light).

$$\frac{n'}{l'} - \frac{n}{l} = \frac{n'-n}{r}$$

$$\frac{1}{-3} - \frac{4}{3l} = \frac{1 - {}^4/_3}{-7.8}$$

$$\frac{1}{l} = -0.282$$

$l = -3.55$ mm (distance of pupil from cornea)

$$m = \frac{h'}{h} = \frac{nl'}{n'l}$$

$$\frac{4.5}{h} = \frac{{}^4/_3 \times 3}{3.55}$$

$h = 3.99$ mm (diameter of pupil)

5.15 FURTHER RELATIONSHIPS

From *Figure 5.17* it will be seen that the lateral magnification may also be expressed in the following forms:

$$\left. \begin{array}{l} m = \dfrac{B'Q'}{AD} = -\dfrac{x'}{f'} \\[2em] \text{and} \\[1em] m = \dfrac{AE}{BQ} = -\dfrac{f}{x} \end{array} \right\} \tag{5.07}$$

where x and x' are the distances of object and image from the first and second principal foci respectively. These distances are called the **extra-focal distances**. From expression (5.07) we have.

$$ff' = xx' \tag{5.08}$$

This is known as **Newton's Relation,** an expression that is frequently useful. From equation (5.06)

$$\frac{h'}{h} = \frac{nl'}{n'l} \quad \text{we have}$$

$$n'h'l = n h l'$$

In the paraxial region, as the angles u and u' (*Figure 5.17*) are very small.

$$\frac{AD}{l} = u, \quad \text{and} \quad \frac{AD}{l'} = u'$$

Hence

$$n' h' u' = n h u = H \tag{5.09}$$

This optical invariant has important applications, and has been credited in various forms to Robert Smith (1689-1768), Helmholtz (1821-1894) and Lagrange (1736-1813). It is usually known as either the **Smith–Helmholtz** or the **Lagrange's Invariant**, H.

From (5.09) we have

$$\frac{u'}{u} = \frac{n\,h}{n'\,h'} = \frac{n}{n'} \times \frac{1}{m}$$

This is called the convergence ratio.

5.16 IMAGE OF A DISTANT OBJECT

All rays from any one point Q on a distant object are parallel on reaching the surface, and, after refraction, reunite in a point Q' in the second focal plane of the surface (*Figure 5.18*). The position of Q' is determined by the chief ray which passes undeviated through the centre of curvature C.

If *w* denotes the angle subtended by the object standing on the axis or its **apparent size**, then

$$F'Q' = h' = f\,tan\,w = f\,w, \text{ when } w \text{ is small} \tag{5.10}$$

This gives another definition of the first focal length which is often useful:

$$f = \frac{h'}{w} = \frac{\text{size of image in F' plane}}{\text{apparent size of object}} \tag{5.11}$$

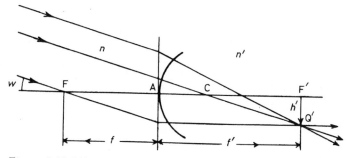

Figure 5.18 Spherical refracting surface. Image of distant object

CHAPTER 5   EXERCISES

1. What is meant by the curvature of a surface? Show that it is equal to the reciprocal of the radius of curvature.

2. In measuring the curvature of a surface with an Aldis spherometer having a distance of 25 mm between the centres of the centre and outer spherical feet, the readings are −0.16 mm on a plane and +2.54 mm on the curved surface. The diameter of the spherical feet is 5 mm. Find the curvature of the surface. What percentage error is introduced by using the approximate formula?

3. The radius of the circle containing the three fixed legs of a spherometer is $1\frac{1}{8}$ inches. The instrument is placed on a convex spherical surface of curvature 3 dioptres. By how much will the central leg have to be raised above the outer legs to be just in contact with the surface?

4. A plano-concave lens has a diameter of 27 mm, radius of curvature 66 mm and central thickness 1.5 mm; what is its edge thickness?

5. An optician's lens measure has its outer points $^3/_4$ inch apart, and a depression of the centre point of $^1/_{32}$ inch gives a reading of +9.25 D on its scale. For what refractive index was it scaled?

6. A lens measure applied to a brass tube of 9 inches external diameter gave a reading of 4.75 dioptres. For what refractive index was it adjusted?

7. A lens made of glass of refractive index 1.523 and having surface powers of +6 D and −11 D is to have a diameter of 40 mm and 2 mm central thickness. What will be the edge thickness of this lens and the least thickness of a plane glass plate from which it can be ground?

8. What must be the distance between the two outer points of a lens measure if a power of +1 D is to give a depression of $^1/_4$ mm to the centre point for a refractive index of 1.53?

9. A pencil of light is made to converge by means of a lens to a point 20 cm from the lens. Find the curvature of a wavefront (a) on leaving the lens, (b) at 1 cm from the lens, (c) at 19.9 cm from the lens, (d) at 21 cm from the lens.

10. A small source of light is placed 75 cm from a lens; what is the curvature of the wavefront at the lens? If the effect of the lens is to add 4 D convergence to the light, find the position at which the light will focus.

11. Where must a small source of light be placed in order that the wavefront of the divergent pencil shall have a curvature of 7 D on reaching a lens. If the lens brings the light to a focus at a point 30 cm beyond the lens, what will be the change in curvature of the wavefront produced by the lens?

12. Show that when a pencil of convergent light is intercepted by a plane parallel plate of glass of thickness t and refractive index n, the position of the focus is moved outwards a distance $t - t/n$.

13. A plane wavefront is incident in air on a curved glass surface of +5 cm radius of curvature, refractive index of the glass 1.5, the line bisecting the wavefront at right angles passing through the centre of curvature of the surface. Find by Huygens' construction the form of the refracted wavefront. What do you notice about its form?

14. Derive the fundamental paraxial equation

$$\frac{n'}{l'} - \frac{n}{l} = \frac{n' - n}{r}$$

for refraction at a spherical surface.
Indicate clearly in your proof that the expression only applies to the paraxial region. Explain the convention adopted as to signs.

15. An object 1 cm high is placed 15 cm from a convex spherical refracting surface of 20 cm radius of curvature, which separates air from glass ($n = 1.52$). Find the position and size of the image. What are the positions of the principal foci of this surface?

16. A spherical refracting surface of radius +8 cm separates air on the left from glass of refractive index 1.5 on the right. A parallel pencil is incident on it, the chief ray of the pencil passing through the centre of curvature. Find by graphical construction the refracted rays corresponding to incident rays at 3, 4, 5, 6 and 7 cm above the chief ray. Note where they cut the chief ray.

17. What is the difference in the apparent thickness of a biconvex lens ($n = 1.5$) having radii of curvature of 8 cm and 20 cm and central thickness 2 cm, the lens being examined first from one side and then from the other?

18. A plano-convex lens is viewed from above (a) with its plane surface uppermost, and then (b) with its convex surface uppermost. In (a) the greatest apparent thickness is 13.3 mm and in (b) 14.6 mm. If the lens is 20 mm thick, what is the refractive index of the material, and what is the radius of curvature of the convex surface?

19. A glass sphere has a diameter of 10 cm, and refractive index 1.53. Two small bubbles in the glass appear to be, one, exactly at the centre of the sphere and the other midway between the centre and the front surface; find their actual positions.

20. A glass paper-weight (n = 1.53) has a picture fastened to its base. Its top surface is spherical and has a radius of curvature of $2\frac{1}{2}$ inches and its central thickness is $1\frac{1}{2}$ inches. Where will the picture appear to be when viewed through the top surface and what will be its magnification?

21. For the purpose of calculation the eye may be considered as a single refracting surface of +60D power, having air on one side and a medium of refractive index $\frac{4}{3}$ on the other. What must be the distance of the retina—the receiving screen—from this surface, if the eye be correctly focused on a distant object? What will be the distance of the object that is focused by an eye in which the distance between refracting surface and retina is 24 mm?

22. Treating the cornea as a single spherical refracting surface of radius +7.8 mm separating air from the aqueous of refractive index $\frac{4}{3}$, find the position and size of the image of the eye's pupil, the latter being 3.6 mm behind the cornea and 4 mm in diameter.
    What effects would be produced on the retinal image in an eye as the pupil is reduced in diameter?

23. A narrow beam of parallel light is incident normally on the flat surface of a glass hemisphere of 10 cm radius. Find where the rays of light are brought to a focus (*a*) when the hemisphere is in air, (*b*) when it is in water. The refractive indices of glass and water may be taken to be $\frac{3}{2}$ and $\frac{4}{3}$ respectively.

24. A piece of glass 2 cm long and of refractive index 1.60 is plane at one end and has a convex spherical surface of radius of curvature 1.875 cm worked on the other end. It is set up with the curved surface facing a distant point source. Find the curvature of the wavefront emerging from the plane surface and the position of the point where the light comes to a focus.

25. Considering the eye as a single refracting surface of +5.56 mm radius of curvature, having air on one side and a medium of refractive index $\frac{4}{3}$ on the other, what will be the size of the image formed of a distant object that subtends an angle of $1°$ at the eye?

26. Explain and prove Young's construction for the path of a ray of light refracted at a spherical surface of radius r separating two media of refractive index n and n'.

Chapter 6
The thin lens

6.1 INTRODUCTION

The lens may be defined as a portion of a transparent substance bound by two polished surfaces, both of which may be curved, or one only may be curved and the other plane. The curved surfaces may be of a number of different forms, spherical, cylindrical, toroidal, or paraboloidal, but the commonest form of lens is that with spherical surfaces, and the work of this chapter will be limited to this form. Such lenses are usually called **spherical lenses**.

The term *lens*—in French *lentille*—is derived from the Latin word for the lentil, from the similarity of the convex lens to the form of the lentil seed. The use of lenses appears to have been known from very early times, lenses having been found that date back to about 1600 BC. These lenses, which are of crystal, were no doubt used as magnifiers.

6.2 FORMS OF LENSES

Thin lenses may be divided into two classes according to the curvature of their surfaces and therefore the effect that they produce on the light. Lenses of the forms shown in *Figure 6.1* have either both surfaces or the steeper of the two surfaces convex.

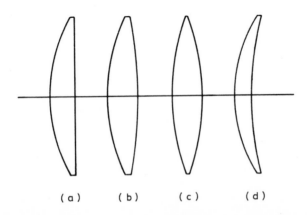

(a)   (b)   (c)   (d)

Figure 6.1 Convex, convergent or positive thin lenses

84

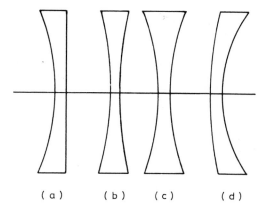

(a)   (b)   (c)   (d)

Figure 6.2 Concave, divergent or negative thin lenses

Such lenses, if *thin* and of higher refractive index than the surrounding medium, tend to produce convergence of the light—a positive effect. They will therefore be termed **convex, convergent** or **positive**. Lenses having both surfaces or the steeper of the two surfaces concave, as shown in *Figure 6.2*, will, under the same conditions, tend to make the light divergent. These will therefore be termed **concave, divergent** or **negative**.

The following terms are applied to the various forms of lenses.

Positive	*Negative*
(1) Plano-convex (*Figure 6.1a*).	Plano-concave (*Figure 6.2a*).
(2) Bi-convex (*Figure 6.1b*).	Bi-concave (*Figure 6.2b*).
(3) Equi-convex (*Figure 6.1c*).	Equi-concave (*Figure 6.2c*).
(4) Convex meniscus (*Figure 6.1d*).	Concave meniscus (*Figure 6.2d*).

The surface towards the incident light will be called the first surface of the lens, and positions in the space occupied by the incident light will be said to be in front of the lens.

Figure 6.3 represents a section through a biconvex spherical lens, the spheres of which its surfaces are portions having centres at C_1 and C_2 respectively. The

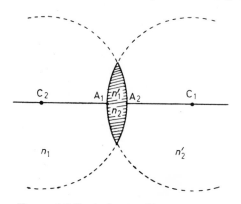

Figure 6.3 Optical axis of lens

straight line C_1C_2 joining the centres of curvature of the two surfaces is called the **optical axis** of the lens. If one surface is plane, it may be considered as a portion of a sphere of infinite radius, and the optical axis will be the line perpendicular to the plane surface passing through the centre of curvature of the other surface. The points A_1 and A_2, where the axis meets the surfaces, are termed the **front** and **back vertex** or **pole** of the lens respectively, and A_1A_2 is the lens thickness. When the lens is so thin that the thickness may be neglected, the common vertex A is the **optical centre**. Lenses are usually 'edged' so that the optical centre coincides with the geometrical centre of the lens aperture. This is not always the case with ophthalmic lenses, such lenses sometimes being 'decentred'. At the optical centre the two surfaces of the lens are parallel and any rays passing through this point are therefore unchanged in direction. The ray passing through the optical centre is the chief ray from any object point.

6.3 REFRACTION BY A LENS

The change in direction of any ray on passing through a lens may be found by considering the refraction at each surface in turn. This may either be calculated from the law of refraction or may be found graphically by one of the methods previously described. It will be noticed that pencils of light on emerging from the lens are no longer homo-centric. As will be expected from the result of refractions at a single curved surface, there will be spherical aberration, the rays through the more peripheral portions of the lens crossing the optical axis closer to the lens than the rays from the more central portions of the lens aperture. Rays close to the optical axis, that is, in the paraxial region, meet the axis after refraction approximately in one point, and we shall assume that for the paraxial region point images are formed of point objects.

6.4 FOCAL POWER

The function of the lens is to change the vergence of the light or the curvature of the wavefronts, thus producing an image—real or virtual—of the object from which the light is coming. This change in reduced vergence that the lens produces is termed its **focal power**. As this depends, in the case of a thin lens, only on the curvature of the surfaces and the refractive indices, the focal power of any one lens is a constant.

The expressions obtained in Chapter 5 for the focal power of a surface may be applied to each surface of the lens in turn, and we have for the paraxial region

First surface

$$\frac{n'_1}{l'_1} - \frac{n_1}{l_1} = \frac{n'_1 - n_1}{r_1}$$

Second surface

$$\frac{n'_2}{l'_2} - \frac{n_2}{l_2} = \frac{n'_2 - n_2}{r_2}$$

$$n_2 = n'_1$$

If the thickness of the lens is neglected, the vergence $1/l'_1$ of the refracted light at the first surface is the vergence $1/l_2$ of the incident light at the second surface, and

$$\frac{n_2}{l_2} = \frac{n'_1}{l'_1}$$

Then from the expressions above for the separate surfaces

$$\frac{n'_2}{l'_2} - \frac{n_1}{l_1} = \frac{n'_1 - n_1}{r_1} + \frac{n'_2 - n_2}{r_2}$$

As n'_2/l'_2 and n_1/l_1 are the reduced vergences of the incident and emergent light respectively, the expression gives the focal power of a *thin* lens with media of any refractive index in the object and image spaces.

Thin lens in air

If the lens is in air, then $n_1 = n'_2 \doteq 1$ and putting

$$n'_1 \doteq n_2 = n$$

we have

$$\frac{1}{l'_2} - \frac{1}{l_1} = \frac{n-1}{r_1} + \frac{1-n}{r_2} = F_1 + F_2$$

where F_1 and F_2 are the surface powers.

$$\frac{1}{l'_2} - \frac{1}{l_1} = (n-1)\left(\frac{1}{r_1} - \frac{1}{r_2}\right)$$

or, dropping the suffixes of l'_2 and l_1

$$\frac{1}{l'} - \frac{1}{l} = (n-1)\left(\frac{1}{r_1} - \frac{1}{r_2}\right) = L' - L \qquad (6.01)$$

$L' - L$ is the focal power, F, of the lens and we have

$$F = (n-1)\left(\frac{1}{r_1} - \frac{1}{r_2}\right) = (n-1)(R_1 - R_2) \qquad (6.02)$$

or

$$\left.\begin{array}{l} F = L' - L \\ L' = L + F \end{array}\right\} \qquad (6.03)$$

For a lens in air L and L' are the vergences at the lens of the incident and emergent light respectively, and the latter expression (6.03) may be stated as follows:

For a lens in air, the final vergence L' of the light is equal to the incident vergence L plus the impressed vergence F.

This is the better form to remember and use because it contains no negative signs.

6.5 CONJUGATE FOCI AND PRINCIPAL FOCI

Applying expression (6.03) to the case of a positive lens, we see that with an object at an infinite distance, the light emerging from the lens is convergent, and a real image is formed. The position of this image is the **second principal focus**, F'. As the object moves up towards the lens, the incident light becomes more and more divergent, and therefore the emergent light becomes less convergent, and the image moves away from the lens until the object is in such a position that the image is at infinity, that is, the emergent light is parallel. The object is now at the **first principal focus**, F. If the object is brought still closer to the lens, the power of the lens is insufficient to overcome all the divergence of the incident light, and the emergent light diverges from a virtual image on the same side of the lens as the object, but farther from the lens than the object is.

With a negative lens, as its effect is to add divergence to the light, a virtual image will always be formed of a real object, and this image will be closer to the lens than the object. The postition of the virtual image when the object is at an infinite distance will be the second principal focus, F'.

When the incident light is convergent, the point to which it is converging may be considered as a *virtual object*. The positive lens will always form a real image of this virtual object closer to the lens, since it adds convergence to the convergent incident light. The negative lens may, however, in this case, form either a real or a virtual image according to whether the convergence of the incident light at the lens is greater or less than the lens power. If the convergence of the incident light at the lens is equal to the power of the negative lens, the emergent light will be parallel, and the image will be at infinity. The position of the virtual object is then the first principal focus, F. With the virtual object closer to the negative lens than the first principal focus—'inside' the first principal focus—a

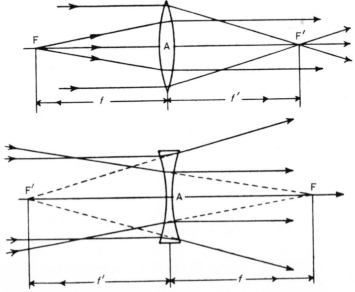

Figure 6.4 Principal foci and focal lengths

real image is formed, and with the virtual object 'outside' the first principal focus the image is virtual. The positions of the principal foci of a positive and a negative lens are shown in *Figure 6.4*.

As with a single surface, the conjugate foci are interchangeable, and therefore, for any one pair of real object and real image positions, except when the lens is situated exactly midway between these positions, there will be two possible positions of the lens, the object distance in one position being equal numerically to the image distance in the other and vice versa.

6.6 FOCAL LENGTHS

The distance AF (*Figure 6.4*), *from* the lens to the first principal focus, is the **first** (or anterior) **focal length**, f, and the distance AF' *from* the lens to the second principal focus is the **second** (or posterior) **focal length**, f'.

When the object is at infinity, $L = 0$, and for a thin lens in air we have

$$L' = F$$

$$l' = f' = \frac{1}{F} \tag{6.04}$$

Also, when the image is at infinity, $L' = 0$, and therefore

$$-L = F$$

$$l = f = -\frac{1}{F} \tag{6.05}$$

Thus, for a lens in air, or in any case where the first and last media are of equal refractive index, the two focal lengths are numerically equal but of opposite sign, the second focal length having the same sign as the power. Except in the case of ophthalmic lenses, which are expressed in focal power, most lenses are specified in terms of their second focal lengths.

Expression (6.03) may also be written

$$\frac{1}{l'} = \frac{1}{l} + \frac{1}{f'} \tag{6.03a}$$

A table of reciprocals or a calculator with a reciprocal key will be found of great convenience in solving problems on conjugate foci.

6.7 CHROMATIC ABERRATION

As the power of a lens depends upon the refractive index and this varies with the colour of the light, it is evident that power of any lens is different for light of different colours.

An axial object point emitting white light will therefore be imaged as a short spectrum, lying along the axis, the violet being focused closest to the lens and the red farthest away, as in *Figure 6.5*. The effect is known as **chromatic aberration**. In order that a lens shall produce good images it is necessary that light of

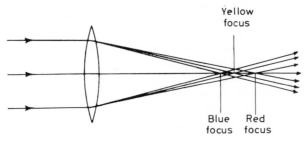

Figure 6.5 Chromatic aberration

different colours shall, as far as possible, be brought to a focus at the same place. This aberration can be largely corrected by combining a positive and a negative lens of suitable powers and made of different types of glass. Such a combination is known as an **achromatic lens**. Chromatic aberration and the principles of its correction are more fully considered in Chapter 13. By suitably choosing the curves and glasses of a combination of lenses, both spherical and chromatic aberration can be largely corrected; the expressions obtained for the paraxial region will then be true for larger apertures.

6.8 IMAGES OF EXTRA-AXIAL POINTS

As with a single refracting surface, the pencil of light from an object point not on the lens axis will, after refraction, no longer be homocentric, even when the lens aperture is small. The light will be affected by various aberrations that are described in Chapter 18. The paths of any rays from an extra-axial object point can be found by calculation from the law of refraction or by one of the graphical constructions.

When the distance of the object from the axis is small as compared with its distance from the lens and when the aperture of the lens is also small, all light reaching the lens from the object will be in the paraxial region, and in this case a reasonably good image of the object will be formed by a single lens. By suitably combining a number of lenses it is possible to correct the oblique aberrations to a great extent, and in the case, for example, of photographic objectives and certain eyepieces, good images are produced of objects at considerable angles from the axis and with lenses of large aperture. When only a small aperture of the

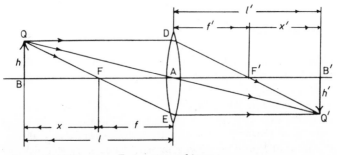

Figure 6.6 Thin lens. Formation of image

lens is to be used, as in the spectacle lens, considerable reduction of the oblique aberrations may be effected by a suitable choice of curves for the two surfaces of a single lens. The expressions obtained for the positions of the images of extra-axial points lying in the paraxial region may also be used in dealing with objects at considerable distances from the axis in the case of corrected systems.

If Q (*Figure 6.6*) is an extra-axial object point of which the lens forms a good image, all the light from the object which passes through the lens will converge to a real image point or diverge from a virtual image point, and the position of this image Q' is fixed if the intersection of two refracted rays can be found. From the properties of the lens that have already been considered the directions of certain rays after refraction are known, and any two of these may be used to find the image position. These known ray directions are as follows:

(*a*) Rays passing through the optical centre of the lens are undeviated—chief rays or principal rays.
(*b*) Rays passing through the first principal focus leave the lens parallel to the axis.
(*c*) Rays incident parallel to the axis pass, after refraction, through the second principal focus.

In *Figures 6.5* and *6.10* three rays QA, QD and QE from a point Q are shown converging after refraction to a real image Q' in the case of the positive lens and diverging as from a virtual image Q' in the case of the negative lens.

6.9 OBJECT OF FINITE SIZE. LATERAL MAGNIFICATION

The position and size of the image of an object of finite size can be determined by finding the image positions for a number of points on the object by the above method. It will be found that, within the paraxial region, for an object perpendicular to the axis an image is formed also perpendicular to the axis, as in *Figure 6.6*. The position of the image is determined from the conjugate foci relation (6.03), and its size as follows:

$$\frac{h'}{h} = \text{lateral magnification of the image} = m$$

From *Figure 6.6*

$$\frac{B'Q'}{BQ} = \frac{AB'}{AB}$$

or

$$\frac{h'}{h} = m = \frac{l'}{l} = \frac{L}{L'} \tag{6.06}$$

The magnification will be positive when the image is erect, that is, when Q' is on the same side of the axis as Q, and negative when the image is inverted.

Example:

When an object is a certain distance from a thin positive lens a real inverted image one-half the size of the object is formed. The object is moved 100 mm nearer the lens and the real image formed is now the same size as the object. What is the focal length of the lens?

$$m_1 = -\frac{1}{2} = \frac{l'_1}{l_1}; \quad l'_1 = -\frac{l_1}{2}$$

$$m_2 = -1 = \frac{l'_2}{l_2}; \quad l'_2 = -l_2$$

$$\frac{1}{f'} = \frac{1}{l'_1} - \frac{1}{l_1} = -\frac{2}{l_1} - \frac{1}{l_1} = -\frac{3}{l_1}$$

$$\frac{1}{f'} = \frac{1}{l'_2} - \frac{1}{l_2} = -\frac{1}{l_2} - \frac{1}{l_2} = -\frac{2}{l_2} = \frac{2}{l_1 + 100}$$

$$\therefore \quad \frac{3}{l_1} = \frac{2}{l_1 + 100}$$

$$l_1 = -300, \quad l'_1 = +150$$

$$\frac{1}{f'} = \frac{1}{150} + \frac{1}{300} = \frac{3}{300}$$

$$f' = 100 \text{ mm}$$

Further expressions for magnification are easily derived from (6.03) and (6.06), thus

$$m = \frac{L}{L'} = \frac{L - F}{L'}$$

$$= 1 - l'F = 1 - \frac{l'}{f'} \tag{6.07}$$

This is an expression useful in problems on optical projection equipment. Also

$$\frac{1}{m} = \frac{L'}{L} = \frac{L + F}{L}$$

$$= 1 + lF = 1 + \frac{l}{f'} \tag{6.08}$$

an expression useful in problems on the camera and magnifier.

Note: In expressions (6.07) and (6.08), l and l' must be expressed in metres when used with F, which is in dioptres.

With any lens there must be a position of object and image such that the image is of the same size as the object but inverted, then $m = -1$, and $l' = -l$. The object and image are then said to be in the **symmetrical planes** of the lens, and the distance between these planes for a thin lens, $l' - l = 4f'$. This is the minimum distance between a real object and real image in the case of a positive lens or between a virtual object and virtual image in the case of a negative lens.

The positions of object and image such that $m = +1$, that is, where object and image are of the same size and on the same side of the axis, are known as the **principal planes** of the lens, and in a thin lens these coincide at the lens. These planes are of great importance in the case of thick lenses and lens-systems and are dealt with in Chapters 9 and 19.

6.10 NEWTON'S RELATION AND THE LAGRANGE INVARIANT

In the same way as for a single refracting surface, the conjugate foci and magnification expressions for a lens may be expressed in terms of the extra-focal distances, x and x'. Thus, from *Figure 6.6* we have

$$m = -\frac{f}{x} = \frac{f'}{x} \tag{6.09}$$

and

$$m = -\frac{x'}{f'} \tag{6.09a}$$

Hence

$$-xx' = f'^2 \tag{6.10}$$

Newton's relation, expression (6.10), is the basis of various methods of measuring the focal lengths of lens systems and thick lenses (section 9.7). It is also used in the Badel lens arrangement (section 20.10). The Lagrange Invariant for a thin lens in air is given by $H = hu = h'u'$ which becomes

$$H = \frac{hy}{l} = \frac{h'y}{l'} \tag{6.11}$$

where y is the maximum ray height at the thin lens.

6.11 IMAGE OF A DISTANT OBJECT

When the object is very distant the rays from any one point are parallel to one another on reaching the lens. The position of the image of a distant object point not on the axis may be determined by finding the point at which two parallel incident rays intersect after refraction. The two most convenient rays to consider will be the chief ray, and the ray incident through the first principal focus and therefore refracted parallel to the axis (*Figure 6.7*). It will be found that the

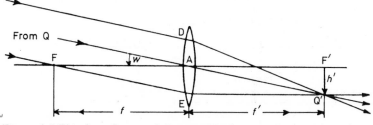

Figure 6.7 Thin lens. Image of distant object

image is formed in the second focal plane, and its position in this plane is therefore determined by the chief ray.

If w is the angle subtended at the lens by a distant object standing on the axis, or the apparent size of the object, then the size of the image,

$$\left.\begin{array}{l} h' = -f' \tan w = f \tan w \\ \quad = fw, \text{ when } w \text{ is small} \end{array}\right\} \tag{6.12}$$

When the object is situated symmetrically about the axis, as is usually the case,

$$h' = 2f \tan \frac{w}{2} \tag{6.12a}$$

where w is the angle subtended by the whole object. When w is small

$$2 \tan \frac{w}{2} = \tan w = w$$

Problems involving the determination of the size of the image of a distant object frequently occur in dealing with the eye, the telescope and the camera.

6.12 MAGNIFICATION OF IMAGES IN THE DIRECTION OF THE AXIS

It is sometimes necessary to consider the size of images of objects lying along the lens axis, or the separation of the images of objects at different distances from the lens. Let $B_1 B_2$ (*Figure 6.8*) be the two ends of an object lying along the axis and $B'_1 B'_2$ the corresponding image; then

$$\frac{1}{l'_1} - \frac{1}{l_1} = \frac{1}{l'_2} - \frac{1}{l_2} = \frac{1}{f'}$$

$$\frac{l'_2 - l'_1}{l'_1 l'_2} = \frac{l_2 - l_1}{l_1 l_2}$$

$$\frac{l'_2 - l'_1}{l_2 - l_1} = \frac{l'_1 l'_2}{l_1 l_2}$$

$l'_2 - l'_1$ is the length of image corresponding to the object of length $l_2 - l_1$, and $(l'_2 - l'_1)/(l_2 - l_1)$ is termed the longitudinal or axial magnification.

When the lateral magnification, $m = l'/l$, is very small, as with a comparatively

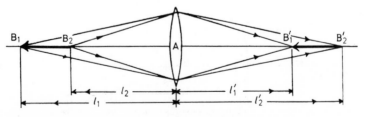

Figure 6.8 Longitudinal or axial magnification

distant object, l'_1/l_1 and l'_2/l_2 are nearly equal, and the longitudinal magnification,

$$\frac{l'_2 - l'_1}{l_2 - l_1} = m^2 \text{ (approx.)} \tag{6.13}$$

Thus, if an image in a photograph, for example, is one-hundredth the size of the object, the 'depth' of the image along the axis is only one ten-thousandth the depth of the object. This fact is of importance in considering depth of focus and allows reasonably good images of objects at different distances being produced on a plane surface, such as the photographic plate.

6.13 PRISMATIC EFFECT OF A LENS

Neglecting aberration, all rays from an axial point are deviated at the lens in such a way that they pass through a single image point on the axis, and each ray therefore suffers an amount of deviation depending on the distance from the axis at which it meets the lens. We may therefore consider the lens as consisting of a series of innumerable prisms, of which the refracting angles, and hence the deviations produced, increase steadily from the centre to the periphery. The refracting angles of these prisms are towards the periphery in the positive lens and towards the centre in the negative lens, while at the optical centre, that is on the axis, the surfaces are parallel, and there is no deviation. The deviation produced on a ray passing through any point on a lens will be termed the **prismatic effect** of the lens at that point, and this increases from the optical centre towards the edge of the lens.

This prismatic effect may be easily seen on looking through a lens at an object. When the eye is looking through the optical centre of the lens, that is, when the line joining the object to the eye passes through the optical centre, the object will appear unchanged in position laterally. On moving the lens at right angles to the axis the object will appear displaced, in the opposite direction to the lens in the case of a positive lens, and in the same direction as the lens in the case of a negative lens, the rays from the object entering the eye having been deviated by the prismatic portion of the lens through which they have passed. This apparent displacement of an object seen through a lens, as it is moved laterally, forms the basis of the neutralisation method commonly used for determining the power of ophthalmic lenses. Spectacle lenses are often 'decentred' before the eye in order to give a prismatic effect in addition to the focusing effect of the lens.

The amount of prismatic effect produced at any point on the lens must depend on the power of the lens and the distance from the axis and may be found as follows:

Any ray BD (*Figure 6.9a*) parallel to the axis, meeting the lens at a distance AD = c from the optical centre, is deviated by the lens (neglecting aberrations) towards the second principal focus F$'$. Then if v is the angle through which the ray is deviated,

$$\tan v = \frac{c}{f'} = cF$$

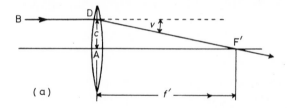

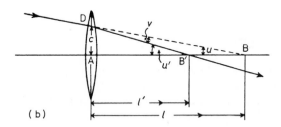

Figure 6.9 Prismatic effect of lenses

We define

$\tan v \times 100$ = prismatic effect P in prism dioptres (section 4.13).

Hence,

$$P \text{ (prism dioptres)} = c \text{ (in cm)} \times F \text{ (in dioptres)} \qquad (6.14)$$

where c is known as the decentration of the ray.

In *Figure 6.9b* a ray incident towards B is deviated by the lens and passes through B'.

Then

$$v = u' - u$$

As the angles are small—paraxial region—

$$u = \frac{c}{l}; \quad u' = \frac{c}{l'} \quad \text{and } v = \frac{c}{f'}$$

hence

$$v = c\left(\frac{1}{l'} - \frac{1}{l}\right) = \frac{c}{f'} = cF$$

Therefore for all paraxial rays meeting a lens at the same height above the axis, the deviation produced by a given thin lens is a constant, and is equal to c/f' radians $= cF$ prism dioptres, when c is in centimetres and F in dioptres of power. The use of centimetres is contrary to recommended practice but remains the normal usage.

6.14 SYSTEMS OF TWO OR MORE LENSES—ELEMENTARY TREATMENT

In practice, single thin lenses are very seldom used and most optical instruments consist of a system of two or more lenses. The most frequent use of a single and comparatively thin lens occurs in the ophthalmic lens, but this is used in conjunction with the eye, which is itself a compound system. Systems of lenses and thick lenses are considered in Chapters 9 and 19 but the position and size of the image formed by a system of thin lenses may be found by applying the results obtained in the present chapter to each of the lenses of the system in turn.

Thin lenses in contact

When a number of thin lenses are placed in contact, their axes being coincident, the vergence of the light on leaving any one lens is the vergence of the light on entering the next. It may be shown that the power of the system is then the sum of the powers of the separate lenses.

$$L_1' = L_1 + F_1 \text{ and } L_1' = L_2$$

$$L_2' = L_2 + F_2 \text{ and } L_2' = L_3$$

$$L_3' = L_3 + F_3 \text{ etc.}$$

$$L_2' = L_2 + F_2 = L_1' + F_2 = L_1 + F_1 + F_2$$

$$L_3' = L_3 + F_3 = L_2' + F_3 = L_1 + F_1 + F_2 + F_3$$

$$L_3' - L_1 = F_1 + F_2 + F_3$$

$L_3' - L_1$ is the power of the complete system, F
Hence

$$F = F_1 + F_2 + F_3 \tag{6.15}$$

In practice, usually not more than two or three lenses in contact can still be considered as a thin lens.

Separated thin lenses

When the lenses of a system are separated, the vergence of the light leaving one lens is changed before reaching the next, and expression (6.15) no longer applies. The position and size of the final image formed by a system of coaxial thin lenses may, however, be found by considering each lens in turn, and treating the image formed by one lens as the object of the next. Thus we have

$$L_1' = L_1 + F_1 \text{ and } l_2 = l_1' - d$$

where d is the distance between the lenses.

$$L_2' = L_2 + F_2$$

Hence l_2' the distance of the image from the second lens may be found.
Also

$$m_1 = \frac{h'_1}{h_1} = \frac{l'_1}{l_1} \text{ and } h'_1 = h_2$$

$$m_2 = \frac{h'_2}{h_2} = \frac{h'_2}{m_1 h_1} = \frac{l'_2}{l_2}$$

hence

$$m = \frac{h'_2}{h_1} = m_1 \times m_2 = \frac{l'_1}{l_1} \times \frac{l'_2}{l_2}$$

The above method is cumbersome where much work is to be done with complicated systems, and the methods of Chapter 9 are preferable.

6.15 GRAPHICAL CONSTRUCTIONS

Many problems involving the determination of the position and size of the image formed by mirrors, lenses and lens systems may be quickly solved by graphical methods, and the student should make himself quite familiar with the constructions used, and practised in the production of accurate drawings.

Choice of scale will be an important factor in obtaining accurate results. Dimensions perpendicular to the axis are usually very small as compared with those along the axis, and these vertical dimensions, such as object and image sizes, may be drawn to a greatly magnified scale, while dimensions along the axis can be drawn in many cases full size or smaller. Angles will therefore be shown magnified, and some that are really equal will appear unequal; but all distances along and perpendicular to the axis will be shown in their relative scales.

As the principles upon which the graphical constructions are based apply only to rays in the paraxial region, that is, to surfaces and lenses of very small aperture, the surface or lens may be represented graphically as a straight line perpendicular to the axis. The diagrams so obtained will not, of course, represent the actual path of the light, but this may easily be put in when the positions of object and image have been found.

The constructions are based on the following facts:

(a) Rays passing through the optical centre of a lens, or the centre of curvature of a surface, are unchanged in direction.
(b) Rays passing through the first principal focus of a lens or surface leave the lens or surface parallel to the axis.
(c) Rays incident parallel to the optical axis pass, after refraction by a lens or surface, through the second principal focus.
(d) All rays that are parallel to one another in the object space intersect after refraction in the second focal plane.
(e) All rays proceeding from a point in the first focal plane emerge from the lens or surface, after refraction, parallel to one another.

The following typical constructions should, with a few exceptions, be self-explanatory:

Positive Lens. Real Object. Real Image (Inverted).	*Figure 6.6*
Positive Lens. Real Object. Virtual Image (Erect).	*Figure 6.10*
Positive Lens. Virtual Object. Real Image (Erect).	*Figure 6.11*
Positive Lens. Distant Object. Real Image (Inverted).	*Figure 6.7*

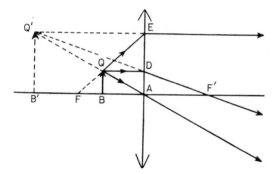

Figure 6.10 Positive lens. Real object. Virtual image (erect)

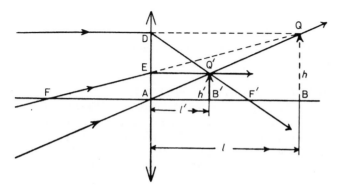

Figure 6.11 Positive lens. Virtual object. Real image (erect)

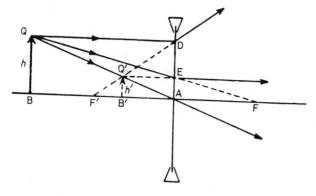

Figure 6.12 Negative lens. Real object. Virtual image (erect)

Negative Lens. Real Object. Virtual Image (Erect).   **Figure 6.12**
*Positive Lens. Image of Axial Object Point, and Construction of the Refracted
Ray for any Incident Ray.*   **Figure 6.13**

    A ray from the given object point B meets the lens at D and the first focal
plane at E. Join EA; the emergent ray DB′ is parallel to EA (rule (*e*) above),
hence B′. This construction also enables the complete path of any portion ST

of an incident ray to be found; it is extremely useful and will be frequently used. An alternative construction is to draw a line from E parallel to the axis meeting the lens at K; join KF'. Then the emergent ray DB' is parallel to KF'.

Negative Lens. Image of Axial Object Point.

The construction is as in *Figure 6.13*.

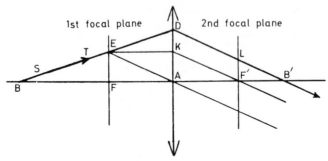

Figure 6.13 Positive lens. Image of axial object, and construction for the refracted ray of any incident ray

Lens System. Two Positive Lenses. Real Object. Figure 6.14.

The commencement of the construction is as in *Figure 6.6*. The rays continuing beyond Q'_1 meet the second lens at D_2, K_2 and E_2, the last two cutting the F_2 plane in G_2 and J_2. Draw the emerging rays $K_2Q'_2$ and $E_2Q'_2$ parallel to G_2A_2 and J_2A_2 respectively. The ray E_1D_2 is parallel to the axis and thus passes through F'_2 intersecting the other two emerging rays in the final image Q'_2.

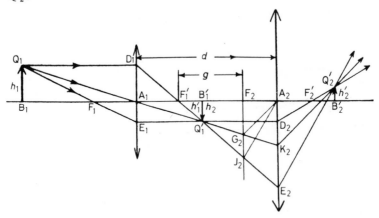

Figure 6.14 Lens system. Two positive lenses, real object

Lens System. Distant Object. Equivalent Lens. Figure 6.15.

The image $F'_1Q'_1$ is formed in the second focal plane of the first lens. It is a virtual object for the second lens. Rays $A_2Q'_1$ and $E_2F'_2$ fix the position of Q'_2. This final image is in the second focal plane of the system. Q'_2P' drawn parallel to the original incident beam fixes P', the position of a thin lens producing the same effect. This is called the **equivalent thin lens** (see Chapter 9).

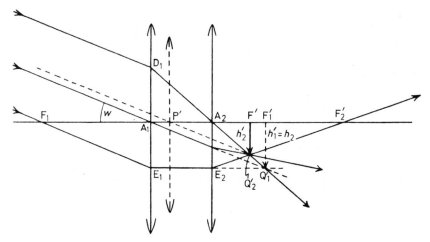

Figure 6.15 Lens system. Distant object. Equivalent lens

6.16 THE Γ-DIAGRAM

When applying graphical methods to a number of centred thin lenses a convenient method of working links the first focal plane of each lens with the lens
position by using form of the greek letter Γ. Drawn to scale in *Figure 6.16* are
three lenses of various powers forming an optical system. Each lens is shown
by a vertical line and linked with this is a horizontal line representing (to scale)

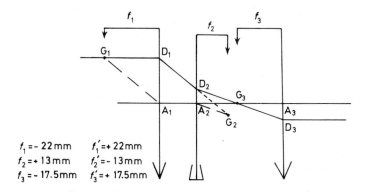

$f_1 = -22\text{mm}$ $f_1' = +22\text{mm}$
$f_2 = +13\text{mm}$ $f_2' = -13\text{mm}$
$f_3 = -17.5\text{mm}$ $f_3' = +17.5\text{mm}$

Figure 6.16 The Γ diagram

the *first* focal length. The incident ray is drawn to meet the first lens at D_1 cutting the first focal plane of that lens at G_1. The refracted ray is simply drawn
parallel to G_1A_1 where A_1 is the optical centre of the first lens. The same
procedure is applied in turn. The value of the hanging Γ format is that it reduces
the possibility of error caused by associating one focal plane with another lens
centre.

6.17 EFFECTIVE POWER OF A LENS

The vergence of the emergent light at A (*Figure 6.17*) immediately on leaving
the lens is $1/l' = L'$. At some place X at a distance d from the lens, the vergence has changed to

$$\frac{1}{l' - d} = \frac{1}{l'_x} = L'_x$$

$$L'_x = \frac{1}{l' - d} = \frac{L'}{1 - dL'}$$

If the incident light is parallel, $L' = F$, and substituting F_x for L'_x in this
case, we have

$$F_x = \frac{F}{1 - dF} \tag{6.16}$$

d, of course, being in metres when F is in dioptres. This is termed the **effective
power** of the lens at the position X. The idea of effective power is very useful
in dealing with systems of lenses, e.g., the spectacle lens and the eye. It will be
seen that the effective power of a positive lens increases, while that of a negative lens decreases, as the position X moves away from the lens (assuming a
distant object). This particularly applies to spectacle lenses in front of the eye.*

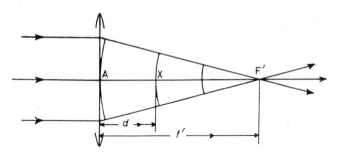

Figure 6.17 Effective power of a lens

6.18 THE MEASUREMENT OF FOCAL LENGTH AND POWER

The effect of a lens is specified by its focal power or focal length, and it is
necessary to be able to measure these quantities, often with considerable accuracy. A large number of methods are available, and a few of the most useful and
typical will be described here.

When the refractive index of the glass is known, the power of a thin lens may
be found from a spherometer measurement of its curvatures. This is done in the
case of the optician's lens measure. (Section 5.2).

*For a full treatment see H. Obstfeld, *Optics in Vision*, Butterworths

Neutralisation

The method commonly used by the ophthalmic optician is that known as **neutralisation**. The power of the lens is determined by finding a lens that, when placed in contact with the lens to be measured, gives a combination having zero power, as judged by observing that a lateral movement of the combination before the eye produces no apparent movement of an object seen through it. If the two lenses are considered as thin, the lens tested is of equal but opposite power to the known lens. The test is carried out with the optician's trial case, which consists of a set of positive and negative lenses of known power increasing by small steps.

6.19 THE OPTICAL BENCH

Most methods involve the use of an **optical bench**. This consists of a horizontal bar, usually of either one or two metres length, graduated in centimetres and millimetres. A number of stands to carry lenses, screens, objects, etc., slide along the bar, their positions being read by an index against the scale. As an object, cross-lines stretched across a frame, or a series of small holes, in the form of a cross, drilled in a metal plate, may be placed in front of a frosted electric lamp. To receive a real image a ground glass screen is used, the image being viewed *through* the glass. Adjustable holders may be provided to carry lenses and systems of various diameters, and it is very convenient to have a lens holder which will carry the standard trial case lenses of 38 mm diameter, a large selection of lenses thus being easily available for experimental purposes. Other bench fittings include a collimator and telescope (see below), and a measuring rod to enable distances between lenses, screens, etc., to be determined when these are not exactly over the index mark.

6.20 OPTICAL BENCH METHODS. POSITIVE LENSES

Collimator method

When a suitable distant object is available the focal length of a thin positive lens may be found directly by receiving the image of the object on a screen, its distance from the lens being the second focal length. In the laboratory, however, a convenient distant object is not usually available, but the same effect may be obtained by means of a piece of apparatus known as a **collimator**. This consists of a positive achromatic lens mounted at one end of a tube with an illuminated object of some form at the other; the object may be a small circular aperture, a narrow slit, cross-lines or a scale, according to the purpose for which the collimator is required. The object is placed at the first principal focus of the lens, so that light from any point on the object is parallel on emerging from the lens, as though coming from a distant object. The collimator is an important piece of apparatus in many optical experiments.

To use the collimator in determining the focal length it is mounted with its axis in line with the bench and the image formed by the lens under test is received on the screen as with an actual distant object.

Telescope method

The position of the first principal focus and hence the focal length of a thin positive lens can be found if we can determine when the light emerging from the lens is parallel. This can be done by means of a telescope. The telescope can be of the astronomical form (see Chapter 10) with cross-lines in the focal plane of the eyepiece. To adjust the telescope the eyepiece is first sharply focused on the cross-lines; the telescope is then directed to a very distant object and focused so that the image of the distant object falls exactly in the plane of the cross-lines. This is checked by noting that there is no parallax between cross-lines and image, that is, there is no apparent movement of one against the other as the eye is moved slightly from side to side. An object, the lens to be tested and the adjusted telescope are now set up in line on the bench, and the distance between lens and object varied until a clear image of the object is seen in the telescope. Since the telescope has been focused for parallel light, the object must now be at the principal focus of the lens, and the focal length can be read on the bench.

Mirror method

A simple method of determining when the light leaving a lens is parallel is by means of a plane mirror placed behind the lens to reflect the emergent light back through the lens. When the object is at the first principal focus of the lens, the parallel light reflected by the mirror must focus, after passing again through the lens, in the plane of the object. If the object is inside the first principal focus, the divergent emergent light will be still divergent after reflection and will therefore focus, after again passing through the lens, outside the principal focus. Similarly if the object is outside the first principal focus, the reflected image will be inside.

Object, lens and plane mirror are set up on the bench, and the reflected image received on a small screen placed in the plane of the object; or a small object, such as a pin, may be used, this being adjusted until it and the reflected aerial image coincide as judged by parallax.

This method of accurately placing an object at the principal focus of a lens by obtaining coincidence of the object and its reflected image is known as **auto-collimation**.

Conjugate foci methods

When the focal length of the lens to be measured is sufficiently short for the object and its real image to come within the length of the bench, the focal length or the focal power may be determined by forming a real image of the object on a screen and measuring the distance l and l' of the object and image from the lens. The focal length or focal power is then found from the conjugate foci expression (6.03) or (6.03a). It should be remembered that in order that a real image of the object shall be received on the bench, the length of the bench must be at least four times the focal length of the lens.

6.21 OPTICAL BENCH METHODS. NEGATIVE LENSES

As the negative lens will not form a real image of a real object, methods of measuring negative lenses are a little more complicated than those described above. In most cases an additional positive lens is used to give convergent light incident on the negative lens, the image formed by the positive lens acting as a virtual object for the negative lens. The following are two simple methods of measuring the focal length or focal power of negative lenses.

Conjugate foci method

A positive lens A_1 (*Figure 6.18*) is arranged to give a real image of an object B on a screen at B_1'. The negative lens to be measured, A_2, is placed between the positive lens and the screen, and the distance $A_2 B_1'$ is then the object distance l_2 for the negative lens. If this distance l_2 is less than the focal length of the negative lens, the light on leaving this lens will be still convergent, and the screen may be moved back to the new position B_2', where a sharp image is again formed. The distance $A_2 B_2'$ is then the image distance l_2' for the negative lens, and the focal power or focal length can be found from expressions (6.03) or (6.03a). Both object and image distances are positive in this case.

This method is also useful for measuring the focal length and power of weak positive lenses. The lens to be measured is placed between the lens A_1 and the first image, as with the negative lens, and the screen must then be moved towards the lens to focus the new image.

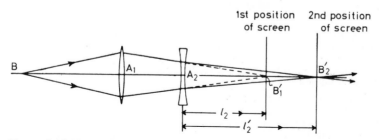

Figure 6.18 Measurement of focal length of negative lens. Conjugate foci method

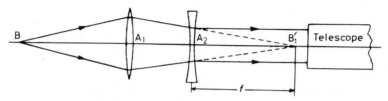

Figure 6.19 Measurement of focal length of negative lens. Telescope method

Telescope method

Using the same arrangement as in the previous experiment, but replacing the screen in its second position by a telescope focused on a distant object, the negative lens may be adjusted until the light emerging from it is parallel (*Figure 6.19*), when a sharp image of the object will be seen in the telescope. The image formed by the positive lens is now at the principal focus of the negative lens, and the focal length can be read off directly on the bench.

Other methods that are suitable also for measuring the focal lengths of thick lenses and systems, are described in Chapter 9.

CHAPTER 6   *EXERCISES*

1. The radius of curvature of one surface of a biconvex lens made of glass having a refractive index of 1.53 is six times that of the other surface. If the power of the lens is 5D, what are its radii of curvature?

2. Distinguish between a real and a virtual image, illustrating your answer by describing the formation of (a) a real image, (b) a virtual image, by a convex lens.

3. Explain what is meant by the focal power and the focal length of a thin lens, illustrating your answer by the case of both positive and negative lenses. A positive lens is to be made of glass having a refractive index of 1.55 and is required to have a focal length of 150 mm, what will be the radii of curvature of its surfaces if the lens is made (a) equi-convex, (b) plano-convex?

4. A thin plano-convex lens is to be made from glass of refractive index 1.62, so that its power will be +5 D. Find the radius of the tool with which the convex surface must be ground.

5. A very thin watch glass, the radius of curvature of which is 4 inches, is placed on a table; water is poured into it to a depth of $1/8$ inch. Find the aperture of the water lens thus formed and its focal power and focal length.

6. Draw, full size, a section through an equi bi-convex lens of 80 mm aperture and surface radii 50 mm. From a point on the axis 40 mm from the front surface draw a ray inclined to the axis at 30°. Trace its path through the lens and find the corresponding image point on the axis. Assume the refractive index to be 1.5.

7. Establish a formula which gives the focal lengths of a thin lens in terms of the positions of an axial object and its image.

A rod AB 15 cm in length lies along the axis of a thin converging lens of focal length 20 cm. If the nearer end of the rod is 25 cm from the lens, find the length of the image and show its position on a diagram.

8. Where must an object be placed in front of a convex lens of 8 cm focal length to give an image:
   (a) 25 cm in front of the lens.
   (b) 25 cm behind the lens.
State in each case whether the image is real or virtual.

9. An object is placed 333 mm from an equal-convex lens and a virtual image is formed 2 metres from the lens. Calculate the radii of curvature of the lens if its refactive index is 1.523.

10. A small object is moved along the optical axis of a positive thin lens of focal length 25 cm. Find the position of the image and the vergence of the emergent pencil (or curvature of emergent wavefront) for each of the following object distances:
   (a) $-\infty$; (b) -200 cm; (c) -100 cm; (d) -50 cm; (e) -25 cm; (f) -10 cm.

11. An object 5 mm high is placed 600 mm in front of a negative lens of focal length 250 mm. Find (a) by calculation, (b) by graphical construction to scale, the position, size and attitude (erect or inverted) of the image.

12. A positive lens forms a real image of a luminous object. If the lens is gradually covered over, what happens to the image? Explain clearly with diagrams.

13. Give a graphical construction for the image formed by a convergent lens. A small object is placed upright on the axis of a convergent lens at a distance of 100 mm from the lens, and it is observed that a virtual erect image of twice the linear dimensions of the object is formed. Determine the focal length and power of the lens.

14. A lamp and screen are 3 feet apart and a + 7D lens is mounted between them. Where must the lens be placed in order to give a sharp image, and what will be the magnification?

15. How does the transverse magnification produced by a lens of focal length f depend on the distance of the object from the lens? Calculate the focal length of the lens required to throw an image of a given object on a screen with a linear magnification of 3.5, the object being situated 20 cm from the lens.

16. Prove for (a) a positive lens, (b) a negative lens, that the magnification of an image is changed by unity if the image distance is changed by an amount equal to either focal length of the lens.

17. What is meant by the magnification of an image? An object is placed 30 mm from a positive lens and a virtual image four times the size of the object is required. Find graphically and by calculation the necessary focal length of the lens.

18. An object is to be photographed three-quarters full size using a lens of 6 inches focal length. What must be the distance of the object from the lens?

19. In photographing a person 6 feet high and standing 12 feet from the camera it is required to obtain an image 4 inches high. What must be the focal length of the lens?

20. An object 1 cm long is placed perpendicular to the optical axis of a positive lens, the lower end of the object being 5 cm above a point on the axis 20 cm from the lens. The lens has a focal length of 8 cm and an aperture of 3 cm. Find graphically the size and position of the image.

21. When a real image is formed 9 inches from a convex lens the magnification is unity. Where will a real image be formed when the magnification is fifty? What is the focal length of the lens?

22. It is required to form a real image 800 mm from an object and magnified four times. Find the position and focal length of the required lens.

23. The distance between an object and its real image formed by a lens is 32 inches. If the image be three times the size of the object, find the position and focal power of the lens.

24. The difference in the positions of the image when the object is first a great distance and then at 5 metres from the anterior principal focus of a convex lens is 3 mm. Calculate the focal length of the lens.

25. Apply carefully a graphical construction to find the position and size of the image of an object 2 mm high formed by a positive lens of focal length 25 cm. The object distance is (a) − 50 cm, (b) − 80 cm. Explain the reasons for the various steps in the construction.

26. With a projection lantern it is required to project an image 20 feet in diameter on a screen 50 feet from the lens. What focal length is required, the diameter of the object being 3 inches?

27. Find the size and position of the image formed by a convex lens of 10 inches focal length, of the following object, viz., an arrow 1 foot 8 inches long lying along the axis of the lens so that its middle point is 2 feet 1 inch from the lens. What would be the size and position of the image if the arrow were turned through 90° about its middle point?

28. When an object is 10.5 inches in front of a lens the virtual image formed is 4.3 inches from the lens. What will be the position of the image when the object is 43 inches in front of the lens?

29. Calculate, and show with a diagram, the position of the image when an object is placed 16 inches in front of a convex lens, the focal length of which is 20 inches. What will be the magnification of the image?

30. A ray of light is incident in a direction inclined at 30° to the axis of a positive thin lens of 20 mm focal length, cutting the axis at a point 50 mm in front of the lens. A point moves along this ray from infinity on one side of the lens to infinity on the other. Find graphically the conjugate ray and trace the movement of the conjugate image point.

31. Light is made to converge to a point 30 cm from a convex lens. When a concave lens is placed 10 cm from the convex lens and in the converging pencil of light, the point to

which the pencil converges is moved to 10 cm farther from the convex lens. Find the focal length of the concave lens.

32. A real inverted image of an object is formed on a screen by a thin lens, and the image and object are equal in size. When a second thin lens is placed in contact with the first, the screen has to be moved 20 mm nearer the lenses in order to obtain a sharp image, and the size of this image is three-quarters that of the first image. Find the focal length of the two lenses.

33. An image formed on a screen by a convex lens is 50 mm long. Without moving either the object or screen, which are 1.5 m apart, a second image can be produced which is 200 mm long. Show how this is possible and find the size of the object and the focal power of the lens.

34. A lens + 6.55D is 10 cm in diameter and is fixed horizontally 24 cm above a point source of light. Calculate the diameter of the circle of light projected on a ceiling situated 8.2 metres above the lens.

35. A distant object subtends an angle of 30°. Give a careful diagram showing the formation of an image of this object by a lens of + 5D power, and from the diagram find the size of the image.

36. A distant object subtends an angle of 10° at the centre of a positive lens of 15 cm focal length and 5 cm aperture, one end of the object being on the lens axis. Show graphically the path of a pencil of light from each end of the object and filling the aperture of the lens. Find from the diagram and by calculation the size of the image formed.

37. A distant object subtends an angle of 5°; what focal length lens is necessary to form a real image 8 mm long?

38. An image of a distant object is formed by a + 2D lens. Find the position of a second lens of + 5D power in order that a real image may be formed half the size of the original image.

39. A distant object subtends an angle of 10°; find the size of the image formed by a + 4D lens. Where must a − 8D lens be placed to form a real image four times the size of the original image?

40. Find the focal length of a thin lens which will form an image of the same size as that formed by the system in question 39, using the same object.

41. A small object stands upright on the axis of a convergent lens A. Prove by means of a diagram, that if another lens is placed in the anterior focal plane of the lens A the size of the image remains unaltered while its position is changed.

42. A converging lens forms an image of a distant object at a plane A and when a diverging lens is placed 1 inch in front of A (between the converging lens and A) a real image is formed 5 inches beyond A. Find the focal length and focal power of the diverging lens.

43. What is meant by chromatic aberration of a lens? The light from a distant small source of light is brought to a focus by means of a thin equi-convex lens with curves of 20 cm radius of curvature. If the refractive indices for red and blue light be 1.515 and 1.524 respectively, find the chromatic aberration of the lens.

44. The refractive index of a certain kind of crown glass is: for yellow light 1.5186, red 1.5161, blue 1.5247. The radii of curvature of a lens made of this glass are + 55 mm and + 87 mm. Find the power and focal length of the lens for each colour.
    What appearances would you expect at the place where the lens focuses a pencil of white light?

45. Parallel rays of white light fall axially on a converging lens of glass. Draw a diagram showing the convergence of the resultant red and blue rays to separate foci, and show that a white screen can be placed in such positions that the image is (1) a white spot fringed with red or (2) a white spot fringed with blue.

46. A thin converging lens of focal length *f* is used to form an image of a luminous object on a screen. Determine the least possible distance between object and screen.
    A lens of 10 cm focal length forms a sharp image of a fixed object on a fixed screen for two positions of the lens which are 15 cm apart. If the distance from object to screen be

doubled, what will be the distance apart of the two positions of the lens for which there will again be a sharp image?

47. A ray of light on passing through a + 5D lens is deviated through the same angle as it would have been when refracted by a prism of 10° refracting angle, $n = 1.62$. At what distance from the optical centre is the ray meeting the lens?

48. When a positive lens is placed close to the eye and moved laterally objects seen through the lens are seen to move in the opposite direction, while with a negative lens the apparent movement of the object is in the same direction as the lens. Explain why this is and show how the effect may be used to find the power of a lens.

49. Two thin watch glasses each of radius of curvature 30 cm are cemented together in the form of a double convex lens. The lens is placed in water, the refractive index of which is $^4/_3$. Calculate the focal length and prove any formula you use.

50. A thin bi-convex lens having curvatures of 3D and 5D and $n = 1.62$ is immersed in a large tank of water. An object is placed in the water one metre from the lens; what will be the position of the image if that is also formed in the water?

51. An object 33 mm high is placed at 444 mm from a lens, which brings the light to a focus at an equal distance beyond the lens, but of this distance 124 mm is travelled in glass of refractive index 1.55. Find the power of the lens and the size of the image.

52. A plane mirror is set up behind a convex lens and a pin is placed in front. By adjusting the distances of the pin and mirror from the lens it is found that in two positions the image of the pin coincides with the pin, the image in one case being erect, in the other inverted. Draw diagrams to illustrate the formation of the image in each case and show how the focal length may be determined.

53. Describe a method of measuring the focal length of a negative lens.

54. Explain exactly how to determine the refractive index of a given concave lens by using an optical bench and its usual fittings.

55. In measuring the focal length of a lens by a telescope method, the telescope was focused on an object 8 feet away instead of on a distant object. With the telescope placed close to the lens a sharp image was obtained when the object was 6 inches away from the lens. What was the focal length of the lens?

56. Describe carefully how you would determine the focal length of a weak positive lens, say of 0.5D power. Illustrate the method by giving a typical set of bench readings, and show how the focal length could be found from these.

57. An illuminated object A is placed at the first principal focus of a 20D positive lens L on the other side of which, at some distance, is a telescope which is adjusted until the object is seen distinctly. When a lens of unknown power is placed at the second principal focus of L, between L and the telescope, it is found necessary to move the object 2 cm further away from L in order to see it distinctly in the telescope. Calculate the power of the unknown lens.

Chapter 7

Reflection at curved surfaces

7.1 INTRODUCTION

It has been shown in Chapter 3 that the effect of a plane reflecting surface is to change the direction of the light without altering its vergence, and an image of an object is therefore formed at the same distance from the surface as the object, but on the opposite side of it. If, however, the surface is curved, the reflected light will be changed both in direction and vergence.

The curved reflecting surface may be either convex or concave to the incident light, and may be of various forms, usually spherical, paraboloidal, or cylindrical. For the present, only the spherical form will be considered.

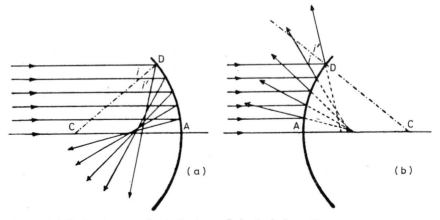

Figure 7.1 Reflection at spherical mirror. Spherical aberration

In *Figures 7.1a* and 7.1b DA represents the principal section of a concave and of a convex reflecting surface; C is the **centre of curvature**, and A, the centre of the aperture, is the **pole** or **vertex** of the surface; the line AC is the **principal** or **optical axis**. Clearly any directions, such as AC and DC passing through the centre of curvature will be normal to the portions of the surface through which they pass. The direction of any part of the light reflected at a curved surface may therefore be found from the law of reflection, the angle of incidence for any ray being the angle which the ray makes with the radius of curvature through the point of incidence. *Figures 7.1a* and *7.1b* show the light from a distant object point on the axis reflected at a concave and a convex surface respectively. It will be seen that the light which comes from the object point is no longer travelling exactly from or towards a point image after reflection; the rays meeting the surface close to the axis will cross the axis farther from the surface than those more remote from the axis. Thus there is spherical aberration, and the form of the reflected beam is very similar to that of the beam after refraction at a spherical surface (section 5.7). Only when the object point is at the centre of curvature of a spherical mirror will the reflected pencil be homocentric, as in this case all the rays meet the surface normally and are reflected along their original paths; object and image are therefore coincident.

7.2 FOCAL POWER

If the light is limited to the paraxial region of the mirror, the light from a point object will, after reflection, be converging to or diverging from an approximate point, and an image of the object will be formed either before or behind the mirror. The effect produced by a curved mirror, and hence the position of the image, may be found, as in the case of the refracting surface and the lens, either by considering the change in vergence or curvature of the wavefronts, or the change in the direction of the rays.

Change in vergence

Figure 7.2 shows light, converging towards a virtual object point B, intercepted by a convex mirror at A, and diverging after reflection from the virtual image B′

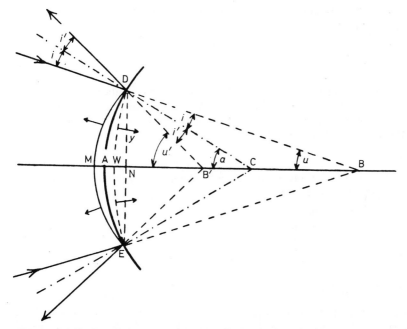

Figure 7.2 Reflection at spherical surface. Change in vergence

DAE represents the mirror, of curvature $\dfrac{1}{r}$ = R.

DWE represents the incident wavefront, of curvature $\dfrac{1}{l}$.

By applying Huygens' construction as for a plane reflecting surface (section 3.2), the exact form of the reflected wavefront can be found. It will be seen that when the mirror aperture is large, the section of the reflected wavefront is not circular, that is, the wavefront is not spherical, and the reflected light does not therefore diverge from a point. If, however, we consider only a very small portion of the mirror near the axis, the reflected wavefront will be approximately spherical with its centre at B'.

In the time that the centre of the wavefront would have travelled a distance AW, the reflected wavefront will have travelled back an equal distance AM, and similarly for all other points on the incident wavefront. DME is therefore the reflected wavefront of curvature $1/l'$, corresponding to the incident wavefront DWE. Then

MA = AW = AN − WN

MN = MA + AN = AW + AN = AN − WN + AN = − WN + 2AN

For small portions of the curves, the sags are proportional to the curvatures; hence

$$\frac{1}{l'} = -\frac{1}{l} + \frac{2}{r} \qquad\qquad (7.01)$$

As the reflected light is travelling in the opposite direction to the incident light, we shall regard

$$L' = -\frac{1}{l'}$$

as the reduced vergence when the light is in air. That is, the reflected light is divergent, negative vergence, L' negative, when l' is positive, and convergent, positive vergence, L' positive, when l' is negative.

Expression (7.01) may then be written

$$L' = L - 2R$$

$L' - L$, the change in vergence produced by the mirror, is the **focal power**, F, of the mirror and, for a mirror in air,

$$L' - L = F = -2R \left.\vphantom{\begin{array}{c}1\\1\\1\end{array}}\right\}$$

or   (7.02)

$$L' = L + F$$

Change in ray path

From the law of reflection, angle i = angle i'. When the angles a, u and u' are small, the arc AD may be considered as a short line of length y perpendicular to the axis and

$$a = \frac{y}{AC}, \quad u = \frac{y}{AB}, \quad u' = \frac{y}{AB'}$$

From the figure,

$$a = i + u, \text{ and } u' = i' + a = i + a$$

Hence

$$a = u' - a + u$$

$$2a = u' + u$$

Substituting the above values for a, u and u'

$$\frac{2}{AC} = \frac{1}{AB'} + \frac{1}{AB}$$

or

$$\frac{1}{l'} = -\frac{1}{l} + \frac{2}{r} \text{ as before} \qquad (7.01)$$

7.3 REFLECTION REGARDED AS A SPECIAL CASE OF REFRACTION

If we apply the sign convention for angles to the case of reflection, we see that the angles of incidence and reflection are always of opposite signs. The law of reflection may therefore be considered as a special case of the law of refraction where $n' = -n$, and $n' \sin i' = n \sin i$ then becomes $i' = -i$, i' being the angle of reflection.

By means of this device any expression derived for refraction at a spherical surface may be applied to reflection at a spherical surface, $-n$ being substituted for n'. Thus the fundamental paraxial expression (5.02)

$$\frac{n'}{l'} - \frac{n}{l} = \frac{n' - n}{r}$$

becomes for reflection

$$-\frac{n}{l'} - \frac{n}{l} = -\frac{2n}{r}$$

or

$$\frac{1}{l'} = -\frac{1}{l} + \frac{2}{r} \qquad (7.01)$$

Similarly the expressions (5.05)

$$\frac{n'}{f'} = -\frac{n}{f} = F = \frac{n' - n}{r}$$

become for reflection in air

$$-\frac{1}{f'} = -\frac{1}{f} = F = -\frac{2}{r} \qquad (cf.\ 7.04)$$

For a mirror in any medium of refractive index n, its focal power

$$F = -\frac{n}{l'} - \frac{n}{l} = L' - L = -\frac{2n}{r} = -2nR \tag{7.03}$$

7.4 CONJUGATE FOCI. PRINCIPAL FOCUS. FOCAL LENGTH

The positions of object and image for any mirror can be found from expressions (7.01) or (7.02), the same sign convention being used as for the refracting surface or lens. It will be seen that the general effect of a concave mirror is similar to that of a positive lens, and the effect of the convex mirror is similar to that of a negative lens, with the addition, however, in each case that the general direction of the light is changed on reflection.

Thus, parallel light from a distant object will, after reflection at a concave mirror, be converging to form a real image in front of the mirror (*Figure 7.3a*); the position of the image is then the **second principal focus**, F′, of the mirror, and its distance AF′ from the mirror, the **second focal length**. As the object moves up towards the mirror, the image moves away, and object and image coincide at the centre of curvature. When the object reaches the **first principal focus**, which in the case of the mirror coincides with the second principal focus, the image is at infinity, and the reflected light is parallel. With the object inside the principal focus, the reflected light will be diverging from a virtual image behind the mirror, and at a greater distance from the mirror than the object, until on reaching the mirror surface object and image again coincide.

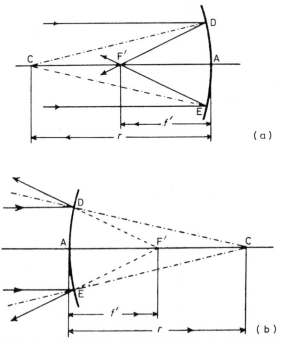

Figure 7.3 Principal focus and focal length of spherical mirror

With a convex mirror, as it makes the light more divergent, a virtual image is always formed with a real object. When the object is very distant, this virtual image is at the second principal focus F′, its distance AF′ from the mirror being the second focal length, f' (*Figure 7.3b*).

When the incident light is converging towards a virtual object, in the case of a concave mirror a real image will always be formed between the principal focus and the mirror. In the case of the convex mirror, however, as with the negative lens, the image may be real or virtual according to the position of the virtual object. When the incident light is converging towards the first principal focus, which coincides with the second principal focus, the reflected light is parallel, and the image therefore at infinity. If the virtual object is between the mirror and the principal focus, a real image is formed in front of the mirror, while if the virtual object is outside the principal focus, the reflected light will be diverging from a virtual image. Object and image coincide at the centre of curvature and at the surface.

The image positions for a number of different object positions, should be found graphically and by calculation.

When $l = \infty$, $l' = f'$ and when $l' = \infty$, $l = f$. Therefore from (7.01) and (7.02)

$$f' = -\frac{1}{F} = \frac{r}{2} \qquad (7.04)$$

The principal focus of a mirror is therefore situated midway between the mirror and its centre of curvature.

7.5 EXTRA-AXIAL POINTS. GRAPHICAL CONSTRUCTION. MAGNIFICATION

When an object point is situated some distance from the axis of the mirror, the reflected light does not pass through a point, even if the aperture of the mirror is quite small, as may be seen by finding graphically from the law of reflection the reflected directions of a number of rays from such a point. If, however, the object is close to the axis and the mirror aperture small, so that the incident light is in the paraxial region, a reasonably good image will be formed. This image must lie somewhere on the ray from the object point that passes through the centre of curvature, for this ray is reflected back along its original direction, and the position of the image can be determined by finding where this ray is intersected by some other ray. A convenient second ray to consider will be one incident parallel to the axis, that after reflection will pass through the principal focus. The image, formed by a mirror, of an object of finite size, may be determined by finding the images of a number of points on the object. The image of a small plane object perpendicular to the mirror axis will be a plane also perpendicular to the axis.

As with the refracting surface and lens, the position and size of the image formed by a curved mirror may conveniently be found graphically. The construction being true only for the paraxial region, the mirror may be represented by a straight line passing through the vertex perpendicular to the axis. Suitable scales may be chosen, as explained in section 6.15. The constructions are based on the following facts:

All rays parallel to the axis will, after reflection, pass through the principal focus and vice versa.

All rays incident through the centre of curvature are reflected back along their original paths.

All rays incident at the vertex are reflected at an equal angle on the opposite side of the axis.

Three examples of these constructions are shown, viz.:

Concave mirror, real inverted image. Figure 7.4.

Concave mirror, object inside principal focus, virtual, erect and enlarged image. Figure 7.5. This illustrates the formation of the image in the ordinary concave shaving mirror.

Convex mirror, virtual, erect and diminished image. Figure 7.6. In the convex driving mirror of the motor car, the diminished image permits of a greater extent of image being seen than would be seen with a plane mirror of the same size.

In *Figures 7.4, 7.5* and *7.6*

$$\frac{h'}{h} = m = -\frac{AB'}{AB} = -\frac{l'}{l} = \frac{L}{L'} \qquad (7.05)$$

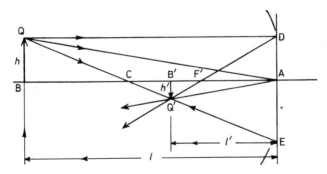

Figure 7.4 Concave mirror (convergent). Real inverted image

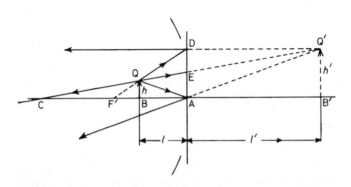

Figure 7.5 Concave mirror (convergent). Virtual erect image

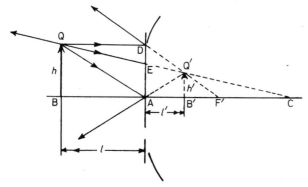

Figure 7.6 Convex mirror (divergent). Virtual, erect and diminished image

The magnification may also be expressed in terms including f' as for a thin lens.

$$\left. \begin{array}{l} m = \dfrac{L}{L'} = \dfrac{L' - F}{L'} = 1 - \dfrac{l'}{f'} \\[3ex] \dfrac{1}{m} = \dfrac{L'}{L} = \dfrac{L + F}{L} = 1 - \dfrac{l}{f'} \end{array} \right\} \qquad (7.06)$$

Other expressions, in terms of object and image distances measured from the centre of curvature or from the principal focus (Newton's relation) are easily derived.

7.6 THE LENS MIRROR

Some light is reflected at each of the surfaces of a lens, and that reflected at the second surface again passes through and is refracted by the first. Mirrors are sometimes produced by silvering the second surface of a lens; examples occur in the Mangin mirror (section 7.9) and the large double convex lens silvered on the second surface that is used in place of the concave shaving mirror. Such a system is known as a **lens mirror** and its general properties are considered in Chapter 19 (section 19.11). When, however, the lens may be considered as thin the focal power of the lens mirror may be found simply as follows:

The light passing through the first surface of power F_1 is reflected at the second surface, which is a mirror having a power $-2nR_2$, and again passes through the first surface. The total change produced in vergence, that is, the power of the lens mirror, is therefore:

$$F_M = 2F_1 - 2nR_2 \qquad (7.07)$$

Example:

A thin biconvex lens of + 5D power is set up on an optical bench. It is found that the image formed by reflection from the second surface coincides with the object at a distance of 100 mm. On reversing the lens, object and reflected image coincide at 125 mm from the lens. Find the curvatures of the surfaces and the refractive index of the glass.

Treating the lens-mirror as a single concave mirror, since object and image coincide the radius of curvature of the equivalent mirror would be – 100 mm in the first place and – 125 mm when the lens is reversed. Therefore the powers of the lens-mirror are + 20D, and + 16D when reversed.

The power of a lens-mirror (from 7.07) is

$$F_M = 2F_1 - 2nR_2 = 2(n-1)R_1 - 2nR_2$$

$$20 = 2(n-1)R_1 - 2nR_2 \quad \text{or} \quad 10 = nR_1 - R_1 - nR_2 \tag{1}$$

$$16 = -2(n-1)R_2 + 2nR_1 \quad \text{or} \quad 8 = -nR_2 + R_2 + nR_1 \tag{2}$$

(In (2) the signs of R_1 and R_2 are changed as the lens is reversed). For the lens

$$F = (n-1)(R_1 - R_2)$$

$$5 = (n-1)(R_1 - R_2) \tag{3}$$

Solving the simultaneous equations (1), (2) and (3) we obtain R_1 = + 3D, R_2 = – 5D, n = 1.625.

7.7 MEASUREMENT OF FOCAL POWER AND FOCAL LENGTH

As the power of a mirror depends only upon its curvature, the power may be determined from a spherometer measurement of curvature. Optical methods are, however, often preferable, especially in the case of surfaces of very short or very long radius, and such methods avoid the possibility of damage to the surface, as may occur when a spherometer is used on a finely polished surface. The optical methods involve the use of an optical bench and fittings, as described in sections 6.20 and 6.21.

Concave mirrors. Radius of curvature. Parallax method

The mirror to be measured is mounted vertically on the optical bench, and a well illuminated small object, such as a pin, is placed in front of it. On moving the object along the bench, a position will be found where the real inverted aerial image formed by the mirror coincides in position with the object. This coincidence is judged by observing that object and image do not appear to separate as the head is moved across the bench, that is, there is no parallax. Object and image must then be at the centre of curvature of the mirror, and the radius of curvature can be measured.

Focal length. Collimator method

When light from a distant object is available, the focal length of the concave mirror can be obtained directly. In the laboratory a collimator will generally be used. The collimator and the mirror are set up on the bench with their axes almost coinciding—it may be necessary to tilt the mirror slightly in order to separate the incident and reflected beams. A small screen covering not more than half the aperture of the mirror, is mounted between them, and adjusted until a well defined image of the collimator object is received on it from the mirror. The screen is then at the principal focus of the mirror, and the focal length can be read off on the bench.

Focal Length. Telescope method

A telescope, previously focused on a distant object, is mounted on the bench, and directed towards the mirror. A small object, such as a pin, well illuminated, is mounted vertically between the mirror and telescope, the head of the pin being as near as possible on the axis of the mirror. On moving the object along the bench, a position will be found where a sharply focused image of the object is seen in the telescope. The light reflected from the mirror must now be parallel, and the object is therefore at the principal focus.

Focal power or focal length. Conjugate foci method

The focal power or focal length of the concave mirror may be determined by the method of conjugate foci, as described for a positive lens. A real image of an object is received on a small screen, and the distances l and l' measured; the power is then obtained from expression (7.02).

Convex mirrors. Radius of curvature

When a convergent beam is intercepted by a convex mirror in such a position that the beam is converging towards the centre of curvature of the mirror, the reflected light will return along its original path. This fact is the basis of a useful method of measuring the radius of curvature of a convex mirror. An illuminated object, positive lens and the mirror to be measured, are set up on the optical bench, and adjusted until a sharp image of the object is formed on the mirror surface, the distance from the lens to this position of the mirror being made greater than the radius of curvature of the mirror. The mirror is then moved towards the lens until an image is formed in the plane of the object; this position is most accurately found by parallax. The distance $A_2 A_1$ between the two positions of the mirror is then the radius of curvature of the mirror (*Figure 7.7*).

The same arrangement can be used to determine the radius of curvature of a concave mirror, the mirror in this case being moved away from the lens to its second position.

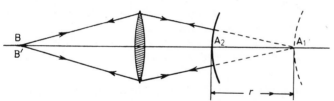

Figure 7.7 Convex mirror. Measurement of radius of curvature

Measurement of surfaces of short radius. Drysdale's method*

A modification of the last method, suggested by Dr. C. V. Drysdale, permits of the measurement of surfaces of very short radius of curvature, such as the surfaces of microscope lenses.

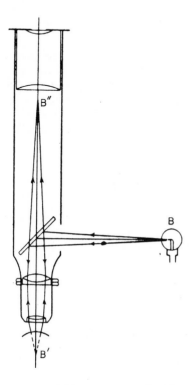

Figure 7.8 Measurement of surfaces of small radius. Drysdale's method

A microscope (*Figure 7.8*), with scale and vernier to read the position of the body, is fitted with a vertical illuminator. This consists of a short length of tube between the end of the microscope body and the objective with an aperture on one side and a small clear glass reflector in the tube at 45° to the axis to reflect light from a source down through the objective. The source B, a small electric lamp, is attached to the microscope tube in line with the aperture and at approximately the same distance from the reflector as is the focal plane of the eyepiece, B″. The surface to be measured is placed on the stage of the microscope, and the microscope is focused on it; some fine dust may be placed on the surface for this purpose. The source is then adjusted until a sharp image of it is focused in the microscope, that is, the image B′ lies on the surface. On moving the microscope down in the case of a convex surface and up in the case of a concave surface, a

*C. V. Drysdale. *On a Simple Direct Method of Measuring the Curvature of Small Lenses*. Trans. Opt. Soc. 2 (1900–01).

second position is found where a sharp image of the source is again focused; the image B′ now lies at the centre of curvature of the surface, and the distance the microscope has moved between the two positions is the radius of curvature of the surface. This method is the principle of a number of instruments now made for the measurement of the curvature of contact lenses.

Convex surfaces of short radius. Ophthalmometer method

If the size of the image, formed by a mirror, of an object of known size at a known distance can be found, the radius or focal length of the mirror can be calculated. Thus in *Figure 7.9*, BQ is an object in front of a convex mirror at A, and B′Q′ is the virtual image formed. Then

$$m = \frac{B'Q'}{BQ} = \frac{h'}{h} = \frac{f'}{-x} = \frac{r}{-2x}$$

$$r = \frac{-2xh'}{h}$$

When the object distance is large as compared with the radius of curvature, the image is formed very close to the principal focus, and x is practically equal to d, the distance between image and object; thus

$$r = \frac{-2dh'}{h} \tag{7.08}$$

This is the principle of the **ophthalmometer** or **keratometer** an instrument that is used to measure the curvature of the cornea of the eye, or more usually, the difference in its curvature in different meridians. A large illuminated object, or two separated small objects are attached to a long focus microscope having a scale in the focal plane of the eyepiece. The microscope is directed towards the surface to be measured, and focused on the virtual image of the object; as the object moves with the microscope in focusing, the distance d between object and image remains constant. The size of the image h' is then found by measuring

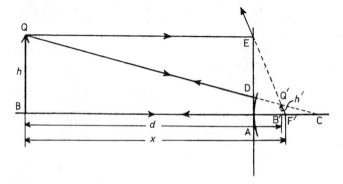

Figure 7.9 Radius of curvature of convex surface of short radius. Ophthalmometer method

its image on the scale in the eyepiece, and h being the size of the object or the separation of the two small objects, r can be found from expression (7.08).

In measurements on the living eye it is impossible, because of movement, to measure the size of the image on a scale, and in the actual ophthalmometer the size of the image is determined by means of a device in the microscope for doubling the image. Various forms of the instrument are made; in some the measurement of the image is made by varying the amount of doubling, while in others the amount of doubling is constant and the object length h is varied to give an image of constant size.

7.8 MEASUREMENT OF CURVATURE OF THE SURFACES OF A LENS

The curvature of the surfaces of a lens may be measured by the above methods, treating the surface as a mirror. The following method based on the principle of the lens mirror may sometimes be useful, particularly in the case of a positive lens with shallow curves.

The lens is placed in front of an illuminated aperture in a white screen, the surface to be measured being turned away from the aperture. The distance between lens and screen is adjusted until an image of the aperture is formed on the screen by means of the light internally reflected from the second surface; then if l is the distance of the screen from the lens, f' the focal length of the lens and r_2 the radius of curvature of the second surface,

$$r_2 = \frac{lf'}{l+f'}$$

The proof of this is left as an exercise.

7.9 MIRRORS OF LARGE APERTURE

Two of the most frequent applications of the curved reflecting surface are the use of the concave mirror for producing a parallel beam, and its use in bringing the light from a distant object to a focus. In the searchlight and the motor headlight a very intense beam of light is required, and it is therefore necessary that the beam shall be approximately parallel; this is obtained by placing a small intense source at the principal focus of a concave mirror. The concave mirror of the reflecting telescope replaces the objective in producing an image of a star or other distant object; thus the mirror in this case must bring parallel incident light to a focus. In each of these cases the aperture of the mirror is large—in the headlight the aperture is often more than twice the focal length—and, as there must be little or no spherical aberration, it is obvious that a spherical mirror cannot be used.

From *Figure 7.1* it will be seen that the effect of the spherical mirror on a parallel incident pencil is too great towards the periphery; and in order that a parallel incident pencil shall be brought to a good focus, or that the light from a small source at the focus shall be reflected as a parallel pencil, the curvature of the mirror must get less and less towards the periphery. In *Figure 7.10*, B is a point source at the focus of a concave mirror of large aperture and, if the reflected beam is to be parallel, the reflected wavefronts must be plane; therefore, the

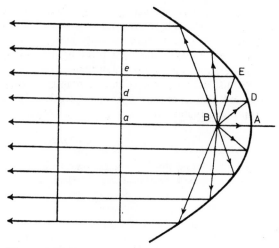

Figure 7.10 Paraboloidal mirror

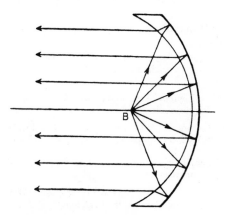

Figure 7.11 Mangin mirror

various paths by which the light travels, BAa, BDd, BEe, etc., must be equal. The curve having this property is the parabola; and mirrors of large aperture are made of paraboloidal form when either the incident or reflected light is parallel.

A similar result is often obtained by using a concave spherical reflector, and placing in front of it a negative lens, the curves being chosen so that the increasing thickness of glass towards the periphery corrects the spherical aberration. In practice, the reflector consists of a suitably designed negative meniscus lens silvered on the convex surface (*Figure 7.11*). This form of reflector, which is frequently used in large objectives, is known as the **Mangin mirror**.

In any system fitted with a paraboloidal or a Mangin mirror only the reflected light from a point on the axis will be truly parallel, but if, as is usually the case, the source is small the light from any point on the source will be reflected as a practically parallel pencil, the various pencils diverging slightly from the axis.

The area illuminated by the reflected light therefore increases as the distance from the lamp is increased, but the light, that would have diverged in all directions, is concentrated in a comparatively small area, and hence a greatly increased illumination of distant objects is obtained. When a mirror is to be used to focus light to a point at some finite distance from the mirror, other than at the centre of curvature, an ellipsoidal mirror must be used. The conjugate points are then situated at the two foci of the ellipsoid. Mirrors of this form are frequently used in the illuminating system of cinema projectors and other optical display equipment.

CHAPTER 7   *EXERCISES*

1. Prove that parallel rays of light are reflected through a single point by a concave spherical mirror provided that only a small portion of the whole sphere, of which the mirror is part, is employed.

What is the effect if a large part of the spherical surface is employed?

2. A small source of light is placed at the principal focus of a concave spherical mirror, the radius of curvature of which is 10 cm and aperture 16 cm. Find graphically the paths of the reflected rays corresponding to incident rays that strike the mirror at points 2, 3, 4, 6 and 8 cm from the axis.

In what manner would the shape of the mirror need to be altered to make the reflected beam parallel?

3. A pencil of light has a convergence of 5.5D on reaching a concave mirror whose focal length is 220 mm. Find the position of the image.

4. The curvature of a concave mirror is 5D. The focal power of another concave mirror is 5D. Explain the difference between the mirrors. In each case find the position of the image of an object 2 metres from the mirror.

5. What is meant by the focal power and focal length of a concave reflecting surface? How are these values related to the curvature of the surface?

6. A parallel pencil of light of 3 cm diameter is incident on a concave mirror of 25 cm radius of curvature. What is the area of the cross section of the reflected pencil 40 cm from the mirror?

How far from the mirror must a second concave mirror of 3D curvature be placed in order that the light may be reflected back along its original path?

7. Deduce the relation between object distance, image distance and focal length for a spherical mirror, directly from the laws of reflection. Describe how the focal length of a convex spherical mirror may be determined experimentally.

8. Prove, by consideration of the light fronts, that for an object in any position the change in vergence produced by a convex mirror is equal numerically to twice the mirror's curvature. What limitations are there to the truth of this statement?

9. How far from a concave spherical mirror must a real object be placed in order that the image shall be (*a*) real and four times the size of the object and (*b*) virtual and four times the size of the object?

10. An object 3 inches long is placed 15 inches from the principal focus of a concave mirror, and a real image 1¹/₄ inches long is formed. Find the curvature of the mirror and the distance of object and image from the mirror.

11. When an object is 31 cm in front of a concave mirror the image is 19 cm in front of the mirror. Where will the image be if the object is a virtual one and 31 cm behind the mirror?

12. Find graphically and by calculation the positions and sizes of the images in the following cases, the object in each case being 3 cm long.

(*a*) Object 10 cm from convex mirror of 10 cm radius.

(*b*) Object 5 cm from concave mirror of 15 cm radius.

13. Draw figures showing the change which takes place in the size and position of the image when an object on the axis of a concave mirror is moved from a great distance until it is in contact with the mirror.

14. The focal length of a concave mirror is 17 inches and a virtual image is required 5 inches from the mirror. Where must the object be placed?

15. The radius of curvature of a convex mirror is 250 mm. Standing on its axis and 600 mm in front of the mirror is an object 5 mm high. Find the position and size of the image (*a*) graphically, (*b*) by calculation. Is the image real or virtual?

16. Find the position of the image of an object which is 9 inches from a concave mirror whose radius of curvature is 12 inches.

17. A concave mirror forms a real image of an object at 20 cm in front of the mirror; if the object is 133.3 cm from the mirror, find (1) the focal power, (2) the curvature, (3) the radius of curvature of the mirror.

18. Explain the formation of an image by reflection of light from a convex spherical mirror.

A small object is placed on the axis of a convex spherical mirror at a distance of 200 mm from it, and the virtual image formed is observed to have half the linear dimensions of the object. Determine the focal length and the radius of curvature of the mirror.

19. A concave mirror has a radius of curvature of 2 feet; show graphically and by calculation where an observer must be situated in order to see a virtual image of his face magnified $2\frac{1}{2}$ times.

20. A convex mirror having a radius of curvature of 25 cm forms, at 20 cm from the mirror, a virtual image of an object 2 cm in height. Calculate the magnification and give a diagram showing the path of the light.

21. When a real inverted image is formed 14 inches from, and in front of a mirror, the magnification produced is two. What will be the position of the object when the magnification is four and the image is erect?

22. A convex mirror for a motor car has a radius of curvature of 18 inches. What will be the size and position of the image of a car 6 feet broad and 5 feet high at a distance of 10 feet from the mirror?

23. An object is placed on the axis of a concave mirror and perpendicular to it. Obtain an expression, in terms of the object distance and the focal length of the mirror, for the ratio of the linear dimensions of the image to that of the object. Represent the result graphically, taking the focal length of the mirror as 10 cm.

24. An object is placed 75 cm from a mirror and a real image is formed 120 cm from the mirror. What will be the focal length of the mirror? If the back surface of the mirror be plane, its diameter 5 cm and its central thickness 3 mm, find its edge thickness.

25. A convex motor driving mirror has a radius of curvature of 15 inches; it is rectangular in shape, its dimensions being 5 inches horizontally, and 3 inches vertically. What extent of object 20 yards from the mirror could be seen by reflection by a driver whose eye is 3 feet from the mirror?

26. The front surface of the cornea of the eye has a radius of curvature of 8 mm; what will be the size and position of the image formed by the reflection of an object 200 mm in front of the cornea and 60 mm long?

27. An object and its image are one metre apart; if the object is four times the size of the image, find the radius of the concave mirror forming the image.

28. A gas flame is 4 feet from a wall, and it is required to throw on the wall a real image of the flame which shall be magnified three times. Find the position and focal length of a spherical mirror to give the required image; give a diagram.

Describe very briefly a method for finding the radius of curvature of a steeply curved mirror, such as a ball-bearing, for example, without touching it.

29. A point source of light is placed 15 cm from a concave mirror of 45 cm radius of curvature and 8 cm diameter. What will be the diameter of the cross section of the reflected pencil at 10, 35 and 100 cm from the mirror?

30. Explain (*a*) the use of convex mirrors as back view reflectors on motor cars, (*b*) the optical arrangement of a motor car headlight.

31. Deduce expressions for the magnification of the image formed by a curved reflecting surface in terms of the distances of object and image from (*a*) the centre of curvature, (*b*) the principal focus.

32. If the sun's disc subtends an angle of 32 minutes at the earth's surface, find the focal length of the concave mirror which would produce a real image of the sun upon a screen, the diameter of the image being 55 mm.

33. A convex mirror and a plane mirror are placed facing each other and 28 cm apart. Midway between them and on the principal axis of the convex mirror a small luminous object is situated. On looking into the plane mirror the first two images of the luminous object are seen. Draw a diagram showing how the two images are formed and calculate the radius of curvature of the convex mirror if the image formed by two reflections is situated at a distance of 38 cm behind the plane mirror.

34. A concave and a convex mirror, each of 20 cm radius of curvature, are placed opposite to each other and 40 cm apart on the same axis. An object 3 cm high is placed midway between them. Find the position and size of the image formed when the light is reflected first at the convex and then at the concave mirror. Give a diagram showing the path of the light from a point on the object to a corresponding point on the image.

35. A concave mirror is immersed in water. The image of an object 30 cm from the mirror is formed at 9 cm in front of the mirror. If both object and image are in the water, what is the radius of the mirror?

36. An object and its image formed by a concave mirror are in front of the mirror and on opposite sides of its centre of curvature. If the object and image are 300 mm and 75 mm respectively from the centre of curvature and the image is nearer to the mirror than the object, find the focal power of the mirror.

37. Two concave mirrors of focal length 20 and 40 cm are turned towards each other, the distance between their vertices being one metre. An object 10 mm high is placed between the mirrors at a distance of 100 mm from the more steeply curved mirror. Find the position and size of the image produced by light which is reflected first from the nearer mirror and then from the farther mirror. Give a diagram showing the path of the light.

38. Explain carefully with diagrams the advantages of a convex over a plane surface for the driving mirror of a motor car.

39. Two small illuminated objects are placed 50 cm apart and 40 cm in front of an eye, the line joining the objects being perpendicular to the visual axis. Find the position and distance between the images reflected from the cornea, the radius of which is 8 mm.

40. An object 5 cm high is perpendicular to and has one end on the axis of a + 5D lens, the object being 20 cm from the lens. At 30 cm on the other side of the lens is placed a concave mirror of 30 cm radius of curvature, its axis coincident with that of the lens. Find graphically and by calculation the position and size of the final image. Show in a diagram the path of the light from each end of the object.

41. Describe two methods of measuring the curvature of a polished concave surface.

42. When an object is held at a distance of 8 cm from one face of a thin lens of glass $n = 1.5$, the image of the object formed by reflection in this face is found to lie in the same plane as the object. If the object is placed at a distance of 30 cm from the lens, the image produced by the lens is inverted and of the same size as the object. Find the radii of the surfaces.

43. A lens made of glass of refractive index 1.5 is set up on the optical bench. With the object at 40 cm it is found that the image formed by the lens coincides with the image produced by reflection at the first surface of the lens. When the object is placed 80 cm from the lens the lens forms a real image which is $2^1/_5$ times the size of the object. Find the radii of curvature of the lens surfaces.

44. A thin convergent lens gives an inverted image the same size as the object when the latter is 20 cm from the lens. An image formed by reflection from the second surface of the lens is found to be coincident with the object at 6 cm from the lens. Find the radius of curvature of the second surface of the lens and, assuming $n = 1.5$, the radius of curvature of the first surface.

45. A concave mirror of focal length 115 mm is placed on a bench and liquid is poured into it until the greatest depth is 0.5 cm. It is found that, on placing a luminous point on the vertical axis of the mirror at a height of 159 mm from the liquid surface, the luminous point and its image coincide. Determine the refractive index of the liquid and give the theory underlying your calculation.

46. A thin converging lens and a convex spherical mirror are placed coaxially at a distance of 30 cm apart, and a luminous object, placed 40 cm in front of the lens, is found to give rise to an image coinciding with itself. If the focal length of the lens is 25 cm, determine the focal length of the mirror. Explain fully each step in your calculation.

47. A parallel beam of light falls normally on the plane surface of a thin planoconvex lens and is brought to a focus at a point 48 cm behind the lens. If the curved surface be silvered, the beam converges to a point 8 cm in front of the lens. What is the refractive index of the glass, and what is the radius of curvature of the convex surface?

48. Derive Newton's expression for the focal length of (*a*) a convex mirror, (*b*) a concave mirror.

49. A concave mirror is formed of a block of glass ($n = 1.62$) central thickness 5 cm. The front surface is plane and the back surface, which is silvered, has a radius of curvature of 20 cm. Find the position of the principal focus and the position of the image of an object 20 cm in front of the plane surface.

50. A + 1.25D equi-convex lens of refractive index 1.53 has one of its surfaces silvered, and parallel light falls axially on the other surface. What will be the effect?

51. A thick glass rod of refractive index n has a convex spherical surface of radius r worked on one end. Of a very distant bright object lying on the axis of the rod an image is formed by reflection and a second image by refraction, at the surface. Show that the distance separating the images is given by

$$\frac{r}{2}\left(\frac{n+1}{n-1}\right)$$

and that the refracted image is numerically

$$\frac{2}{(n-1)}$$

times as large as the reflected image.

52. A convex lens of focal length f is placed at a distance of $4f$ from a concave mirror of radius of curvature f, and an object is placed midway between the two. Compare the sizes of the images formed by the light refracted through the lens (*a*) directly, and (*b*) after one reflection at the mirror.

53. Two small lamps A and B are set up 200 mm apart facing symmetrically a polished steel ball of diameter 20 mm. The perpendicular distance from the midpoint of AB to the nearer surface of the ball is 25 cm. If this ball is replaced by a second ball of 22 mm diameter, its nearer surface occupying the same position as that of the first ball, by how much must the separation of A and B be changed in order that the separation of the reflected images shall remain unaltered?

54. A ray of light travels parallel to, and at a distance y from, the axis of a concave spherical mirror of radius r. Find the distance from the centre of curvature of the point at which the ray intersects the axis of the mirror after reflection. Show that the spherical aberration of this ray is given by

$$\frac{r}{2}\left(\frac{r}{\sqrt{r^2 - y^2}} - 1\right).$$

55. A ray of light is incident at an angle θ on a concave spherical mirror parallel to a radius CA where C is the centre of curvature and A a point on the mirror surface. Show that the reflected ray intersects CA at a distance $r(1 - \frac{1}{2} \sec \theta)$ from A; r is the radius of curvature of the mirror.

If this reflected ray strikes the mirror again and the second reflected ray passes through the end B of the diameter BCA of the spherical surface of which the mirror forms part, what must be the angle θ?

Chapter 8

Cylindrical and sphero-cylindrical lenses

8.1 INTRODUCTION

The commonest form of surface used for optical purposes is spherical. Such surfaces can be produced comparatively easily to a high degree of accuracy by mechanical means, and the lenses of almost all optical instruments have spherical or plane surfaces. It has been shown in previous chapters that good images will not be formed by refraction at a single surface or a spherical lens when the aperture is at all large, but it is possible to obtain good images with a suitable combination of plane and spherical surfaces. Such a combination of surfaces is more easily and accurately manufactured than the aspherical surfaces required to obtain the same result with fewer surfaces. The increasing use of plastics in place of glass has made possible the manufacture of more complicated forms of surface, and such surfaces are widely used in producing simpler optical systems.

There are, however, a number of cases where the use of a lens is required for other purposes than the direct production of a good image of an object. The most frequent example of this is the use of the lens to correct the astigmatic eye, which has different powers in different meridians and so does not form a sharp image on the retina. The correcting lens must therefore be one having corresponding different powers and hence different curvatures in various meridians.

8.2 THE CYLINDRICAL SURFACE

The simplest form of 'non-spherical' surface is cylindrical. In practice, usually only one surface of a lens will have a cylindrical curve, the other being plane or spherical.

The cylindrical surface (*Figure 8.1*) has zero curvature in the section DOO'D', the plane of which is parallel to the axis MN of the cylinder. In the section EAE'G the surface has its maximum curvature R, and the curve is circular. These two directions DD' and EE' are the **principal meridians** of the surface. Along any intermediate oblique meridian, such as HH', the curve will be elliptical, and the curvature will have a value somewhere between zero and the maximum R.

If only a small portion of the surface is considered, the curve in an oblique meridian may be regarded as circular, and the curvature may be found as follows: For a given sag AG the section in the principal meridian of maximum curvature has an arc EE', while the section in an oblique meridian for the same sag has an arc HH' and

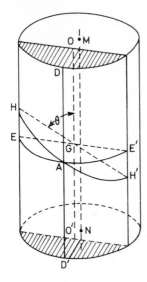

Figure 8.1 Cylindrical surface

$$HH' = \frac{EE'}{\sin \theta}$$

where θ is the angle that the oblique section makes with the axis. For small portions of the curve the curvatures of two curves having the same sag will be inversely proportional to the square of their chords (section 5.2); therefore, if R_θ is the curvature in a meridian at an angle θ with the axis and R the maximum curvature,

$$\frac{R_\theta}{R} = \left(\frac{EE'}{HH'}\right)^2$$

hence

$$R_\theta = R \sin^2 \theta \qquad (8.01)$$

Similarly, the curvature at A along a meridian at right angles to HAH', and therefore making an angle $(\theta + 90°)$ with the axis, is given by

$$R_{(\theta+90°)} = R \sin^2 (\theta + 90°)$$
$$= R \cos^2 \theta \qquad (8.02)$$

Hence

$$R_\theta + R_{(\theta+90°)} = R (\sin^2 \theta + \cos^2 \theta) = R$$

that is, *the sum of the curvatures along any two directions of a cylindrical surface intersecting each other at right angles is equal to the maximum curvature.* These theorems are a special case of a more general theorem concerning curved surfaces due to the mathematician Euler (1707–1783).

8.3 REFRACTION BY CYLINDRICAL LENS. LINE FOCUS

Light from a luminous point B (*Figure 8.2*) in front of a thin cylindrical lens continues unchanged in divergence in the axis meridian after leaving the lens. In any other meridian the light is refracted by an amount depending on the power in that meridian, and in the figure the light in the meridian at right angles to the axis is shown converging to B'_H. The refraction in the oblique meridians is such that all the light from the object point on leaving the lens will pass through an **astigmatic** or **line focus** at B'_H. The position of this line focus can be found from the usual conjugate foci expression taking the maximum power of the lens; and its direction, as will be seen from the figure, is parallel to the cylinder axis.

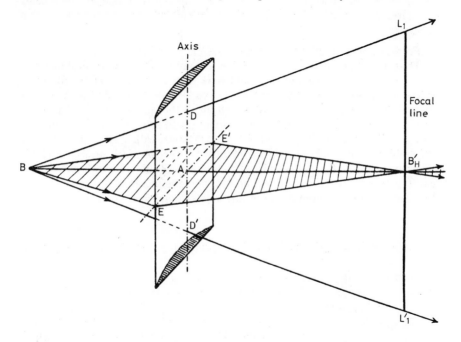

Figure 8.2 Refraction by cylindrical lens. Line focus

Between the lens and B'_H the light is also travelling as though it had come from a virtual line focus at B perpendicular to the cylinder axis, and an eye placed close behind the lens sees the luminous point B drawn out into a line perpendicular to the axis. This is the principle of the **Maddox Rod**, an important device used in sight-testing for disassociating the images seen by the two eyes in order to test for muscle imbalance. The Maddox Rod is a glass cylinder of about $1/8$ inch diameter mounted in an opaque disc. The disc is placed before one eye with the rod in line with the pupil, and the eye then sees a distant small source as a line perpendicular to the axis of the rod. In practice, a glass disc ground with a series of parallel cylindrical corrugations generally takes the place of a single rod.

8.4 SPHERO-CYLINDRICAL LENSES

An eye that has *regular astigmatism* requires for its correction a lens which the power varies from a minimum in one meridian to a maximum in the meridian at right angles. In a great many cases, however, the minimum power required will not be zero, as would be given with a cylindrical lens. The required lens could be made up in the following different forms:

(1) Each surface cylindrical with the axes of the cylinders at right angles—*crossed cylinders*.
(2) One surface spherical and the other cylindrical—*sphero-cylinder*.
(3) One surface of a form having different curvatures in different meridians— toroidal or toric curve, the other surface usually being spherical—*sphero-toric*.

Forms (2) and (3) are those generally used in practice.

Thus, if a thin lens is to have a power of, say + 5D in the vertical meridian, and + 3D in the horizontal, it would have in the crossed-cylinder form, one cylindrical surface of + 5D power with its axis horizontal and another cylindrical surface of power + 3D with its axis vertical. In the sphero-cylindrical form the lens could be made in two ways. One surface could be spherical with a power of + 3D and the other surface cylindrical of + 2D power with axis horizontal to give the additional power in the vertical meridian; or the spherical surface could have a power of + 5D, and the cylindrical surface a power of − 2D with its axis vertical. The toric form of this lens will be considered in section 8.8. The changing of the form of a lens of given power in this way is known as transposition.*

8.5 REFRACTION BY SPHERO-CYLINDRICAL LENS. ASTIGMATIC BEAM

In considering the refraction by a lens of the form described in the previous paragraph, we may regard the sphero-cylindrical lens as typical of any lens having different powers in the principal meridians, and the general effect of such a lens will be the same in whichever form it is made. The principal meridians of the sphero-cylindrical lens will be the meridian containing the axis of the cylinder and the meridian perpendicular to the axis.

Figure 8.3 shows the form of the pencil from a luminous point B on the optical axis after passing through a sphero-cylindrical lens having positive power in each of its principal meridians, the axis of the positive cylindrical surface in this case being vertical. Light passing through the vertical section DD' is affected only by the power of the spherical surface, and converges to B'_v; in sections parallel to DD' the light converges to points on a horizontal line $C_2 C'_2$ passing through B'_v. In the horizontal section EE' the power is the sum of the powers of the spherical and cylindrical surfaces, and the light in this section converges to B'_H, sections parallel to EE' converging to points on a vertical line $C_1 C'_1$ through B'_H. A pencil of this form, where the light originating at a point passes through two perpendicular lines, is called an **astigmatic pencil**.

*Fuller particulars of transposition will be found in EMSLEY and SWAINE *Ophthalmic Lenses*.

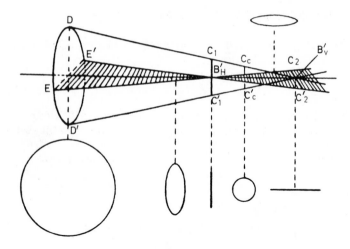

Figure 8.3 Astigmatic pencil

Cross-sections of such a pencil are shown in *Figure 8.3*. If the aperture of the lens is circular, the cross-section of the emergent pencil is at first an ellipse with its major axis parallel to the cylinder axis, narrowing down to a line parallel to the cylinder axis at B'_H. The cross-section then becomes circular at $C_c C'_c$, after which it becomes elliptical with the major axis perpendicular to the cylinder axis and degenerates into a line again at B'_V. The circular cross-section at $C_c C'_c$ where the pencil will have its smallest cross-sectional dimension, is known as the **circle of least confusion**.

The properties of the astigmatic pencil were investigated by the mathematician Sturm (1838), and the pencil of the form shown in *Figure 8.3* is known as Sturm's Conoid. The distance between the focal lines is known as the **interval of Sturm**.

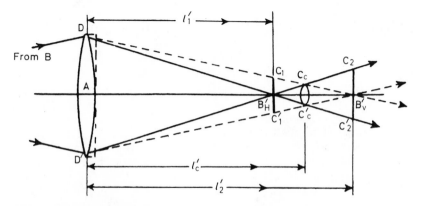

Figure 8.4 Astigmatic pencil. Positions and lengths of line foci

8.6 POSITIONS AND LENGTHS OF LINE FOCI, ETC.

The positions of the line foci are found by applying the conjugate foci expression $(L' = L + F)$ to each meridian in turn.

Lengths of line foci

To find the lengths of the line foci and the position and diameter of the circle of least confusion, it is convenient to represent the sections of the pencil in the two principal meridians in one plane, as in *Figure 8.4* where the section in a horizontal meridian is shown in full lines and in a vertical meridian in broken lines.
    From *Figure 8.4*

$$\left.\begin{array}{c} \dfrac{C_1 C'_1}{DD'} = \dfrac{B'_H B'_v}{AB'_v} = \dfrac{l'_2 - l'_1}{l'_2} \\[4mm] \text{or} \\[4mm] C_1 C'_1 = DD' \left(\dfrac{L'_1 - L'_2}{L'_1}\right) \\[6mm] \dfrac{C_2 C'_2}{DD'} = \dfrac{B'_H B'_v}{AB'_H} = \dfrac{l'_2 - l'_1}{l'_1} \\[4mm] \text{or} \\[4mm] C_2 C'_2 = DD' \left(\dfrac{L'_1 - L'_2}{L'_2}\right) \end{array}\right\}$$

(8.03)

(8.03a)

Circle of least confusion

$$\frac{C_c C'_c}{DD'} = \frac{l'_c - l'_1}{l'_1} = \frac{l'_2 - l'_c}{l'_2} \tag{8.04}$$

$$l'_2 l'_c - l'_1 l'_2 = l'_1 l'_2 - l'_1 l'_c$$

$$l'_c = \frac{2 l'_1 l'_2}{l'_1 + l'_2}$$

or

$$L'_c = \tfrac{1}{2}(L'_1 + L'_2) \tag{8.05}$$

giving the position of the confusion circle.
    Equation (8.04) may also be written

$$z = C_c C'_c = DD' \left(\frac{L'_1 - L'_2}{L'_1 + L'_2}\right) \tag{8.06}$$

where z is the diameter of the confusion circle.

8.7 IMAGE OF OBJECT OF FINITE SIZE

The optical axis of a sphero-cylindrical lens is a line passing through the centre of curvature of the spherical surface and perpendicular to the cylinder axis. In the case of a thin lens the point on the lens through which the optical axis passes is the optical centre.

As with a spherical lens, any ray passing through the optical centre will be undeviated and will be a chief ray from the object point from which the ray started. Since one ray from every point of the object passes undeviated through the sphero-cylindrical lens, the image formed when the aperture is quite small, will correspond in its general shape to the object, and its size may be found in the same way as with the image formed by a spherical lens. Each object point is, however, imaged as a line, parallel to one or other of the principal meridians, the length of the line being dependent upon the aperture of the lens and the distance between the two focal lines. The form of the image will therefore only agree with that of the object when the length of each line focus is quite small as compared with the size of the image.

The nature of the image formed by a sphero-cylindrical lens with its principal meridians horizontal and vertical is shown in *Figure 8.5*. It should be noticed that lines sharply defined in the image are perpendicular to the meridian focusing in the image plane. The form of the image produced by astigmatic beams is of considerable importance in the case of the astigmatic eye.

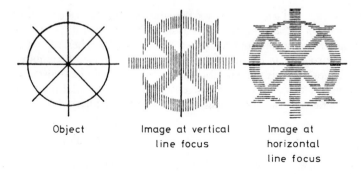

| | Object | Image at vertical line focus | Image at horizontal line focus |

Figure 8.5 Image formed by sphero-cylindrical lens

8.8 TORIC LENSES

When the astigmatic lens is required in meniscus form, as is often the case with ophthalmic lenses, it will be necessary for one surface of the lens to be of a form having different curvatures in its principal meridians. Such a surface is the **toroidal** or **toric** surface, which takes its name from the *torus*, the architectural term for the moulding at the base of an Ionic column.

A toric surface, two forms of which are illustrated in *Figure 8.6*, is generated by the revolution of the arc of a circle about an axis OO' which lies in the plane of the circle but does not pass through its centre. Common examples of a toric surface occur in the motor tyre, anchor ring and capstan.

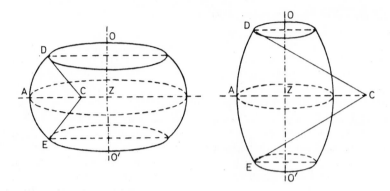

Figure 8.6 Toric surfaces

A straight line AZ, which passes through the centre of the circle perpendicular to the axis of revolution will be the optical axis of the surface. One principal meridian of the surface will be a section containing the axis of revolution, and the other, a section through the optical axis and perpendicular to the axis of revolution; in the figure these meridians are vertical and horizontal respectively. As with the cylindrical surface, the principal meridians are the directions of maximum and minimum curvature. The principal meridian of the toric curve along which the curvature, and hence the power, is numerically the smaller is know as the base curve.

By the use of a toric surface, any given astigmatic lens can be made up in an infinite number of forms. The toric surface will have the required difference in power in the principal meridians, and the other surface will be spherical and of the necessary power to give, in conjunction with the toric surface, the required power of the lens. For example the lens given in section 8.4 could be transposed into toric form with a toric surface of + 6D power in the horizontal meridian and + 8D power in the vertical, the second surface then being spherical of − 3D power.

CHAPTER 8   *EXERCISES*

1. An optician's lens measure divided for a refractive index of 1.523 is placed on a cylindrical surface of 100 mm radius of curvature. What will be the reading when the line containing the three legs of the measure is (*a*) perpendicular, (*b*) parallel, (*c*) at 30° to the cylinder axis?

2. A sphero-cylindrical lens has surface powers of + 3D.S. and + 2D.C. (*n* = 1.523). It is edged circular 45 mm diameter and its minimum edge thickness is 1 mm. Find its maximum edge thickness and its central thickness.

3. A thin lens has cylindrical surfaces of equal curvature, the axes of the cylinders being at right angles; show that the effect of the lens is the same as that of a spherical lens with a curvature equal to either of the cylinders.

4. Show that the sum of the curvatures along any mutually perpendicular directions of a cylindrical surface is equal to the maximum curvature.
   Find the sphero-cylinder equivalent to the combination
   + 4.0 D.C. ax 80°/ − 2.0 D.C. ax 170°

5. A sphero-cylindrical lens has surface powers of + 4.50D sphere and − 2.75D cylinder, and is made of glass of refractive index 1.622. Find the radii of curvature of the tools with which the surfaces are ground.

6. A point source is placed 0.5 m from a cylindrical lens of + 5D power with axis vertical and 40 mm diameter. Find the position, length and direction of the line focus.

If a + 3D spherical lens be placed in contact with the cylindrical lens, what will be the positions and lengths of the line foci and the position and diameter of the circle of least confusion?

7. A small source of light is placed 0.75 m from a sphero-cylindrical lens + 6D.S./ −3.5D.C. ax. V, diameter 50 mm. Find the position and lengths of the focal lines.

8. A point source of light is situated 70 cm from the following lens + 6D.S./− 2.25D.C. ax. V 40 mm diameter. Find the positions, lengths and directions of the line foci and the position and diameter of the circle of least confusion. What will be the appearance of the image in the plane of each of the focal lines when the point source is replaced by an illuminated circular object 5 mm diameter?

9. The lens + 10.00 D.S./− 4.00 D.C., the aperture of which is circular and of 60 mm diameter, is set up with a distant luminous point on its optical axis. The focal lines are found to be respectively horizontal and vertical, the horizontal one being the nearer to the lens. What is the shape and size of the patch of light formed on a vertical screen fixed 125 mm behind the lens? What lens placed in contact with the above lens will convert the combination into a sphere the principal focus of which is formed on the screen?

10. Two thin cylindrical lenses, one of power + 5D axis vertical and the other + 4D axis horizontal and 45 mm diameter, are placed in contact 0.5 m from a point source of light. Find the dimensions of the patch of light on a screen parallel to the plane of the lenses and (*a*) 0.2m, (*b*) 0.4 m, (*c*) 0.6 m from the lenses.

11. A narrow illuminated slit 15 mm long is placed 166.7 mm from a + 10D cylindrical lens of 9 mm diameter. Find the nature and size of the 'image' when the slit is (*a*) parallel, (*b*) perpendicular to the cylinder axis.

12. A sphero-cylindrical lens has powers + 5.0D.S./− 2.0D.C. ax. V. It is required to make a lens equivalent to this with one surface plane; what powers must be worked on the other surface? What type of surface is this? Give a diagram.

13. A lens made of glass of refractive index 1.523 has a toric surface the powers of which are + 10D in the vertical meridian and + 6D in the horizontal; the other surface is plane. Find the radii of curvature of the surfaces when a lens of the same power is made up in the crossed cylinder form.

14. A polished metal cylinder 2 inches in diameter stands on a table with its axis vertical and at a distance of 10 feet from a window 6 feet broad. What would be the breadth of the image seen by reflection from the cylinder?

15. A small crossline chart (the limbs of the cross being horizontal and vertical) and a white screen are set up vertically on an optical bench one metre apart. The lens + 5.0 D sph./− 1.0 D cyl. axis V is moved along the bench between them. How far from the chart must the lens be placed in order that (*a*) the horizontal line (*b*) the vertical line of the chart shall be in focus on the screen?

16. The lens

$$\frac{- 4.0 \text{ D sph.}}{+ 9.0 \text{ D cyl. ax. } 30°/+ 12.0 \text{ D cyl. ax. } 120°}$$

is set up facing a small source of light. The astigmatic difference, or interval of Sturm, is found to be $16\frac{2}{3}$ cm. How far from the lens is the source situated? For what real position of the source is the astigmatic difference a minimum?

17. An object B and a screen B′ are placed 625 mm apart. Between them is a combination consisting of the following two thin lenses, of aperture 20 mm each, placed in contact:

Lens (1) $\qquad \dfrac{+ 16.00 \text{ S}}{-5.75 \text{ C. ax. H}/- 6.25 \text{ C. ax. V}}$

Lens (2) $\qquad \dfrac{- 5.75 \text{ S}}{+ 5.50 \text{ C. ax. H}/+ 6.00 \text{ C. ax. V}}$

Find a position for the combination such that a sharp and diminished image of B will be formed on the screen.

A point source is placed at B and the lens (2) is rotated through 90°. What will now be the shape and size of the appearance on the screen?

Chapter 9

Lens systems and thick
lenses — elementary

9.1 INTRODUCTION

Most optical instruments consist of a number of lenses separated by air spaces, or, as in the eye, of a number of media of different refractive indices separated by curved surfaces. In most cases the systems are coaxial or centred, the centres of curvature of the surfaces being on a common axis.

The work of this chapter will be limited to coaxial systems of thin lenses in air, and to single thick lenses having air on each side.

It has been shown in section 6.14, that the position and size of the image formed by a system of thin lenses may be determined by calculating through the system, treating the image formed by one lens as the object for the next. Such a method, while simple in principle, becomes tedious where a number of image positions have to be found for the one system, and efforts were made, notably by Euler, Lagrange, Möbius and Gauss, to simplify the treatment of the lens system. As a result of the work of these investigators, particularly that of Gauss (1777-1855), it is possible to express the effect of any lens system in terms of an equivalent thin lens or surface. Once the power and positions of this equivalent lens, relative to the system have been found, any further work is greatly simplified, as the simple formulae already found for the thin lens and surface can be applied to the system.

9.2 EQUIVALENT FOCAL LENGTH AND POWER. PRINCIPAL POINTS

In *Figure 6.15* it is seen that a thin lens of a certain focal length placed at P' would produce an image of a distant object of the same size and in the same position as that produced by the system. The focal length P' F' of this thin lens, which will theoretically replace the system, is termed the **equivalent focal length** of the system, and is usually the most important quantity in the specification of optical systems, such as photographic and microscope objectives and eyepieces. The reciprocal of this distance, usually in metres, is the **equivalent power**. The equivalent focal length of a system may therefore be defined as *the focal length of a thin lens, which will produce of a distant object an image of the same size as is produced by the system.*

We must now consider the position of this equivalent lens with respect to the system. *Figure 9.1a* shows the path of a ray $B_1 D_1$ incident parallel to the axis through the thin lenses at A_1 and A_2. The effect is the same as if the incident ray had been refracted at H' by the equivalent lens placed at P'. F' is the second principal focus of this lens, or of the combination of A_1 and A_2. In the lower half of the figure, a ray FE_1 is shown incident from a point F, such that after refraction by the system, it emerges parallel to the axis. The effect is the same as if the incident ray had been refracted at H by the equivalent lens placed at P. F is the first principal focus of this lens or of the combination. It will be found that FP = P'F' or $f' = -f$ where f and f' are the first and second equivalent focal lengths respectively of the system.

The path of the ray from F may be found graphically by starting from B_2 and proceeding from right to left.*

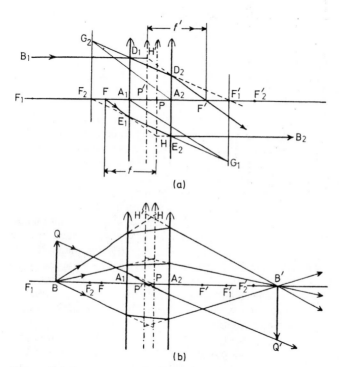

Figure 9.1 Lens system. Equivalent lens and principal planes (a) Distant object, distant image (b) Near object and image

In *Figure 9.1b,* the paths of a number of rays from a near object BQ have been traced through the same system (the construction lines are not shown in the figure), and again the effect of the system is seen to be the same as though the incident light had entered the equivalent lens when it was in the position PH, and had left when the lens was in the position P' H'. The effect of the system is

---

*In both cases the path of the ray after the first lens is directed toward the principal focus of the first lens. This ray path is extended to intersect the focal plane of the second lens in G_2 (G_1) and the final ray path is drawn parallel to $A_2 G_2$ ($A_1 G_1$) as was explained in section 6.15. *Figure 9.1* is an important diagram and will repay careful study.

therefore that of a single thin lens with focal lengths PF and P′ F′ situated at two different positions, at P with respect to the incident light, and at P′ with respect to the emergent light. These two positions are called the **principal planes** of the system, and the points P and P′, where these planes intersect the axis, are the **principal points**.

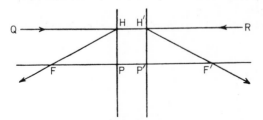

Figure 9.2 Principal or unit planes

PH and P′H′ (*Figure 9.2*) represent the principal planes of a system with its principal foci at F and F′. A ray QH′ from an infinitely distant axial point may be considered as refracted once at H′, and the refracted ray passes through F′; similarly a ray RH may be considered as refracted at H and passing through F. Considering QH and FH as two rays intersecting in an object point at H then, as H′F′ and H′R are the two conjugate rays, H′ is the image of the object point H formed by the system. Hence *the principal planes are conjugate planes in which the magnification is* +1. The principal planes are sometimes known as **unit planes**.

Two further points on the axis may be found for any system which has the property that any incident ray directed towards the first leaves the system as though from the second, and with its direction unchanged. These are the **nodal points**, N and N′ (Moser 1844; Listing 1845), and are useful in considering the size of the image. When the object and image spaces have the same refractive index, the principal and nodal points coincide as in *Figure 9.3*, and are then sometimes known as the **equivalent points**.

There are therefore six points on the axis of a lens system or thick lens that together are known as the **cardinal points** of the system; these are the two principal foci, F and F′, the two principal points, P and P′, and the two nodal points, N and N′. The significance of the first four points as a means of simplifying the theoretical treatment of optical systems was first suggested by Gauss (1840), and these points are sometimes named after him.

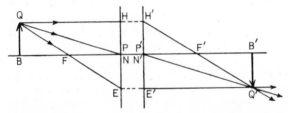

Figure 9.3 Lens system. Principal and nodal points. Graphical construction of image

9.3 GRAPHICAL CONSTRUCTION OF IMAGE FORMED BY A LENS SYSTEM

When the positions of principal foci and principal points are known, the position and size of an image formed by a centred lens system, however complicated, may be found easily by the graphical methods described for a thin lens (section 6.15). Thus in *Figure 9.3* an object BQ is situated in front of a lens system, the principal foci and principal points of which are at F, F′, P and P′ respectively. Two rays from the object point Q are sufficient to determine the position and size of the image. These rays are one leaving Q parallel to the axis, and the other ray passing through the first principal focus F. The first of these meets the first principal plane at H and, from the properties of the principal planes, must, whatever refractions it undergoes in the system, travel in the image space as though from a point H′ on the second principal plane, H′ being at the same distance from the axis as H. Since the incident ray is parallel to the axis, the emergent ray must pass through the second principal focus F′.

The incident ray passing through the first principal focus F must emerge parallel to the axis, as though from a point E′ on the second principal plane. The distance of E′ from the axis will be the same as that of E, the point at which the ray met the first principal plane. The intersection of these two emergent rays determines the position of the image Q′.

9.4 EXPRESSIONS FOR EQUIVALENT POWER, BACK VERTEX POWER, ETC. TWO THIN LENSES IN AIR

In addition to the equivalent focal length, the distances of the first and second principal foci from the front and back vertices of the system respectively will often be required; these distances are called the **front** and **back vertex focal lengths**, f_V and f'_V. The reciprocals of these distances, with the sign changed in the case of the front vertex focal length, are known as the **front** and **back vertex powers**, F_V and F'_V or F.V.P. and B.V.P. These vertex powers are of considerable importance in dealing with spectacle lenses. In *Figure 9.1*

$$\text{F.V.P. or } F_V = -\frac{1}{f_V} = -\frac{1}{A_1 F}$$

$$\text{B.V.P. or } F'_V = \frac{1}{f'_V} = \frac{1}{A_2 F'}$$

In *Figure 9.4* parallel light is incident on the first lens of a system of two thin lenses in air. The effective power at A_2 of the first lens A_1

$$= \frac{F_1}{1 - dF_1} \quad \text{from equation (6.16)}$$

This is the L value for the second lens (L_2).

Then the vergence of the light on emergence from the second lens A_2

$$= L_2 + F_2 = \frac{F_1}{1 - dF_1} + F_2$$

$$= \frac{F_1 + F_2 - dF_1 F_2}{1 - dF_1} = F'_V = \frac{1}{f'_V} \tag{9.01}$$

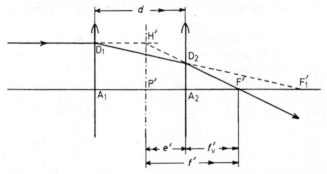

Figure 9.4 Lens system, equivalent and back vertex focal lengths

The equivalent lens of focal length P′F′ placed at P′ will produce the same deviation of the ray as is produced by the system. To find its power, let $A_1 D_1 = 1$ cm, then the prism effect at $D_1 = F_1$ prism dioptres (from equation 6.14). The deviated ray from D_1 meets the second lens at D_2 and

$$A_2 D_2 = \frac{f'_1 - d}{f'_1} = 1 - dF_1$$

Hence, the prism effect at D_2

$$= F_2 (1 - dF_1)$$

and the total prism effect

$$= F_1 + F_2 (1 - dF_1) = F_1 + F_2 - dF_1 F_2$$

This would be the prism effect produced by the equivalent lens at H′, P′H′ being 1 cm.

Hence the equivalent power

$$F = F_1 + F_2 - dF_1 F_2 \tag{9.02}$$

and

$$f' = -f = \frac{1}{F} = \frac{f'_1 f'_2}{f'_1 + f'_2 - d} \tag{9.02a}$$

Furthermore, from equation (9.01)

$$F'_V = \frac{F}{1 - dF_1} = \frac{1}{f'_V} \tag{9.03}$$

$$f'_V = \frac{f'_2 (f'_1 - d)}{f'_1 + f'_2 - d} \tag{9.03a}$$

Similarly, by considering a ray incident through the first principal focus, we obtain

$$F_V = \frac{F}{1 - dF_2} = -\frac{1}{f_V} \tag{9.04}$$

$$f_V = -\frac{f'_1\,(f'_2 - d)}{f'_1 + f'_2 - d}$$ (9.04a)

The distance e of the first principal point *from* A_1 and the distance e' of the second principal point *from* A_2 may be found when f, f', f_V and f'_V are known, for

$$e = f_V - f$$ (9.05)

and

$$e' = f'_V - f'$$ (9.06)

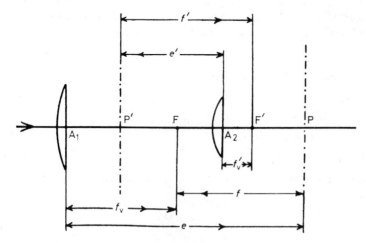

Figure 9.5 Huygens' eyepiece. Position of cardinal points

Example
Find the equivalent focal length and the positions of the principal foci and principal points of a Huygens eyepiece, consisting of two thin lenses of focal length, $f'_1 = + 45$ mm and $f'_2 = + 18$ mm, the lenses being separated 31.5 mm.

$$F_1 = \frac{1000}{45} = 22.22\text{D}, \quad F_2 = \frac{1000}{18} = 55.56\text{D}$$

$$F = F_1 + F_2 - dF_1F_2$$

$$= 77.78 - (0.0315 \times 22.22 \times 55.56)$$

$$= 38.86\text{D}$$

Therefore $f' = + 25.73$ mm and $f = -25.73$ mm.

$$F_V = -\frac{1}{f_V} = \frac{F}{1 - dF_2}$$

$$= \frac{38.86}{1 - (0.0315 \times 55.56)} = -51.19\text{D}$$

$$f_V = +19.27 \text{ mm}$$

$$F'_V = \frac{1}{f'_V} = \frac{F}{1 - dF_1}$$

$$= \frac{38.86}{1 - (0.0315 \times 22.22)} = +129.5D$$

$f'_V = +7.72$ mm

The first principal point is therefore situated $f_V - f = +45.0$ mm from the first lens, and the second principal point at $f'_V - f' = -18.0$ mm from the second lens. The positions of the various points are shown to scale in *Figure 9.5*.

The following further expressions are sometimes useful in dealing with lens systems, and students should derive these for themselves as exercises.

$$ff' = xx' \qquad (9.07)$$

which is Newton's relation (see section 6.10) and

$$F = -g F_1 F_2 \qquad (9.08)$$

where g is the distance from the second principal focus of the first lens to the first principal focus of the second lens.

$$F = F_2 (1 - aF_1) \qquad (9.09)$$

where $a = d + f_2$, the distance from the first lens to the first principal focus of the second lens.

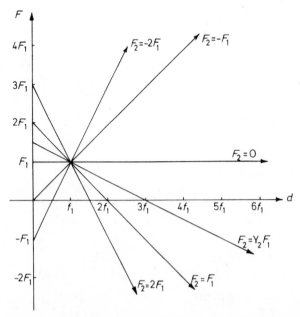

Figure 9.6 Equivalent power of a lens system in terms of the power and focal length of the first lens

The formula for equivalent power (9.02) yields a simple graph to show the effect of varying the value of d if the axes are graduated so that F is measured in multiples of F_1 and d is measured in multiples of f_1' .Thus *Figure 9.6* shows that when the power of the second lens is equal to that of the first, the power of the system equals $2F$ when they are in contact. As the separation d is increased the equivalent power reduces. When the second lens is at the second principal focus of the first it has no effect on the location of the image and so the equivalent power is equal to that of the first lens no matter what the power of the second. For most combinations of two lenses there exists a separation distance for which the equivalent power is zero. Such lens systems are said to be **afocal**.

Other values of F_2 are shown in *Figure 9.6*. Notice how two positive lenses have reducing equivalent power with increasing d while the converse applies if F_2 is negative. What happens if F_1 is negative? For what values of F_1 and F_2 are afocal systems impossible?

Similar graphs may be drawn for F_V and F'_V which do not yield straight lines. The student is encouraged to calculate these as a valuable exercise.

9.5 THICK LENS IN AIR

The thick lens, with air on either side of it, may be dealt with in a similar manner to the lens system above.

The focal powers of the two surfaces will be, from equation (5.04)

$$F_1 = (n - 1)R_1$$

$$F_2 = (1 - n)R_2$$

where n is the refractive index of the lens and R_1 and R_2 the curvatures of the surfaces.

A ray BD_1 (*Figure 9.7*) parallel to the axis is refracted by the first surface towards F_1', the second principal focus of the first surface, and after refraction at the second surface crosses the axis at F'. F' is the second principal focus of the thick lens. $P'H'$ is the position of the equivalent lens with respect to the

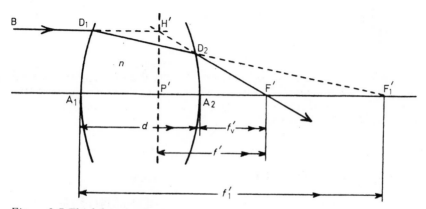

Figure 9.7 Thick lens in air

emergent light, that is, the second principal plane and $P'F'$ the second equivalent focal length.

As before, taking $A_1 D_1$ as 1 cm, the prismatic effect due to the first surface = F_1 prism dioptres, and the prismatic effect at D_2 due to the second surface = $A_2 D_2 \times F_2$.

$$A_2 D_2 = \frac{f'_1 - d}{f'_1} \quad \text{and} \quad f'_1 = \frac{n}{F_1} \quad \text{(from equation (5.05))}$$

$$\text{or } A_2 D_2 = 1 - \frac{d}{n} F_1$$

Therefore the total prismatic effect produced

$$= F_1 + \left(1 - \frac{d}{n} F_1\right) F_2$$

$$= F_1 + F_2 - \frac{d}{n} F_1 F_2$$

This would be the prism effect produced by the equivalent lens at H', $P'H'$ being 1 cm.

Hence the equivalent power is given by

$$F = F_1 + F_2 - \frac{d}{n} F_1 F_2 \tag{9.10}$$

$$A_2 F' = P'F' \times \frac{A_2 D_2}{P'H'} = \frac{1}{F}\left(1 - \frac{d}{n} F_1\right)$$

Hence

$$\frac{1}{A_2 F'} = \frac{1}{f'_V} = F'_V = \frac{F}{1 - \frac{d}{n} F_1} \tag{9.11}$$

Similarly

$$-\frac{1}{f_V} = F_V = \frac{F}{1 - \frac{d}{n} F_2} \tag{9.12}$$

It should be noticed that the expression for the thick lens are of the same form as those for two thin lenses in air, the value d in the latter being a special case of the reduced separation d/n. Graphs similar to those suggested in the previous section may be drawn.

9.6 POSITION AND SIZE OF IMAGE

In *Figure 9.8* a ray BH incident on the first principal plane PH of a lens system or thick lens is refracted along $H'B'$ from the second principal plane $P'H'$. This ray has been deviated through the angle $KLH' = v$ and from the figure

$$v = u' + u$$

or, as u is a negative angle, taking signs into account because *Figure 9.8* is not an all-positive diagram

$$v = u' + (-u)$$

$$u = \frac{c}{PB}, \quad u' = \frac{c'}{P'B'}$$

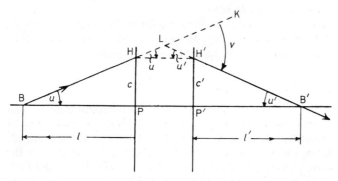

Figure 9.8 Lens system. Prismatic effect. Position of image

As the heights of intercept in the two principal (unit) planes must be equal, $c' = c$, and

$$v = \frac{c}{P'B'} - \frac{c}{PB}.$$

The total prismatic effect in prism dioptres produced on a ray at 1 cm from the axis is equal to the equivalent focal power of the system (section 9.4), or the deviation in radians $= c/f'$

Hence $v = \dfrac{c}{P'B'} - \dfrac{c}{PB} = \dfrac{c}{f'}$

and

$$\frac{1}{P'B'} - \frac{1}{PB} = \frac{1}{f'}$$

Representing the distances PB and P'B', the object and image distances measured from the principal planes, by l and l' we have for a system or thick lens

$$\frac{1}{l'} - \frac{1}{l} = \frac{1}{f'} = F$$

Thus, the conjugate foci expression obtained for a thin lens may be used for a lens system or thick lens, if object and image distances are measured from the first and second principal points respectively.

Also, since a ray from any object point passing through the first nodal point emerges from the second nodal point undeviated by the system, the various expressions for the magnification of the image formed by a thin lens are applicable to a system if object and image distances are measured respectively from the first and second nodal points.

9.7 THE MEASUREMENT OF FOCAL LENGTH AND POWER

The measurement of the vertex focal lengths and thence the vertex powers of a system or thick lens may be made by determining the positions of the principal foci by the methods described in Chapter 6.

The determination of the equivalent focal length is, however, less simple, owing to the difficulty of finding directly the positions of the principal points. The following are a few of the methods applicable to lens systems.

Newton's method—Positive systems

The system to be measured is set up about midway along an optical bench, and the positions of the principal foci found, using either a telescope or a collimator (section 6.20). An illuminated object is then placed at some distance x outside the first principal focus, and an image is received on a screen. The distance of the image from the second principal focus is x', and the focal length is found from the expression $f' = \sqrt{-xx'}$. Most accurate results will be obtained when x and x' are nearly equal.

Magnification methods

Methods depending on the measurement of the magnification of the image formed by a system are often very convenient. Providing the image is sufficiently free from distortion, the focal length can be obtained quickly and accurately by such methods.

Two magnifications

An illuminated object of known size, such as a rectangular aperture, is set up on the axis of the lens to be tested. The image is focused on a screen, and its size measured. This is repeated with the object in a different position. From the magnification m_1 and m_2 thus found and the distance moved by the object, $l_2 - l_1$, or by the image, $l'_2 - l'_1$, the focal length may be calculated.

$$f' = \frac{l'_2 - l'_1}{m_1 - m_2} = \frac{l_1 - l_2}{\dfrac{1}{m_1} - \dfrac{1}{m_2}} \tag{9.13}$$

This method has been applied by Abbe (1840-1905) to the measurement of the focal lengths of microscope objectives. The two positions of object and image are obtained by altering the tube-length of the microscope, and the size of the image is measured with a micrometer eyepiece, a stage micrometer being used as object.

Two magnifications, negative systems

The two magnifications method may also be applied to negative systems by using an auxiliary positive lens to form a real image of the object (*Figure 9.9*). The negative system is inserted in the convergent light, and the image formed by the positive lens becomes a virtual object for the negative system. By measuring the magnification of the final image with the system being tested in two different positions, A_1 and A_2, the focal length can be found from equation (9.13).

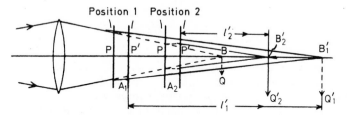

Figure 9.9 Negative system. Measurement of focal length. Two magnifications method

The foco-collimator

If two distant objects subtend a known angle w at the lens, then their images formed in the second focal plane of the lens are separated by a distance

$$h' = 2f \tan \frac{w}{2}$$ from equation (6.12a)

and

$$f = \frac{h'}{2 \tan \dfrac{w}{2}}$$

The distant objects may conveniently be replaced by two collimators fixed at an angle w between their optical axes. The method is a simple and rapid one for

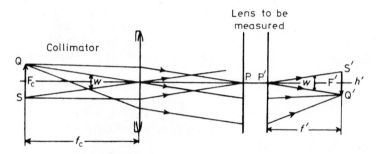

Figure 9.10 The foco-collimator

the measurement of photographic objectives, which have a flat field reasonably free from distortion.

The angle *w* may be supplied by one collimator only if it is of good enough quality over a reasonable field. Such collimators are sometimes called *foco-collimators*. The principle of their use is shown in *Figure 9.10*. The object QS is placed at the first focal plane of the collimator and the image Q'S' formed by the lens under test is measured to determine the magnification. Sometimes QS is a scale so that the focal length of the lens system may be read off directly by comparing the image with a standard length.

Rotation method—Nodal points

The following method provides a means of determining the positions of the nodal points, and since, for a system in air, these coincide with the principal points, the equivalent focal length of the system can be measured.

In *Figure 9.11* light from a distant luminous point is incident on a lens system having its nodal points at N and N', and an image is formed at F'. On rotating the system about the second nodal point N', no movement takes place in the image, for from the property of the nodal points, any ray from the distant object point directed towards the first nodal point in its new position must still emerge along N'F'. If the system is rotated about any other point, N' and therefore F' move.

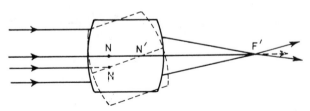

Figure 9.11 Focal length of lens system. Rotation method

The lens to be measured is set up on the optical bench in a holder, commonly called a **nodal slide**, that can be rotated about a vertical axis, the lens sliding in the holder so that any point on its axis can be brought over the axis of rotation. Parallel light from a collimator is focused on a screen, and the position of lens and screen adjusted until a rotation of the lens produces no movement of the image. The distance from the axis of rotation to the screen is then the equivalent focal length.

This method can also be used with the telescope method (section 6.20). An object is brought up towards the lens until its image is clearly seen in a telescope focused for parallel light, and the lens is adjusted, as above, so that the image remains stationary on rotating the lens. Owing to the magnification given by the telescope, this method will usually be more accurate than that of receiving an image on a screen.

The principle of the stationary image with the lens rotating about its second nodal point has been applied to panoramic cameras.

CHAPTER 9   *EXERCISES*

1. The following lenses, all assumed to be thin, are placed in contact. What will be the focal power and focal length of the combination?

(*a*) + 4D,

(*b*) − 6 inches focal length,

(*c*) + 4 inches focal length,

(*d*) − 2.62D.

2. Deduce an expression for the focal length of two thin lenses in contact.
A convex lens produces a real image of an object at a distance of 40 cm from the lens. A concave lens, the focal length of which is 15 cm is placed 20 cm from the convex lens on the side away from the object. Find the new position and nature of the image.

3. Two equal + 4D meniscus lenses of spectacle glass (*n* = 1.523) with − 6D concave surfaces are placed together with these concave surfaces facing one another and the space between them is filled with water. Find the power of the complete glass-water lens (neglect thickness).

4. A 10D plano concave lens of refractive index 1.57 lies on a horizontal plane sheet of glass, with its concave surface uppermost and is filled up level with water, having a refractive index of 1.33. What will be the power of the combination?

5. Two thin convex lenses of focal length 50 mm and 70 mm respectively are separated by 16 cm. Show by graphical construction that (within the paraxial region) incident rays parallel to the axis at all incident heights intersect their conjugate refracted rays always in a plane perpendicular to the axis.

6. Two thin lenses of powers + 20D and + 25D respectively are separated by a distance of 15 cm. Find graphically the focal length and the positions of the principal points, P and P′, of the combination. Three rays at different inclinations are incident on the combination at the axial point P. Trace them through the two component lenses, showing that they emerge from the system through P′ parallel to their original directions.

7. Show with a diagram what are meant by the principal and nodal points of a lens. Explain how these positions are used in finding the position and size of an image formed by a lens.

8. A distant object subtends an angle of $10°$; find the size of the image formed by a + 6D lens. Where must a − 10D lens be placed to magnify the image four times?

9. A distant object subtends an angle of $20°$. Find graphically the size and position of the image of this object formed by a lens system consisting of a + 6D followed by a + 10D lens, the lenses being coaxial and 5 cm apart. What must be the position of an object in front of the first lens in order that its image formed by the system may be at infinity?

10. What are meant by the principal, focal and symmetrical planes of a thick lens or lens combination? Illustrate your answer by a diagram.

11. Two cylindrical lenses, one of + 4D and the other of − 6D, are set up parallel to one another, with their axes vertical and with an interval of 12 centimetres between them. If a point source of light is placed at 20 cm from the positive lens, where will the image be, and what will be its character?

12. Two thin positive lenses of focal lengths 4 cm and 3 cm respectively, each of 2 cm aperture, are mounted coaxially 2 cm apart. An object 2 cm high is placed with its foot on the common axis and 8 cm in front of the 4 cm lens.
On a careful diagram drawn exactly full size, construct the course of the pencil of light from the extremity of the object, through the lenses, to the final image. From your diagram give the size of this image and its distance from the second lens.

13. A screen is placed 14.5 mm behind a + 60D lens (assumed to be thin). Find the size of the blurred image formed on the screen of a distant object that subtends an angle of $20°$. What must be the power of a second lens, place 12 mm in front of the first, so that a *sharp* image of the same object shall be focused on the screen?

14. Derive Newton's relation in the case of a system consisting of two thin separated lenses.

15. A parallel pencil of light falls axially on a + 5.5D spherical lens combined with a − 2D cylindrical lens with axis horizontal. What lens placed 8 inches behind this combination will restore the emergent light to parallelism in both meridians?

16. The image of a distant object subtending an angle of $10°$ is formed by a lens combination consisting of a + 4D and a + 2D lens separated 100 mm. Find the size and position of the image and the equivalent focal length of the system.

What will be the effect on the size and position of the image and on the equivalent focal length of turning the system round so that the + 2D lens is towards the object?

17. An object 20 mm in height is placed in front of an optical system of lenses. An inverted image 12 mm in height is formed on the opposite side of the system. If the object were placed 100 mm nearer the system the image would be formed at infinity. Find the focal length of the system.

18. Two positive thin lenses of focal length 80 mm and 60 mm respectively are coaxial and are separated by 35 mm. Find by graphical construction the equivalent focal length of the combination and the positions of its principal and focal points. Mark these positions clearly on your drawing. Check the results by calculation.

19. Find the equivalent focal length of a lens system consisting of a + 4D and a − 4D lens separated 15 cm. What will be the position and size of the image formed by this system of an object placed 1 metre in front of the first lens?

20. A spectacle lens of power + 10D is mounted coaxially with and 5 cm in front of a second lens of power − 8D. Find the power of the combination and the position of the principal points. Of what well-known optical system is this a model?

21. What are meant by the equivalent focal power and principal points of a lens system? What will be the size of the image of a distant object which subtends an angle of $8°$ at a lens system consisting of two lenses of + 10 inches and − 5 inches focal length respectively, separated by 7 inches?

22. Find the positions of the focal and principal points of a Ramsden eyepiece consisting of two lenses of 2 in focal length separated 4/3 in.

23. Two positive thin lenses of focal lengths 20 cm and 50 cm respectively are separated by a distance of 100 cm. Find the power of the resultant system and the positions of its principal and focal points. Discuss the effect of replacing the original lenses by two of the same focal lengths but of opposite sign. Give a diagram.

24. A lens system is formed of two lenses having power of + 6D and − 10D respectively, separated by 10 cm. Find the position and size of the image of a distant object which subtends an angle of $10°$.

25. A lens system consists of a + 4D lens and a + 3D lens separated by 10 cm. Find the equivalent focal length and the positions of the principal points. An object 3 cm long is situated 50 cm in front of the first lens. What will be the position and size of the image produced by the combination?

26. Find the equivalent focal power and the position of the principal points of a system consisting of a + 6D and a − 8D lens separated 8 cm. Use the values obtained to find the position and size of the image of an object 3 cm long placed 50 cm from the first lens.

27. Two thin positive lenses of focal lengths f'_1 and f'_2 are separated by a distance d. What is the value of d for the focal power of the combination to be zero? When the lenses are in this position, find a relation between the incident height of a ray parallel to the axis in the object space and the height from the axis of the conjugate emergent ray.

28. The positions of object and image such that the magnification is − 1 are known as the symmetrical points. Show that the distances of these from the principal points are respectively $2f$ and $2f'$.

29. A Huygens eyepiece consists of two thin lenses having focal lengths of $2\frac{1}{2}$ inches and 1 inch respectively, separated by $1\frac{3}{4}$ inches. Find the equivalent focal length and the positions of the principal points. Show in a diagram, drawn to scale, the positions of these points and the focal length.

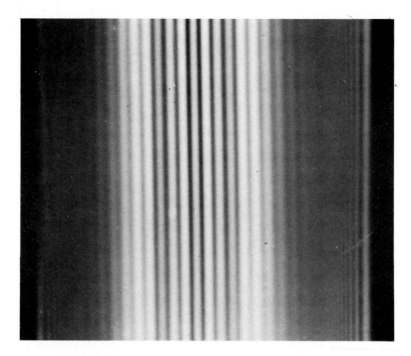

Plate 1   Fringes formed when slit, illuminated with monochromatic light, is viewed via a Fresnel Bi-prism (see section 14.6)

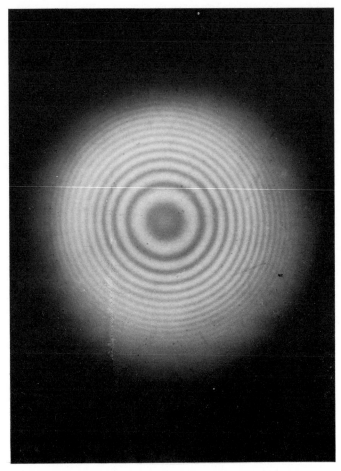

Plate 2   Newton's Rings with monochromatic light. The light centre to the ring system is due to the two glasses not being quite in contact (see section 14.10)

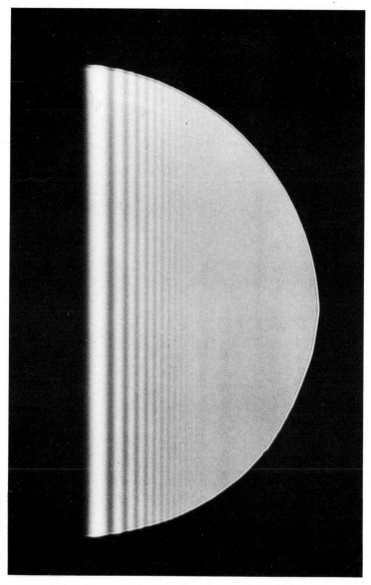

Plate 3   Shadow of a straight edge showing Fresnel Diffraction effects (see section 15.7)

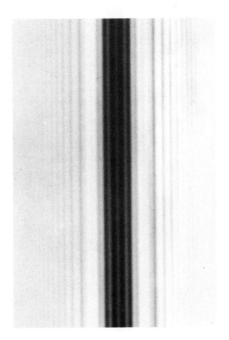

Plate 4   Shadow of a narrow wire showing Fresnel Diffraction effects (see section 15.7)

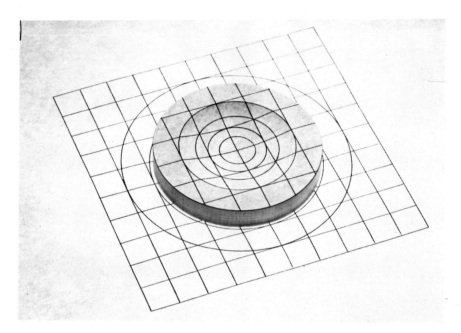

Plate 5   Transmission of an image by a faceplate of optical fibres (see section 4.16)

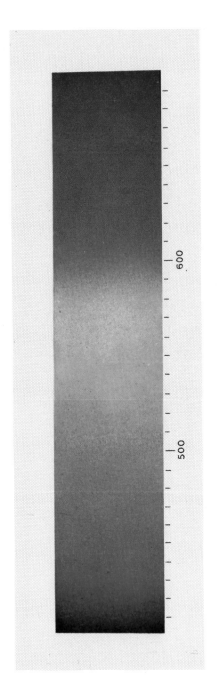

Plate 6b Newton's Rings with white light by reflection. The dark disc at the centre indicates optical contact (see section 14.10). Glass surfaces by Mr. D. Jones, Pilkington P-E Ltd, North Wales

Plate 6a Normal Visual Spectrum. The figures indicate the wavelength in nanometres (nm) (see sections 11.1, 12.7, 13.6, 15.14)

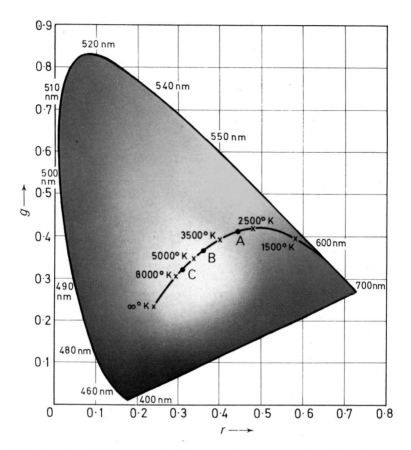

Plate 6c   The CIE Chromaticity Diagram. The letters A, B *and* C *indicate the colour and colour temperature values of the standard illuminants (see sections 13.6, 13.25). Based on the colour master published in* Lamps and Lighting *(Edward Arnold) and reproduced by courtesy of Thorn Lighting Ltd, Research and Engineering Laboratories*

In these pictures it is not possible with printing inks to reproduce exactly the colours of the optical effects described. The particular illuminant used when viewing will also affect their appearance. The intention is to indicate the major effects of the spectral range of light

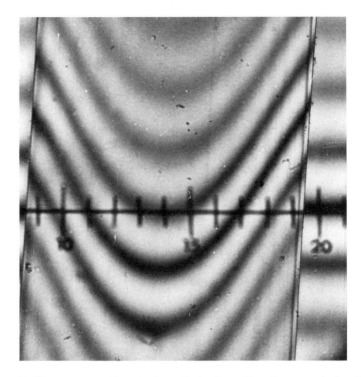

Plate 7   Measurement of strain by polarization. The glass sample has been toughened by heat treatment. Residual strain in the glass displaces the fringes upwards for tension and downwards for compression (see section 16.8)

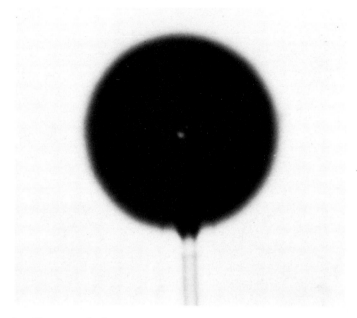

Plate 8a Photograph showing diffracted light forming a bright spot at the centre of a circular shadow (see section 15.4)

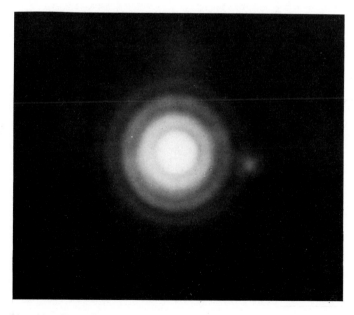

Plate 8b Airy Disc. Pattern of light formed by a perfect optical system of circular aperture as the image of a point object (see section 15.10)

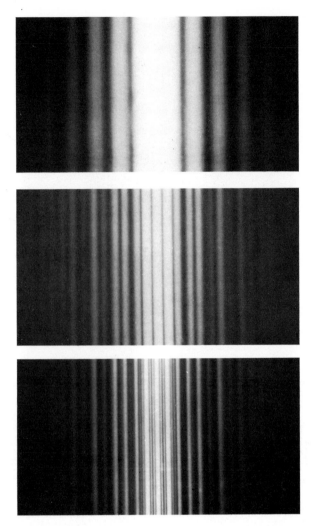

Plate 9   Fraunhofer Diffraction with (a) one, (b) two and (c) four slits (see sections 14.6, 15.9, 15.13)

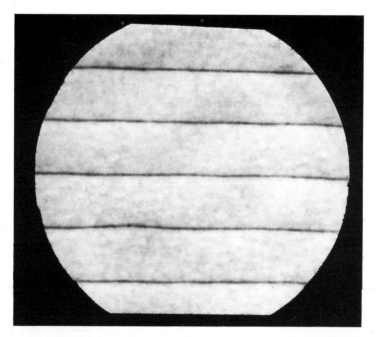

Plate 10a   Multiple-beam interference fringes (see section 14.7)

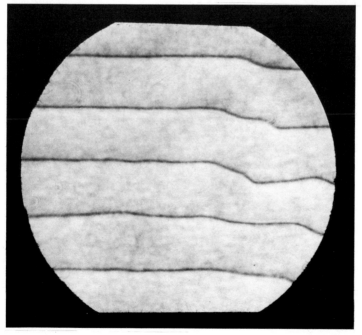

Plate 10b   Multiple-beam intereference fringes showing region of inhomogeneity (see section 14.12). (Plates 10a and 10b by courtesy of Pilkington P-E. Ltd,)

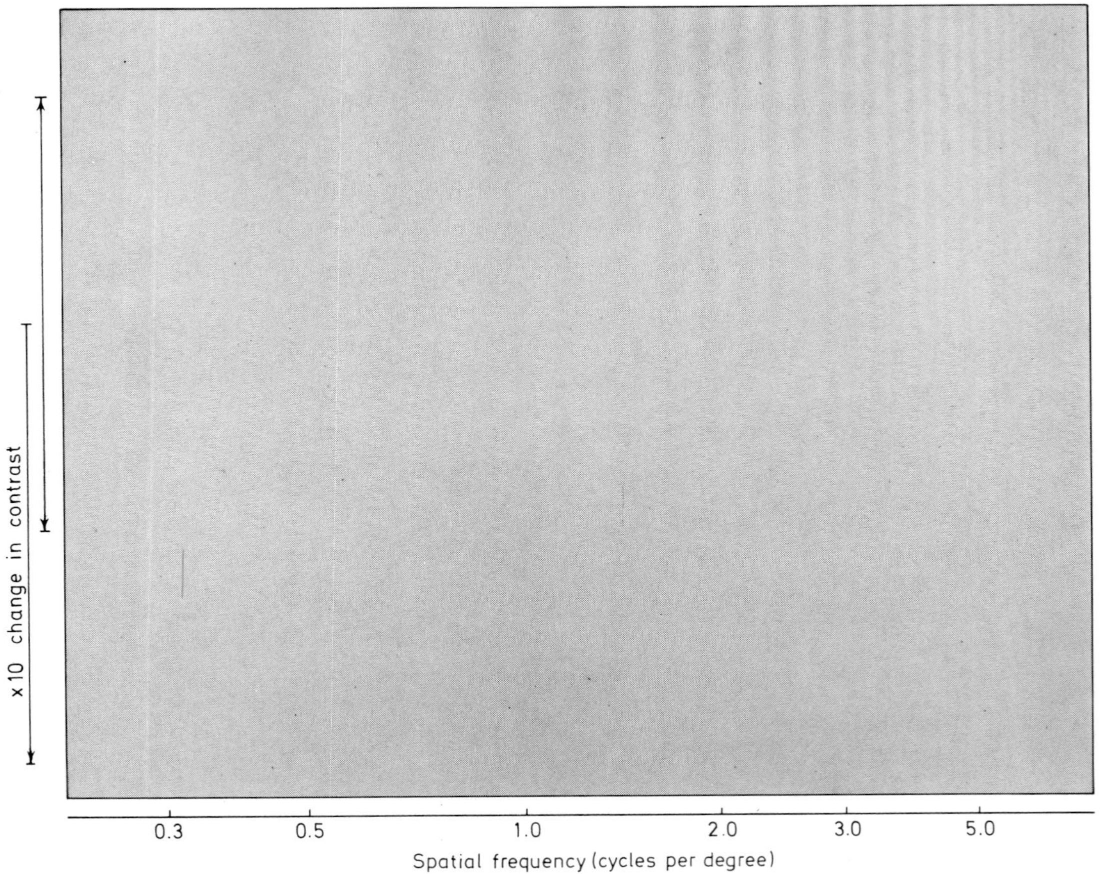
x10 change in contrast
0.3
0.5
1.0
2.0
3.0
5.0
Spatial frequency (cycles per degree)

Plate 11 Double-swept contrast/frequency grating. In this plate the contrast modulation of the grating changes by a factor of 10 (one log unit) for every 60 mm along the y-axis although variations in printing can affect this. The angular frequency values along the x-axis are correct for arm's length viewing at 600 mm (see section 20.9). The original of this plate was supplied by Professor Arden of the Institute of Ophthalmology, London. Permission for its use here is gratefully acknowledged

30. Show that the equivalent focal power F of a system of two thin lenses of focal power F_1 and F_2 is given by the expression

$$F = - gF_1 F_2$$

where g is the distance from the second principal focus of the first lens to the first principal focus of the second lens.

31. Show that in a system of two thin lenses

$$F = F_2 \ (1 - aF_1)$$

where $a = d + f_2$, the distance from the first lens to the first principal focus of the second lens.

32. When the distance of a real image from a lens system is increased by 5 cm the magnification is increased from 0.5 to 0.65. What is the focal length of the system?

33. Show that for a thick lens in air

$$F = (n - 1) \left(\frac{1}{r_1} - \frac{1}{r_2} \right) + \frac{(n - 1)^2}{n} \cdot \frac{d}{r_1 r_2}$$

34. A positive meniscus lens of central thickness 7.60 mm is made of glass of refractive index 1.520, the back surface power being $- 6D$. What power must be given to the front surface in order that the back vertex power of the lens shall be $+ 9.7D$?

35. Find the equivalent power, back vertex power and position of the second principal focus of a meniscus lens (in air) with the following constants:

Radius of first surface $= + 50$ mm
Radius of second surface $= + 250$ mm
Central thickness $= 15$ mm
Refractive index $= 1.6$

36. A bi-convex lens of glass $n = 1.52$ has curvatures of 5 inches and 8 inches radius respectively and a central thickness of $1^1/_2$ inches. Find the size and position of the image of an object 2 inches long placed 15 inches from the first surface of the lens.

37. A slab of glass 3 inches long, $n = 1.5$ has its end faces ground convex so that each has a radius of curvature of $^1/_2$ inch. What are the optical properties of the thick lens so made?

38. A spherical electric bulb 3 inches in diameter is immersed in water; what will be the equivalent focal length of the optical system thus produced?

39. A $+ 10D$ equiconvex lens is inserted into a thin glass box with plane parallel sides which just contains it, and the space is filled with water of refractive index 1.33. If the power of the combination is $+ 4D$, what is the refractive index of the lens and the curvature of its surfaces?

40. Find the equivalent focal length and principal points of a sphere of glass of which the refractive index is 1.54 and the diameter 2.7 inches.

41. Describe and explain the principle of one method of determining the equivalent focal length of a lens system. Give a diagram.

42. In measuring the focal length of a lens system by the foco-collimator, the collimator used had a lens of 10 inches focal length and a graticule in the focal plane with two marks 0.35 inches apart. The size of the focused image formed by the lens being tested was a quarter of an inch. Find the focal length of the lens.

43. A scale is placed at the first principal focus of a positive lens A_1 and is focused by A_1 and a second lens A_2; show that the image is formed at the second principal focus of A_2 and that the magnification is equal to the ratio of the focal lengths of the lenses. Is the image erect or inverted?

44. Define the term 'focal length' as applied to a lens of which the thickness cannot be neglected.

Describe briefly an experimental method of finding the difference between this 'focal length' and the distance between the apex of the lens and its principal focal point. Prove any formula employed.

45. An optical system consists of a + 10D lens followed at a distance of 5 cm by a − 8D lens. Find the power of the whole system, and hence the size of the image of a very distant object subtending an angle of 0.05 radians in the field of view. How would the distance of the image from the last lens, and the size of the image, change if the system were reversed so that the light enters the negative lens first?

46. With the aid of a diagram explain why the planes passing normally to the optical axis through the principal points of a lens system are called the *unit planes*.

The powers of the first and second surfaces of a lens of total power F, in air, are F_1 and F_2; the lens has central thickness d metres and refractive index n.

Show that the principal points P_1 and P_2 of the lens are separated by the distance

$$P_1 P_2 = d \left(1 - \frac{F_1 + F_2}{nF} \right)$$

47. A compound lens of power + 32D consists of two thin positive components of equal power separated by an air space; it forms a real image of a real object, the image being 1.5 times larger than the object. The distance from object to image is found to be 125.21 mm.

Find the separation of the principal points and show their positions on a diagram. Also find *either*

(*a*) the power of the individual components

or   (*b*) the separation of these components.

48. Two coaxial thin positive spherical lenses are separated by 20 mm. The first lens has power + 5D and the back vertex power of the combination is + 50D.

Keeping the separation and the power of the second lens unchanged, by how much must the power of the first lens be altered in order to increase the back vertex power of the combination by one-third of a dioptre?

49. The surface powers of a bi-convex lens of refractive index $n = 1.50$ and central thickness $d = 18.75$ mm are respectively $F_1 = 5.0$ and $F_2 = 8.0$ dioptres. Calculate the positions of its principal points, P and P'.

An expression for the separation of the principal points is

$$PP' = \frac{(n-1)\,ad}{na - d}$$

where a is the distance $C_1 C_2$ separating the centres of curvature of the lens surfaces. Show that this expression gives the correct numerical result for the lens in question.

Chapter 10

The principles of
optical instruments

The more important optical instruments may be divided into two classes: (*a*) those which are used directly with the eye as aids to vision, as, for example, the microscope and telescope, and (*b*) projection instruments, such as the camera and the slide projector.

However, the eye itself may be represented as a projection instrument having a lens system and a mosaic of detectors. In this chapter a very brief description of the eye is given in terms of paraxial optics and its features which relate to its use with aids to vision. A fuller treatment of the eye as an optical instrument is given in Chapter 20.

10.1 THE EYE

Figure 10.1 represents a horizontal section of the human right eye. The greater part of the globe of the eye is almost spherical, about one inch diameter, and is formed by the tough opaque outer coat, known as the **sclera**, S. The front portion of the globe is the more steeply curved and transparent **cornea**, C, which is the chief refracting surface. The cornea is about $^1/_2$ mm thick, and its surfaces, in the normal eye, are approximately spherical. The average radii of curvature of the cornea are 7.7 mm for the front surface and 6.8 mm for the back; its refractive index is 1.38. The space immediately behind the cornea, known as the anterior chamber, is filled with a fluid, the **aqueous humour**, of almost the same refractive index as water. The cornea and aqueous humour together therefore form a positive lens of about 43D power.

Behind the anterior chamber is the double convex **crystalline lens**, L, composed of layers of transparent fibres. The curvature, refractive index and hardness of the layers increase towards the centre of the lens, and the whole is enclosed in the thin lens capsule. Under the action of the **ciliary muscle**, C.M., the lens changes in form, and the power of the system is adjusted to focus images of objects at different distances on the receiving screen. The radii of curvature of the front and back surfaces of the crystalline lens when the eye is focused on a distant object are about 10 mm and 6 mm respectively; its thickness is about 3.6 mm and its refractive index varies from about 1.41 at the centre to 1.38 for the outer layers. The space behind the crystalline lens is filled with a transparent jelly-like medium, the **vitreous humour**, having the same refractive index as the aqueous.

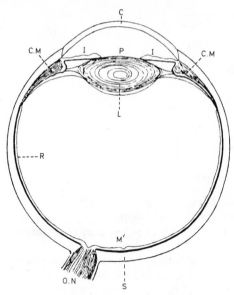

Figure 10.1 Horizontal section of the right eye (× 2.5 natural size)

Directly in front of the crystalline lens in the **iris**, I, the diaphragm, which controls the amount of light entering the eye. This has a circular aperture, P, varying in diameter from about 2 to 8 mm, known as the **pupil**.

Although the centres of the various surfaces and the pupil do not lie exactly on a common axis, we may, for most purposes consider the eye as a centred system, and the approximate positions of the focal, principal and nodal points are shown in *Figure 10.2*.

The innermost coat of the globe in contact with the vitreous humour is the **retina**, R. This is the receiving screen upon which sharp images of external objects must be formed, and from the closely packed nerve endings of which the stimulus is conveyed along the **optic nerve**, ON, to the brain. A little to the temporal side of the end of the optic axis is a small area of the retina known as

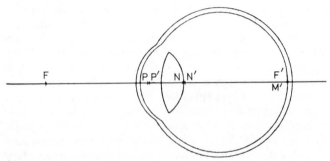

Figure 10.2 Cardinal points of the eye

the **macula lutea**, M′, or yellow spot, having a small depression at its centre, known as the **fovea centralis**. It is here that vision is most distinct, and the image of the object on which the attention is fixed falls on this portion of the retina. The portion of the retina at which the optic nerve enters is insensitive to light and forms the 'blind spot'. Between the retina and the sclera is the **choroid**, a coat which contains the blood vessels which nourish the eye and a layer of pigment cells which absorbs any light passing through the retina, and prevents any light that may pass through the sclera from reaching the retina.

10.2 ACCOMMODATION

As the eye can see distinctly, in turn, objects both far and near, it must be provided with some adjustment whereby images of objects at different distances are brought to focus on the retina. This focusing adjustment, which is known as **accommodation**, is effected by a change in the form of the crystalline lens produced by the action of the ciliary muscle. When the eye accommodates to see a near object, the ciliary muscle contracts and relaxes the suspensory ligament attached to the lens capsule, allowing the front surface of the lens to bulge forward to a steeper curvature and therefore of increased power. At the same time, as can easily be seen by watching an eye, the iris contracts, reducing the size of the pupil. The amount of change in power produced—the **amplitude of accommodation**—is greatest in the case of young persons, being at the age of 10 about 14D, which means that a normal eye at this age can see objects distinctly as close as 7 cm. This power of accommodation gradually becomes less with advancing years, owing probably to a hardening of the crystalline lens, until it is no longer possible to see objects distinctly at the distance required for reading and other close work without the aid of lenses. This condition is known as **presbyopia**.

10.3 EMMETROPIA AND AMETROPIA. FAR AND NEAR POINTS

When the optical system and the axial length of the eye are such that light from a distant object is focused exactly on the retina when accommodation is completely relaxed, the eye is said to be **emmetropic** (*Figure 10.3a*) and the condition is that of emmetropia. When this is not the case the eye is **ametropic**. The three types of ametropia are **myopia, hypermetropia** or hyperopia and **astigmatism**, the first two usually being due to abnormalities in the axial length of the globe and the last to want of sphericity in one or more of the refracting surfaces, usually the cornea.

In myopia, or short sight, the refracting power or the axial length is too great and light from a distant object focuses in front of the retina (*Figure 10.3b*). Such an eye will therefore not be able to see distant objects clearly without the aid of a correcting lens. Objects at some closer distance will, however, be clearly seen with accommodation relaxed. In hypermetropia, or long sight, the refracting power or the axial length is too small, and light from a distant object focuses behind the retina (*Figure 10.3c*). The uncorrected hypermetropic eye with accommodation relaxed, will not see any object clearly, but, if the amount of error be not too great, the additional power required to focus the light on the retina may be provided by accommodation. In this way the hypermetropic eye

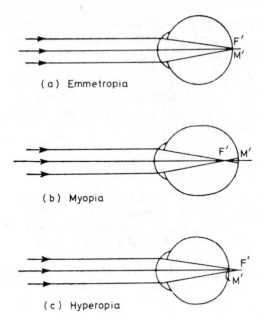

(a) Emmetropia

(b) Myopia

(c) Hyperopia

Figure 10.3 The eye. Emmetropia and ametropia

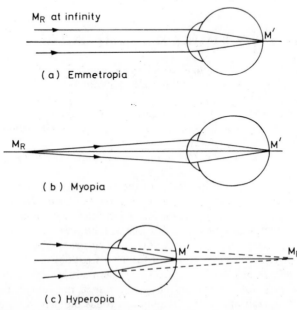

(a) Emmetropia

(b) Myopia

(c) Hyperopia

Figure 10.4 The far point of the eye

may have quite good vision for distant objects. In astigmatism the refracted beam from a point object is astigmatic and focuses through two lines (Chapter 8); these line foci may occupy any positions with respect to the retina, but in no case will an object be seen distinctly.

The object point conjugate to the retina, when accommodation is completely relaxed, is known as the **far point**, *punctum remotum*, M_R, of the eye. This will be (*a*) at infinity in the case of an emmetropic eye (*Figure 10.4a*), (*b*) a real object point in myopia (*Figure 10.4b*) and (*c*) a virtual object point behind the eye in hypermetropia (*Figure 10.4c*). The vergence the incident light must have at the cornea in order that the light may focus on the retina, when accommodation is relaxed, is termed the **static refraction** of the eye.

The conjugate point to the retina, when accommodation is exerted to its fullest extent, that is, the nearest position at which an object can be seen distinctly, is the **near point**, *punctum proximum*, M_P, of the eye. The position of this will depend on the amplitude of accommodation available and the type and amount of ametropia; thus for a given amplitude the near point will be closer in myopia and more remote in hypermetropia (Hyperopia) than in emmetropia.

10.4 THE CORRECTION OF AMETROPIA

Ametropia may be corrected by placing before the eye a lens that will give the light incident from a distant object the necessary vergence for it to be focused on the retina. As will be seen from *Figure 10.5*, the necessary condition for correction will be that the power and position of the lens shall be such that the second principal focus of the lens shall coincide with the far point of the eye. Negative lenses will therefore be used to correct myopia and positive lenses hypermetropia, while in astigmatism, cylindrical or sphero-cylindrical lenses are required. In practice, the amount of ametropia is usually expressed as the power of the lens required for correction when placed at the ordinary spectacle distance, about 12 mm from the cornea.

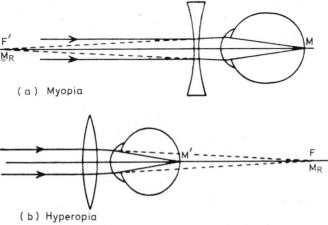

(a) Myopia

(b) Hyperopia

Figure 10.5 Correction of ametropia

10.5 VISUAL ANGLE AND APPARENT SIZE

The apparent size of any object or image seen by the eye depends upon the size of the retinal image, which in turn depends upon the angle subtended by the object or image at the nodal point of the eye. Objects which subtend the same **visual angle** will have the same apparent size, although their actual sizes may be very different if they are at different distances from the eye. Thus a disc $^1/_4$ inch diameter held at arm's length will appear to cover the moon, that is, the disc at that distance and the moon have the same apparent size. The principle of magnifying instruments, such as the magnifier, microscope and telescope, is to produce an image that shall subtend a larger visual angle than that subtended by the object seen directly.

10.6 APPARENT MAGNIFICATION OR MAGNIFYING POWER. THE MAGNIFIER OR SIMPLE MICROSCOPE

An object situated at the near point will subtend the largest visual angle and will therefore have the largest apparent size with which it can be clearly seen by the unaided eye. If it is required to see the object apparently still larger an optical aid must be used. If a positive lens of focal length shorter than the near point distance is placed close before the eye, the object can be brought up to, or a little within, the first principal focus of the lens, and the eye will be able to focus the parallel or divergent light leaving the lens. In *Figure 10.6* the light leaving the lens, as though from the virtual image $B'Q'$ enters the eye and, providing the image is not closer to the eye than the near point, it will be seen distinctly. The size of the retinal image and therefore the apparent size of the object seen through the lens will depend upon the angle w'.

As the distance of the near point differs in different eyes it is necessary in considering the magnification of instruments to adopt some standard value for this distance. This conventional distance is called the **least distance of distinct vision**, q, and is taken as -10 inches or -250 mm measured *from* the eye.

The **apparent magnification** or **magnifying power**, M, produced by a lens may be expressed as the ratio of the angle, w', subtended at the nodal point of the eye by the image to the angle, w, subtended at the nodal point by the object when at the least distance of distinct vision from the unaided eye. Usually the angles are expressed in terms of their tangents. It follows that the magnifying

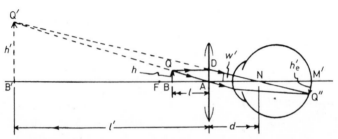

Figure 10.6 Image seen through a lens. Apparent magnification or magnifying power

power is also the ratio of the size of the retinal image obtained with the aid of the lens to the size of the retinal image of the object when at the least distance of distinct vision from the unaided eye.

In *Figure 10.7* the object to be magnified is placed at the first principal focus of a short focus positive lens, and the emmetropic eye with accommodation relaxed sees the erect image at infinity subtending an angle w'. Since the image is at infinity, the distance of the lens from the eye has no effect upon magnification, but in order to obtain the largest field of view the lens should be as close to the eye as possible.

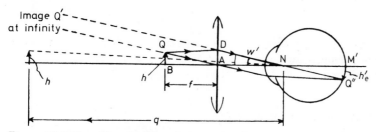

Figure 10.7 Magnifier. Simple microscope or loupe

$$M = \frac{\tan w'}{\tan w} = \frac{w'}{w} \text{ when the angles are small} \tag{10.01}$$

$$\tan w' = \frac{h}{f'}, \quad \tan w = \frac{h}{-q}$$

Therefore

$$M = \frac{-q}{f'} = \frac{250}{f' \text{ (mm)}} = \frac{10}{f' \text{ (in)}} = \frac{F \text{ (D)}}{4} \tag{10.02}$$

When the magnifier is used with the image at any finite distance the magnifying power may be calculated by finding the angles w and w' taking into account the position of the eye. If the eye is assumed to be close to the lens, the use of the magnifier with the image at the near point means that

$$L' = \frac{1}{-250 \text{ mm}} = -4 \text{ D} = L + F$$

Thus the magnifying power, M', under these conditions is given by

$$M' = \frac{h}{l} \times \frac{-250 \text{ mm}}{h} = \frac{L}{-4} = \frac{-4 - F}{-4} = 1 + M \tag{10.02a}$$

Typically, magnifiers are used with the image between infinity and -250 mm. The image location, expressed as dioptres vergence at the eye is known as the **dioptric setting** of the magnifier. A common setting is -2D at which the magnifying power is $M + \frac{1}{2}$.

While the simple microscope possesses the distinct advantages of simplicity and erect image, the necessary short working distance and small diameter of the

 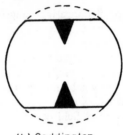 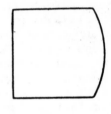

(a) Achromatic   (b) Coddington   (c) Stanhope

Figure 10.8 Types of magnifiers

lens in the case of high powers limit the greatest useful magnification to about 15. Types of magnifiers are shown in *Figure 10.8*. Further discussion of magnifier design is given in section 18.9.

10.7 THE COMPOUND MICROSCROPE

The compound microscope consists essentially of two positive lenses, the first of which, the **objective**, forms a real and magnified image of the object, and the second, the **eyepiece**, is used as a magnifier to examine the aerial image formed by the objective. In practice both objective and eyepiece will be systems of lenses.

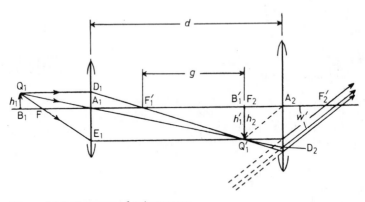

Figure 10.9 Compound microscope

Figure 10.9 shows the path of light from one end of a small object B_1Q_1. A real inverted aerial image h'_1, having a lateral magnification m_1, is formed at B'_1 by the objective A_1. If the microscope is used with the unaccommodated emmetropic eye, B'_1 coincides with the first principal focus F_2 of the eyepiece A_2 which, as a magnifier, gives a magnifying power

$$M_2 = \frac{-q}{f'_2}$$

The total magnifying power M therefore equals $m_1 \times M_2$. From expression (6.10)

$$m_1 = \frac{-g}{f'_1}$$

where g (distance from F'_1 to F_2) is called the **optical tube length**.
Therefore

$$M = \frac{-g}{f'_1} \times \frac{-q}{f'_2} = \frac{-g}{f'_1} \times \frac{250}{f'_2 \text{ (mm)}}$$

(10.03)

In practice, microscope objectives are usually specified in terms of their focal lengths, either in inches or millimetres, and are generally designed for an optical tube length of 160 mm. Eyepieces are marked in terms of their magnifying power, $-q/f'$.

10.8 THE TELESCOPE

Telescopes are used for the following three purposes:

(1) To produce an image of a usually distant object which will subtend a larger visual angle than that subtended by the object, that is, to give an apparent magnification or magnifying power.

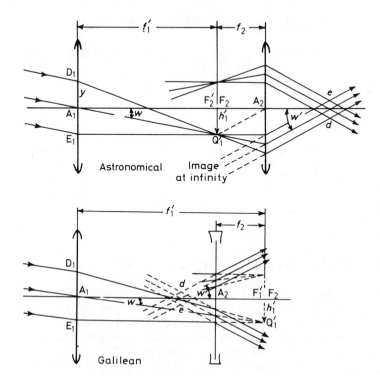

Figure 10.10 Astronomical and Galilean telescopes

(2) As an instrument of large aperture, collecting sufficient light to enable faint distant point sources of light, such as stars, to be seen, and of high resolving power (see Chapter 15). This is the purpose of the large astronomical telescope.

(3) To determine accurately a definite line relative to a certain part of an instrument, that is, as a sighting device, as in surveying instruments, spectrometer, etc.

The telescope consists, in principle, of two lenses, the first—the objective—being a positive lens which forms a real inverted image of the distant object. This image is observed through a second lens—the eyepiece—which acts as a magnifier. The eyepiece may be either positive (**astronomical telescope**), or negative (**Galilean telescope**). In practice, the objective will be an achromatic combination (see Chapter 13) and the eyepiece of one of the forms described in section 10.13.

The path of the light from a point on a distant object through the two forms of telescope is shown in *Figure 10.10*. An unaccommodated emmetropic eye sees an image at infinity subtending an angle w' instead of the angle w subtended by the object. As will be seen from the figure, this image will be inverted in the astronomical telescope and erect in the Galilean. When the telescope is in normal adjustment for an unaccommodated emmetropic eye and a distant object, both incident and emergent light are parallel, and the objective and eyepiece are separated by a distance equal to the sum of their second focal lengths. Such a system is known as an **afocal** or **telescopic system**.

10.9 MAGNIFICATION OR MAGNIFYING POWER OF TELESCOPES

The magnification of the telescope is expressed as the ratio of the angle subtended by the image seen through the telescope to the angle subtended by the object at the eye, without the telescope. When the telescope is in 'infinity adjustment', as in *Figure 10.10*

$$\frac{\tan w'}{\tan w} = \frac{\dfrac{h'_1}{f'_2}}{-\dfrac{h'_1}{f'_1}} = -\frac{f'_1}{f'_2} = M \tag{10.04}$$

The magnification will be negative in the astronomical and positive in the Galilean telescope.

10.10 ENTRANCE AND EXIT PUPILS. RAMSDEN CIRCLE

It will be seen from *Figure 10.10* that the light filling the effective aperture D_1E_1 of the objective will, after passing through the instrument, pass through, or, in the case of the Galilean telescope, appear to come from a circular area de which is the image of the objective formed by the eyepiece. This is the **exit pupil**, **Ramsden circle** or **eye ring**. The objective is the **entrance pupil**.

The magnification may be expressed in terms of the diameters of the entrance and exit pupils. From *Figure 10.11*

$$M = \frac{f'_1}{f'_2} = -\frac{2y_1}{2y_2} \qquad\qquad (10.05)$$

The magnification of a telescope is conveniently determined experimentally by a measurement of the diameters of objective and Ramsden circle.

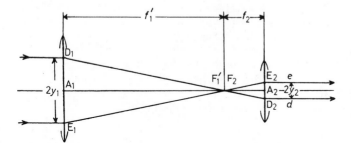

Figure 10.11 Astronomical telescope. Magnification as ratio of pupils

The exit pupil has an important bearing on the field of view of the telescope as will be readily seen from *Figure 10.10*; thus in the astronomical telescope, the exit pupil being real, it may be made to coincide in position with the pupil of the eye, and oblique pencils from the edges of a wide field can enter the eye. In the Galilean telescope, however, the exit pupil being virtual, the pencils from different points of the field diverge from the axis on leaving the eyepiece, and only those within a narrow angle can enter the eye pupil. The narrow angular field of view obtained is the chief drawback of this form of telescope.

Entrance and exit pupils and their significance will be more fully dealt with in Chapter 17.

10.11 GRATICULES

If a scale engraved on a thin parallel glass plate is placed in the common focal plane of objective and eyepiece, it will be seen in focus at the same time as a distant object, and such an arrangement may be used to measure the angle subtended by an object. Such scales, which are frequently used in telescopes and binoculars, especially those used for military purposes, are known as **graticules**. The divisions of the scale are made of such a length that they subtend the required angle, say $1/2$ degree, at the nodal point of the objective and, since object and image subtend the same angle at the nodal points, the angle subtended by any object can be read off directly on the scale.

10.12 REFLECTING TELESCOPES. CATADIOPTRIC SYSTEMS

Owing to the great difficulty and expense of producing objectives of very large aperture, the largest astronomical telescopes are of the reflecting form in which the objective is replaced by a surface silvered concave mirror. The mirror, which is 'figured' to a paraboloidal form to correct spherical aberration, forms a real image. This is observed, usually after further reflection, by the eyepiece in the same way as in the refracting telescope. The image is, of course, free from chromatic aberration, and it was due to the apparent impossibility of producing an achromatic lens that Newton was led to construct the reflecting telescope. Various forms of the reflecting telescope are shown in *Figure 10.12a*.

A combination of reflecting and refracting optics may be used for camera objectives. Such systems are called **catadioptric** and two examples are given in *Figure 10.12b*. Both these designs use spherical mirrors in a Cassegrain relationship. The spherical aberration is corrected in the case of the Schmidt system by a specially figured aspheric plate (much exaggerated in the figure) and in the Maksutov system by a deep meniscus negative lens. All axial reflecting systems block out some of the light. The ratio of the diameter of the obscuring region to that of the full aperture is called the **obscuration ratio**. Variations and combinations of these, and Mangin Mirrors (section 7.9), are used in night-vision objectives, telescope objectives and even some types of magnifier.

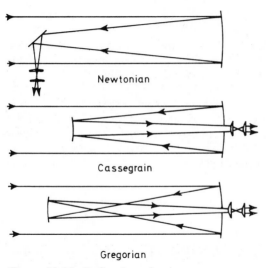

Newtonian

Cassegrain

Gregorian

Figure 10.12a Reflecting telescopes

10.13 EYEPIECES

We have considered the eyepiece of the microscope and telescope as a single lens, but for several reasons it will generally be necessary to replace this single lens by a system of separated lenses. It will be seen from *Figure 10.10* that in order that light from a large field of view may pass through a single lens eyepiece this must be of large diameter. But the lens cannot have a large diameter with the short focal length required for magnification. The difficulty may be overcome by using two separated positive lenses as the eyepiece. The use of two separated lenses will also permit of more control of the aberrations, particularly those affecting oblique pencils, and an example of this is discussed in section 13.12.

If a positive lens A_F (*Figure 10.13*) is placed in the plane of the image $B'_1 Q'_1$ formed by the objective of the microscope or telescope, the image will be un-

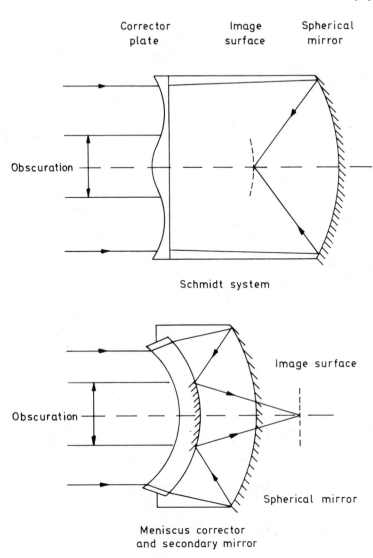

Schmidt system

Maksutov system

Figure 10.12b Catadioptric systems

changed in position and size. The pencils of light from each point on the image will however be deviated towards the axis and will pass through a second lens A_E of smaller diameter than would otherwise be required. The paths of the rays without the lens A_F in place are shown by broken lines in the diagram. The first lens A_F is known as the **field lens** and the second A_E, nearer the eye, as the **eye lens**. The field lens is frequently used in optical systems where large fields are required. The introduction of the field lens will also bring the exit pupil closer to the eyepiece, and this may be an advantage, particularly with low power eyepieces, as the eye may be brought close to the eyepiece and extraneous light cut off by means of an eyecup. In the arrangement shown in *Figure 10.13* any imperfections of the field lens or dust on its surfaces would be seen magnified in the plane of the image. To avoid this, in most practical eyepieces, the field lens is placed a short distance either in front of or behind the image.

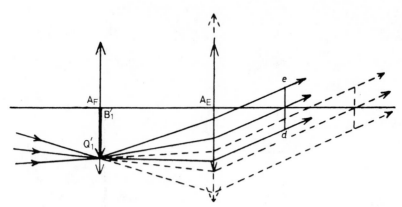

Figure 10.13 Effect of eyepiece field lens

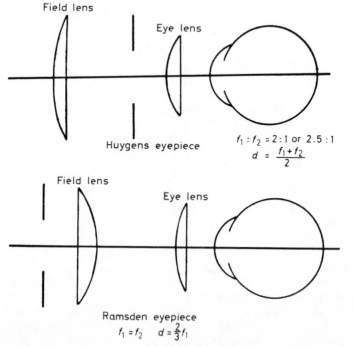

$$f_1 : f_2 = 2 : 1 \text{ or } 2.5 : 1$$
$$d = \frac{f_1 + f_2}{2}$$

Huygens eyepiece

Ramsden eyepiece
$$f_1 = f_2 \quad d = \tfrac{2}{3} f_1$$

Figure 10.14 Huygens and Ramsden eyepieces

The two chief forms of eyepiece are the **Huygens** and the **Ramsden** (*Figure 10.14*). The eyepiece most commonly used in the microscope and the astronomical telescope is the Huygens. This consists of two plano-convex lenses placed with their curved surfaces towards the incident light and separated by a distance equal to half the sum of their focal lengths. The ratio of the focal lengths of field lens to eye lens varies somewhat, but is usually 2:1 in the microscope eyepiece and 2.5:1 in that of the telescope. When a microscope or telescope is to be used for making measurements by means of a scale or graticule placed in the first focal plane of the eyepiece, the Huygens eyepiece cannot be used. It will be seen from *Figure 9.5* that the first principal focus of the eyepiece lies between the two lenses. Thus the image of a scale placed in this position is formed by the eye lens alone whereas the image to be measured is formed by the complete eyepiece, and therefore these two images are not equally affected by any distortion given by the lenses.

To overcome this difficulty the eyepiece must have its first focal plane in front of the field lens, so that the scale and the image formed by the objective are viewed through the complete eyepiece. The most usual eyepiece of this type is the Ramsden, which consists of two plano-convex lenses of equal focal length placed with their curved surfaces facing one another. To obtain the best colour correction the lenses, as in the Huygens eyepiece, should be separated by half the sum of their focal lengths, but as this brings the field lens into the focal plane, any dirt or imperfection on this lens will be sharply focused. The lenses are therefore separated by two-thirds the focal length of either.

10.14 ERECTING EYEPIECE. TERRESTRIAL TELESCOPE

A microscope or a telescope having either a Huygens or a Ramsden eyepiece gives an inverted image of an object. This is not of serious importance in the microscope or in the telescope used for astronomical or measurement purposes, but when the telescope is to be used for observing ordinary terrestrial objects, an erect image is essential. In order to take advantage of the large field of view given by the astronomical telescope some means must be incorporated in the telescope to erect the image. The method of erecting the image by means of reflection, as in the prism binocular, has been described in Chapter 3. A further method is by the use of an **erecting eyepiece**.

If a short focus positive lens A_2 (*Figure 10.15*) is placed in such a position that the image formed by the telescope objective is outside its first principal

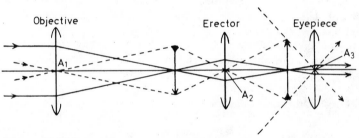

Figure 10.15 Principle of erecting eyepiece

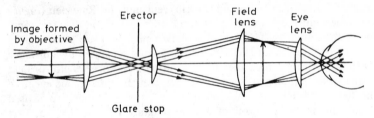

Figure 10.16 Four-lens erecting eyepiece

focus, a real inverted image is formed of the first image, that is, a real erect image of the original object. This image is observed with the ordinary eyepiece A_3. In practice, the erecting eyepiece usually consists of four plano-convex lenses, as in *Figure 10.16*, the first two forming the erecting system and the second two a modified Huygens eyepiece. The chief drawback to the terrestrial telescope with erecting eyepiece is its length, the length of the eyepiece alone being usually about ten times its focal length. By moving the erector along the axis, the size of the second image may be varied. This is the principle of the variable power telescope.

10.15 THE PHOTOGRAPHIC OBJECTIVE

This is essentially a positive lens forming a real image on a flat film. The requirements of such a lens are that it should form, on a plane surface, a well-defined image free from distortion and subtending a large angle, and that the aperture of the lens should be as large as possible. These difficult conditions can only be met by the use of carefully designed systems of a number of lenses. The modern photographic objective—the **anastigmat**—will give a good image subtending an angle of $50°$ or more with an aperture diameter which can be as large as the focal length as described in Chapter 18.

Photographic objectives are specified in terms of equivalent focal length. When a large image of a distant object and therefore a long focal length is required, the long camera necessary with the ordinary type of objective can be avoided by the use of the **telephoto system**. This consists of a positive lens followed by a negative lens of shorter focal length, the lenses being separated by a sufficient distance to give a positive combination. Such a system will have its principal points outside the system in front of the positive lens, and its equivalent focal length is therefore considerably longer than its back vertex focal length. By suitably choosing the lenses and their separation an equivalent focal length several times the vertex distance can be obtained.

For the purpose of controlling the illumination of the image and the depth of focus (see Chapter 17) the photographic objective is fitted with an adjustable aperture, usually placed between the lenses of the combination. The value of the aperture is expressed as the ratio of the focal length to the diameter of the effective aperture,* and is known as the **stop number** or **f number**. It is given as

*The effective aperture is the image of the actual aperture as seen from the front of the system, that is, the entrance pupil. (See Chapter 17.)

$f/8$, $f/16$, etc., where the effective aperture has a diameter $^1/_8$, $^1/_{16}$, etc., of the focal length.

It can be shown that the illumination of the image varies as the area of the aperture, that is, as the square of its diameter and inversely as the square of the focal length. The exposure required under any given conditions therefore varies directly as the square of the f/No. In order to simplify the calculation of exposures the marked apertures are arranged in a series in such a way that each one necessitates an exposure double that required with the preceding larger one, the series being as follows:

	$f/1.4$	$f/2$	$f/2.8$	$f/4$	$f/5.6$	$f/8$	etc.
Exposure required	1	2	4	8	16	32	

10.16 PROJECTION SYSTEMS

Projection systems are also used to form a real image on a screen, but differ from the camera in that the object is usually more or less transparent, the image being formed by the light from a source passing through the object; also the image is usually many times larger than the object. In order that light from each point of the object may pass through the objective or projection lens the source must subtend at least as large an angle at the objective as that subtended by the object. This would entail the use of a source of large area and therefore of low luminance, that is, the amount of light from each part of the source would be small. By using a **condenser**, a large aperture short focus lens or system, between the source and object, an image of the source may be formed close to the objective and this will subtend a sufficiently large angle even when the source is small.

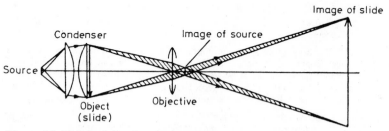

Figure 10.17 Principle of optical projection

The path of the light through the projector is illustrated in *Figure 10.17*. The source should be of high luminance, but small enough for the whole of the beam from the condenser to pass through an objective of reasonable aperture. The focal length of the objective is governed by the size of the image required, and the distance of the screen—the 'throw,' and the focal length of the objective governs that of the condenser, as can be seen from the figure. Further reference to the action of a condenser will be found in section 17.8.

CHAPTER 10   EXERCISES

NOTE: In the following exercises the lenses—objectives, eyepieces, etc.—are to be considered as thin.

1. Explain briefly the following terms, with diagrams
   Myopic eye.
   Hypermetropic eye.
   Crystalline lens.
   Far point.
Why is a negative lens used to correct a myopic eye and a positive lens a hypermetropic eye for distance? Give diagrams.

2. State what is meant by (a) the lateral magnification of an image, (b) the apparent magnification of an image as seen by an eye.

3. An object $^1/_4$ inch long is placed 6 inches from a positive lens of 8 inches focal length; what will be the visual angle subtended by the image as seen by an eye 4 inches from the lens?

4. What must be the position of an object in order that it may be seen distinctly through a + 10D lens placed 5 cm in front of an eye, the eye being accommodated for a distance of 40 cm? If the lens has a diameter of 1 cm, what length of object can be seen through the lens?

5. Show fully how a positive lens produces an apparent magnification of an object seen through it. What will be the magnification produced by a + 15D lens 50 mm from an emmetropic eye when the eye is (a) unaccommodated, (b) accommodated 4D?

6. An object $^1/_4$ inch long is placed 3 inches from a positive lens of 4 inches focal length; what will be the apparent magnification of the image as seen by an eye 2 inches from the lens?

7. Show with a carefully drawn diagram how a magnified image is formed with a compound microscope. Explain exactly what is meant by the magnification of a microscope and derive an expression for calculating the magnification.

8. State exactly what is meant by the magnification of a microscope. What will be the magnification of the microscope having an objective of $^2/_3$ inch focal length, an eyepiece of $1^1/_4$ inch focal length and a tube length of 6 inches?

9. A microscope has an objective of $1^1/_2$ inches focal length and an eyepiece of 2 inches focal length, the distance between the lenses being $6^1/_2$ inches; what will be the magnification of the microscope?
A scale is placed in the first focal plane of the eyepiece and the length of the image of a certain object measured on this scale is found to be 0.16 inch, find the actual length of the object.

10. The objective and eyepiece of a compound microscope, each considered here as a thin lens, have focal lengths of 2 inches and 1 inch respectively and are separated by 9 inches. Calculate where the object must be to give distinct vision to an emmetrope, accommodated for a point 10 inches from the eyepiece. Draw a careful diagram, full size, showing the passage of a pencil of rays through the instrument from one point of the object.

11. A compound microscope has an objective of 15 mm focal length and an eyepiece of 30 mm focal length, the lenses being 180 mm apart. What will be the position of the object and the magnification of the microscope when focused for an emmetropic person with accommodation relaxed?
Find the equivalent focal length of the microscope as a complete system and hence the magnification considering the system as a magnifier.

12. Explain why a compound microscope should consist of a short focus objective and a short focus eyepiece.

13. The microscope of question (11) is used to form a real image on a photographic plate 0.5 m from the eyepiece (photomicrography). How far and in which direction must the microscope be moved with respect to the object in question (11) and what will be the lateral magnification of the image?

14. Give a diagram showing the path of light through a compound microscope and deduce the formula giving the magnification of the instrument.

15. Two thin convex lenses when placed 10 inches apart form a compound microscope whose apparent magnification is 20. If the focal length of the lens representing the eyepiece be $1^1/_2$ inches, what is the focal length of the other?

16. Give a careful diagram showing the path of light through an astronomical telescope. A telescope 12 inches long is to have a magnification of eight times. Find the focal length of objective and eyepiece when the telescope is (a) astronomical, (b) Galilean.

17. Explain, with diagrams of the path of the light, two types of telescope. State the advantages and disadvantages of each type.

18. Explain, with a diagram of the path of the light, the optical principle of the astronomical telescope. What is the disadvantage of this type of telescope? Describe a method by which this disadvantage may be overcome.

19. Show by means of a diagram how the magnification is produced in the case of a Galilean telescope. A Galilean field glass magnifying five times has an objective of 180 mm focal length; find the focal length of the eyepiece and its distance from the objective when the field glass is focused for a myope of 4D.

20. A + 5D cyl. and a + 20D cyl. both with axis vertical, are separated by a distance of 25 cm. How will distant objects appear when viewed through the combination, the eye being close to the stronger lens?

21. An astronomical telescope is to be designed to give a magnification of X 8, a $^3/_4$ inch Ramsden eyepiece being used (separation = $^2/_3$ focal lengths); the aperture of the objective is to be one-sixth its focal length. Treating the objective and eyepiece lenses as thin, show their positions on a scale diagram and trace accurately through the instrument a pencil from one extremity of a distant object which subtends an angle of 5°. Measure the magnification given by the diagram.

Discuss the diagram as drawn: diameter of tube, sizes of eye and field lenses, angular field, etc.

22. A horizontal telescope contains a pair of horizontal cross-wires one-tenth of an inch apart. The telescope is focused on a vertical staff 10 feet away from the object glass, which has a focal length of 10 inches. Find the length on the staff apparently intercepted between the wires.

23. Describe the construction and action of a simple telescope, and show how by adding (a) lenses, (b) prisms, an erect image may be obtained instead of an inverted one.

24. The objective of a reading telescope has a focal length of 10 inches and the eyepiece 2 inches. The telescope is focused on an object 40 inches away, and the image is formed at the distance of most distinct vision (12 inches). Draw a diagram showing the path of rays through the telescope. Determine the magnification and the length of the telescope.

25. A Galilean telescope has an objective of 120 mm focal length and magnifies five times when used by an emmetrope to view a distant object. What adjustment must be made when the telescope is used by (a) a myope of 10D, (b) a hypermetrope of 5D? Find the magnification in each case.

26. A biconvex lens has surface powers of 5D and 20D respectively, is made of glass of refractive index 1.5 and the separation of the surfaces is 375 mm. Find the positions of the principal points of this lens.

What kind of system is this and what properties does it possess?

A mark is made near the axis on the 5D surface. Find the position and magnification of the image of this produced by the 20D surface.

27. A Huygens eyepiece consists of two lenses having focal length of $2^1/_2$ inches and 1 inch respectively, separated $1^3/_4$ inches; find the equivalent focal length and the positions of the principal foci. What will be the magnification of a telescope having an objective of 20 inches focal length with the above eyepiece.

28. Explain the function of the eyepiece in a telescope or microscope. Why is an eyepiece usually formed on two separated lenses? Describe the Huygens and Ramsden eyepieces giving, with reasons, examples of the use of each form.

29. Explain the principle of the telephoto lens. What is its particular advantage?

30. A telephoto lens is made up of two single thin lenses of focal lengths + 20 cm and − 8 cm respectively, separated by a distance of 15 cm. Find (a) by calculation, and (b) graphically, the focal length of the combination and the position of its principal points. What should be the camera extension and what will be the size of the image of a distant object that subtends an angle of 2°?

31. What is the significance of the f/No. of a photographic lens? Describe the system of numbering which is commonly adopted and explain the reasons for its choice.

32. A camera is fitted with a lens of 6 inches focal length. Three of the marked apertures have effective diameters of 1.09, 0.75 and 0.375 inches respectively. Express these as f/Nos. and find the exposures necessary with the first and last, the correct exposure which the 0.75 inch aperture being 2 seconds.

33. The finder on a camera consists of a 6D concave lens having a rectangular aperture of 25 × 20 mm and a peep-hole 75 mm behind it; what extent of object at 10 m distance can be seen in this finder?

34. Explain the principle of the projection lantern. What are the conditions to be satisfied in obtaining the brightest possible image on the screen?

Chapter 11
Photometry

11.1 INTRODUCTION

In the study of geometrical optics which has been developed in the previous chapters we were very concerned with the *direction* of light rays. In the study of **photometry** the *amount* of light is our first concern although obviously the direction is important when the amount in different places is being calculated. In section 1.1 the energy given out by the heated poker included heat as well as light. The heat energy does not stimulate the retina of the eye and is therefore not included in photometry which is solely concerned with the visual sensation produced by the radiated energy.

The visual sensation produced will vary greatly in two respects with the wave-length of the radiation. It has already been stated in section 1.7 that different wave-lengths give rise to the sensation of different colours, and this is considered more fully in Chapter 13. Also the intensity of the sensation produced by equal amounts of emitted energy will vary throughout the visible spectrum. From the spectrum of white light (Plate 6a) it will be evident that the brightest portion is situated in the yellowish-green and that the brightness falls off rapidly towards both the red and violet ends. Assuming a source that is emitting equal energy at all wave-lengths and plotting the luminosity, that is the amount of visual sensation, against the wave-length, we obtain the relative luminosity or visibility curve shown in *Figure 11.1*. The form of this curve varies somewhat with different normal observers and also with different values of luminance, particularly when the luminance is low. The curve shown, which is the result of a large number of observations, has been adopted as the standard luminosity curve for ordinary values of luminance, that is over about ten candela per square metre. It will be seen that the greatest visual effect occurs for a wave-length of 555 nm (see section 12.7) in the yellowish-green portion of the spectrum. This curve is often called the V_λ **curve**. It is also called the **Photopic curve** and is the relative sensitivity of the cones in the retina. Another curve, similar but shifted towards the blue, relates to vision at low luminance.

The eye is incapable of making an absolute measurement of the amount of light entering it; we can look at two sources and estimate that one appears 'brighter' than the other if there is sufficient difference between them, but cannot form a reliable judgment as to by how much they differ. The eye can, however, decide with a fair degree of accuracy whether two adjacent surfaces appear *equally* bright; and this is the basis of all practical visual photometric measurements.

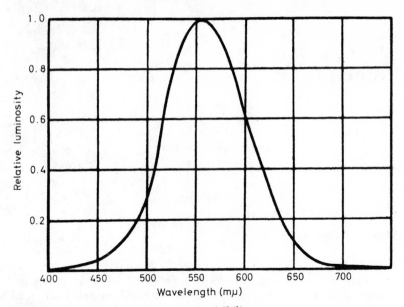

Figure 11.1a Relative luminosity or visibility curve

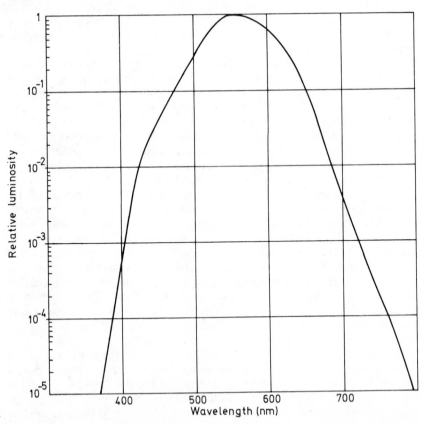

Figure 11.1b The overall visual response plotted with a logarithmic ordinate scale

According to the law of Weber (1834) the smallest perceptible difference of apparent brightness or luminosity is a constant fraction of the luminosity. This fraction, known as *Fechner's Fraction* (1858), is, over a large range of luminosities, about one per cent and the eye can therefore distinguish between two adjacent surfaces that differ in luminance by this amount. This is a reasonable approximation to the curves of *Figure 20.5*.

In photometry we have to differentiate between the following quantities:

(1) The amount of light emitted by a point source, the *luminous flux* and *luminous intensity*.
(2) The amount of light received on a unit area of a surface in a given position; that is, the *illumination* or *illuminance* of the surface.
(3) The amount of light *emitted* or *re-emitted* per unit area of a surface; that is the *luminance* of the surface. This was formerly called the brightness of the surface.

In discussing these measurements it should be remembered that in practical photometry it is rarely necessary to strive for accuracies better than ± 1%, Fechner's Fraction. The exposure meters on cameras for instance may only be accurate to ± 10%. However, the range of the measurements can be over many orders of magnitude and it is often crucially important to eliminate *stray* light from the experimental area. A better description of the response of the eye is given in *Figure 11.1b* where a log scale is used in order to cover the large range of response.

11.2 LUMINOUS FLUX AND LUMINOUS INTENSITY

There is a great difference in the amount of light emitted by different sources. Most practical sources are *incandescent* sources which emit energy over a wide range of wave-lengths, mainly as heat in the infra-red part of the spectrum (see Chapter 13). This is wasted energy as far as the emission of light is concerned and the *luminous efficiency* (see section 11.10) is very important to lamp manufacturers. In considering the amount of *light* given out by a source, the energy emitted must be evaluated or *weighted* according to its ability to stimulate visual sensation as given by the relative luminosity curve (*Figure 11.1*). The rate of flow of *light* from a source is the **luminous flux**, Φ, of the source, and the unit of this is the **lumen** (lm), which is defined later.

In no practical source is the distribution of flux uniform in all directions; hence it is necessary to define the amount of light radiated with respect to a given direction. In no practical source is the amount of luminous flux uniform over all parts of the source; hence it is necessary to define the amount of light radiated with respect to a point on the source or to assume that the whole source is contained at a single point. This last assumption works very well in the majority of cases (see sections 11.3 and 11.4) and the two requirements above can be met by considering the flux emitted by a point source into a cone of very minute solid angle constructed round a given direction. The concentration or density of this flux per unit solid angle is termed the **luminous intensity**, **I**, of the source in the given direction. The unit of luminous intensity is the **candela** (cd) formerly the candle, and luminous intensity is often referred to as **candle-power**. If Φ is the flux in lumens emitted within a cone of solid angle w steradians, theoretically in this case infinitely small, then Φ/w = I lumens per steradian = I candelas, in the given direction.

NOTE: The unit solid angle, or *steradian*, is the solid angle of a cone that, having its apex at the centre of a sphere, cuts off an area of the sphere's surface equal to the square of its radius (*Figure 11.2*). The value of any solid angle in steradians is equal to the area of the sphere's surface included in the angle divided by the square of the radius of the sphere. The surface area of a sphere of radius r being $4\pi r^2$ the solid angle for a complete sphere is 4π steradians.

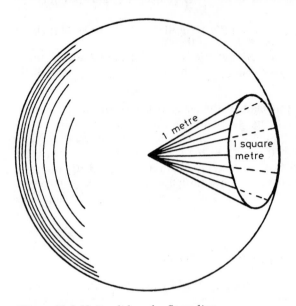

Figure 11.2 Unit solid angle. Steradian

It can be shown that in a cone having a half plane angle θ the solid angle $w = 2\pi (1 - \cos\theta) = \pi \sin^2\theta$ steradians when θ is small.

While no source will be a point as required by the definition of luminous intensity, in many cases the size of the source is negligible at the distances at which it is used. (See section 11.3)

11.3 ILLUMINATION, ILLUMINANCE

A surface receiving light is said to be illuminated, the **illuminance**, E at any *point* of it being defined as the density of the luminous flux at that point, or the flux divided by the area of the surface, when the latter is uniformly illuminated. The metric unit of illuminance is the **lux** (lx) or **lumen per square metre** ($lm.m^{-2}$), the illumination of a surface, normal to the direction of the light, one metre from a source of one candle-power. The imperial unit is the **lumen per square foot** (lm/ft^2), the illumination of a surface normal to the direction of the light one foot from a source of one candle-power. (These units were formerly called the metre-candle and foot-candle respectively).

If Φ is the flux from a point source at the centre of a sphere of radius d, the illumination E on the surface of the sphere, from the definition, will be

$$E = \frac{\Phi}{\text{area of sphere}} = \frac{\Phi}{4\pi d^2}$$

$\dfrac{\Phi}{4\pi}$ is the candle-power I, and therefore,

$$E = \frac{I}{d^2} \qquad\qquad (11.01)$$

This is the law of inverse squares (section 1.3). It should be clearly understood that expression (11.01) applies only to the illumination received on a surface *normal* to the direction of the incident light from a *point* source, where I is the candle-power of the source in the particular direction.

It can be shown by applying the inverse square law to each part of a source larger than a point that the error introduced by assuming it is a point source will be less than 1% if the distance, d, is greater than 10 times the largest dimension of the source.

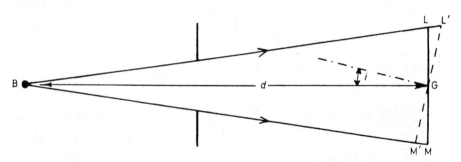

Figure 11.3 Cosine law of illumination

Figure 11.3 represents the section of a pencil of light from a point source B of candle-power I falling on a surface LM perpendicular to the pencil. If Φ is the flux in the pencil of solid angle w, then the illuminance of the surface at distance d from the source is given by

$$E = \frac{\Phi}{\text{area of LM}} = \frac{\Phi}{wd^2} = \frac{I}{d^2}$$

If the surface is rotated about G into a new position L'M' the same flux is distributed over the larger area, and the illumination of the surface is now

$$E' = \frac{\Phi}{\text{area of L'M'}}$$

As the illumination refers to any one point on a surface, we may consider the area L'M' as very small compared with its distance from the source, and then,

$$\text{area of L'M'} = \frac{\text{area of LM}}{\cos \text{LGL}'} = \frac{\text{area of LM}}{\cos i}$$

where i is the angle of incidence.

Therefore

$$E' = \frac{\Phi \cos i}{\text{area of LM}} = \frac{I}{d^2} \cos i \qquad (11.02)$$

Thus, *the illuminance of a surface varies as the cosine of the angle of incidence*; this is known as the cosine law of illumination. The law of inverse squares and the cosine law are the two fundamental laws of photometry.

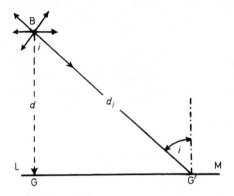

Figure 11.4 Illumination of a surface

In practical applications of photometry we frequently require to find the illumination at different points on a plane surface, as for example, the surface of a street, an illuminated test chart, etc. *Figure 11.4* represents a source B of a candle-power I above a plane surface LM; the illuminance at any point G' on the surface will be

$$E = \frac{I_i \cos i}{d^2_i}$$

where i is the angle of incidence and I_i the candle-power of the source in that direction. From which we have

$$E = \frac{I_i \cos^3 i}{d^2} \qquad (11.02a)$$

$$= \frac{I_i d}{d^3_i} \qquad (11.02b)$$

where d is the perpendicular distance of the source from the surface.

Example:

A room 15 ft by 12 ft is illuminated by a 100 candle-power lamp suspended from the centre of the ceiling and 8 ft above the floor. Find the illumination on the floor (*a*) directly under the lamp and (*b*) in a corner of the room, assuming the candle-power to be 100 in both these directions.

(*a*)   I = 100 c.p. d = 8 ft

$$E = \frac{I}{d^2} = \frac{100}{64} = 1.56 \text{ lumens per ft}^2$$

(*b*) Distance GG′ (*Figure 11.4*) from point on floor directly under lamp to corner of the room.

$$GG' = x = \sqrt{7.5^2 + 6^2} = \sqrt{92.25} \text{ ft}$$

Distance d_i from lamp to corner

$$d_i = \sqrt{d^2 + x^2} = \sqrt{64 + 92.25} = 12.5 \text{ ft}$$

Hence from (11.02b)

$$E = \frac{100 \times 8}{12.5^3} = 0.41 \text{ lumens per ft}^2$$

Example:
Light from a point source of 50 candle-power, which radiates uniformly in all directions, falls on a + 1D spherical lens of 50 mm aperture placed 500 mm from the source. What fraction of the total flux from the source is incident on the lens? What will be the illumination in lumens per square metre on a screen held one metre beyond the lens, on the side opposite from the source, and perpendicular to the optical axis of the lens?

At 500 mm from the source the total flux is spread over a sphere of 500 mm radius, the area of which will be $4\pi \times 500^2$. The area of the lens aperture is $\pi \times 25^2$. Therefore the fraction of the total flux received by the lens

$$= \frac{\pi \times 25^2}{4\pi \times 500^2} = \frac{1}{1600}$$

As the source is 500 mm from a + 1D lens a virtual image of the source is formed one metre in front of the lens. Therefore the diameter of the patch of light on the screen will be twice the diameter of the lens and its area four times the area of the lens.

Total flux emitted from the source = $4\pi I$ = 200π lumens. Flux received by the lens is 1/1600 of total flux

$$= \frac{\pi}{8} \text{ lumens}$$

Area of patch of light on screen = $\pi \times 50^2$ mm^2. Therefore, neglecting any loss of light at the lens,

$$\text{illumination of screen} = \frac{\pi}{8 \times \pi \times 50^2} = \frac{1}{20000} \text{ lm/mm}^2$$

$$= 50 \text{ lm/m}^2 = 50 \text{ lux}$$

Alternative solution:

$$\text{illumination on lens} = \frac{I}{d^2} = \frac{50}{500^2} \text{ lm/mm}^2$$

Area of illuminated patch is four times that of lens and therefore illumination is one-quarter; hence

$$\text{illumination of screen} = \frac{50}{500^2 \times 4} = \frac{1}{20000} \text{ lm/mm}^2 = 50 \text{ lux}$$

11.4 LUMINANCE

Illumination refers only to the amount of light *received* by a surface regardless of the nature of that surface, but in many cases we are more concerned with the amount of light *emitted* in a given direction by a surface. The light emitted per unit area of a surface is the **luminance, B,** of the surface. It will depend, in the case of a non-self-luminous body, both upon the illumination and the fraction of the incident light diffusely reflected by the surface. This latter is known as the **reflection coefficient** or albedo of the surface. Luminance may also refer to a diffusely transmitting substance, such as the diffusing globe of a lamp, or to an actual source, such as a flame or a fluorescent lamp.

In its perception of luminance the eye interprets the flux to be coming from an area projected at right angles to the direction of vision,* and the luminance of a surface is defined as follows: *The luminance in a given direction of a surface emitting light is the luminous intensity measured in that direction divided by the area of this surface projected on a perpendicular to the direction considered.* Luminance is expressed in candelas per unit area of surface, the metric unit being the **nit** (nt), one candela per square metre and the imperial unit, more sensibly, being merely candelas per square foot. The reflection coefficients for a number of surfaces are given in Appendix 1.

Another method of expressing luminance assumes that the surface is a perfect diffuser, i.e. one that appears equally bright from whatever direction it is viewed. This will mean that the luminous intensity per unit area of apparent or projected area is constant for all angles. Since for a plane surface the projected area is proportional to the cosine of the angle between the direction of viewing and the normal to the surface, the luminous intensity of the surface must vary in the same proportion. This is stated in the *cosine law of emission* of Lambert (1727-1777) as follows: *For a perfectly diffusing surface the candle-power per unit area of the surface in any direction varies as the cosine of the angle between that direction and the normal to the surface, so that the surface appears equally bright whatever be the direction from which it is viewed.* No surface completely satisfies this requirement, but a few, such as a coating of magnesium oxide, sand-blasted opal glass and scraped plaster of paris, have diffusion closely approaching the ideal. Many surfaces closely resemble the perfect diffuser when the angle of incidence is small and the direction of viewing is nearly normal. Such surfaces are said to be 'matt'.

Treating the surface as a perfect diffuser, luminance is expressed in terms of the total luminous flux in lumens emitted by a unit area (actual not projected) of surface. A surface emitting or reflecting one lumen per square centimetre has a luminance of one **lambert**, but a luminance of one-thousandth of this value,

---

*The sun or a lamp surrounded by a spherical diffusing globe appears to the eye as a flat disc.

the **millilambert**, is a more generally useful unit. A unit, now outmoded, is the **foot-lambert**, the luminance of a surface emitting a flux of one lumen per square foot. One foot-lambert equals 1.076 millilamberts. The corresponding metric unit is the **apostilb** (asb), the square metre being the unit of area.

It can be shown that for a perfect diffusing surface the luminance in lumens per unit area is π times the brightness in candelas per unit area, i.e. a surface having a luminance of one candela per square foot has a luminance of π foot-lamberts.

As the illumination, E, of a surface is the flux in lumens received by a unit area of the surface and of this flux a fraction, r, the reflection factor, is reflected, the luminance of an illuminated perfectly diffusing surface will be given by rE lumens per unit area or rE/π candelas per unit area.

Some typical approximate values of luminance are given below.

Clear blue sky	10^4 candelas per m^2 or 3 $\times$ 10^3 mL
Overcast sky	10^3 candelas per m^2 or 3 $\times$ 10^2 mL
Heavily overcast sky	10^2 candelas per m^2 or 30 mL
Candle	4 $\times$ 10^3 candelas per m^2
Tungsten lamp filament	2 $\times$ 10^6 to 3 $\times$ 10^7 candelas per m^2
Arc lamp crater	3 $\times$ 10^8 candelas per m^2
Sun	1.6 $\times$ 10^9 candelas per m^2
Moon	2.6 $\times$ 10^3 candelas per m^2
Fluorescent lamp	10^4 candelas per m^2

A further measurement of luminance, recognising that photometry is based on visual parameters, makes allowance for the variations in diameter of the pupil of the eye. The concept of *retinal illuminance* from an extended source assumes that the retina is a smooth surface. The value is given by the equation:

$$E = \frac{Bt \; \alpha \cos \theta}{k}$$

where

B = luminance in direction of viewing (nits)

t = transmittance of eye

α = area of pupil (m^2)

θ = angle of incidence of principle ray through the eye

k = constant equal to the area of the retinal image (m^2) divided by the solid angle of the visual field (steradians)

As $\cos \theta$ is commonly unity and t and k are constants, the retinal illumination may be found by multiplying the number of nits by the area of the pupil in square millimetres. This gives a value in **Trolands**, a unit often used in visual experimentation.

11.5 STANDARD SOURCES OF LIGHT

Because the measurements of photometric quantities are largely comparisons
between sources it is necessary to define a standard source in the same way that
a metre or yard is a standard length. The original standard of luminous intensity
was the **International Candle**. This, despite careful specification, was far from
constant in its intensity and its colour differed considerably from that of modern
sources. The candle as originally specified is now only of historical interest,
although the term *candle-power* still survives. The candle was replaced by
various flame lamps, the chief of which was the Vernon Harcourt lamp burning
pentane vapour with a hollow cylindrical flame and having an intensity of 10
candle-power. Later the standard was maintained with a series of specially constructed electric carbon filament lamps kept at the standardising laboratories of
the various countries.

In 1948 an entirely new primary standard was adopted. This consists of a
very small cylinder containing molten platinum. The surface of the platinum at
the temperature of the 'freezing point' of platinum has a constant luminance.
One sixtieth of the luminous intensity from one square centimetre of this surface
is the candela, which now replaces the candle as the unit of luminous intensity.

As working standards, specially constructed electric filament lamps are now
generally used. In these the filament is mounted in a single plane being perpendicular to the photometer bench. The glass bulb is considerably larger than
would be used in the ordinary way in order to minimise the blackening caused
by the deposition of metal particles on the glass, and the filament is 'aged' by
being run for at least 100 hours before being standardised. In use the lamp is run
at a constant current and voltage slightly below that at which it was 'aged'.

11.6 COMPARISON PHOTOMETERS

Photometry is concerned with the visual effects of emitted radiations so the
primary instrument of measurement is the human eye. As this can only judge
differences in luminous intensity qualitatively it must be assisted if quantitative
results are to be obtained. The method adopted consists in arranging that the
two sources produce equal luminance on two similar adjacent surfaces, each
surface being illuminated by one only of the sources. A simple arrangement is
shown in *Figure 11.5*.

The diffusing surfaces at P lie at equal angles to the light paths but can both
be seen simultaneously. If E_1 and E_2 are the illuminations on the two surfaces

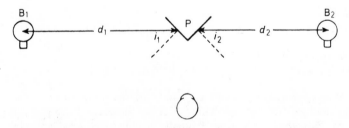

Figure 11.5 Photometer principle

produced by sources B_1 and B_2, whose candle powers are I_1 and I_2 respectively, then

$$E_1 = \frac{I_1 \cos i_1}{d_1{}^2} \quad \text{and} \quad E_2 = \frac{I_2 \cos i_2}{d_2{}^2}$$

where i_1, i_2 are the angles of incidence of the light on the surfaces. If now d_1 or d_2 is changed the eye can observe the surfaces and judge when they are equally bright, that is, when no difference exists and the system is *balanced*. Under these circumstances

$$\frac{I_1}{d_1{}^2} = \frac{I_2}{d_2{}^2} \quad \text{as} \quad i_1 = i_2 \qquad (11.03)$$

Thus the relative candle-powers of the two sources are given by the ratio of $d_1{}^2$ to $d_2{}^2$.

In order that the judgement of equality shall be as accurate as possible the two surfaces must be seen simultaneously and have a sharp dividing edge. Various methods have been devised to obtain these conditions. An early arrangement merely consisted of a rod in front of a screen. When the two sources are properly located the two shadows on the screen are just touching and are equally dense. The ratio of intensities is then given by the ratio of the square of the source-to-screen distances as in the expression (11.03).

A more complex system replaces the V-screen of *Figure 11.5* with a screen of opaque white paper normal to the line between the lamps having at its centre a spot made translucent by treating the paper with oil or wax. Unequal illumination will show a dark central spot on a bright surround or vice versa. Although the precise derivation of the brightness is a little complex the balance position when the spot disappears still occurs when the illuminations are equal and the expression (11.03) applies.

A more complex and sensitive approach is that of the **Lummer-Brodhun Photometer** shown in *Figure 11.6*. The light from each source passes through an

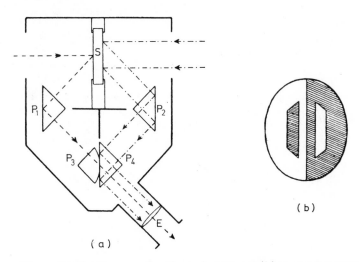

(a)

(b)

Figure 11.6(a) Lummer-Brodhun photometer (b) contrast pattern

aperture in the metal case and falls on either side of a white diffusing screen S; this is formed of plaster of Paris or of two sheets of ground white opal glass with a plate of metal between them. Two right-angled totally reflecting prisms P_1 and P_2 reflect the diffused light from the screen surfaces to the comparison prism P_3 and P_4. The prism P_3 has its hypotenuse face ground to a curve except for a small circular area in the centre, which is plane and polished, and this is pressed into optical contact with the plane hypotenuse of the prism P_4.

Thus light from the left hand side of the screen reaches the eye via the central part of P_3P_4, while the surrounding area receives light from the right hand side. The edge of the central area may be sharp but will disappear when a balance is obtained as before. A more complex working of the prism P_3 can produce a more complex pattern to be erased at balance. One such pattern is given in *Figure 11.6b* and systems incorporating this are known as **contrast photometers.**

These systems give adequate accuracies provided the two sources are of identical colour. When a colour difference is apparent an alternative method can be used in which the two illuminated surfaces are viewed sequentially rather than side-by-side. Any difference in illumination now shows up as a flicker and the basis of this method is that flicker due to colour difference disappears at a lower speed of alternation than flicker due to luminosity difference so that the judgement of equal luminance is not affected by the presence of a colour difference.

Such a system due to **Guild** is shown in *Figure 11.7*. D is a disc which can be rotated at various speeds. D, T and S are coated with diffusing material such as magnesium oxide. The eye is placed at E and sees the inside of the tube T uniformly illuminated to 2.5 ft-lamberts while the hole A subtends an angle of $2°$.

As the disc is rotated the eye sees the aperture field alternately illuminated by the light from S and D and when the disc is rotating slowly the field will appear to flicker even when the luminance at S and D is equal. On increasing the speed

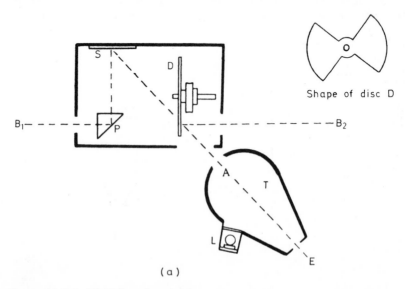

Shape of disc D

(a)

Figure 11.7 Guild flicker photometer

of rotation the flicker becomes less pronounced and at a high enough speed will disappear. The test source B_2 and the photometer are fixed and measurements are made by moving the comparison lamp B_1 while reducing the speed of rotation of the disc until flicker almost disappears. The luminance of S and D is then equal and the relative intensities of B_1 and B_2 can be found as with other photometers.

To obviate any difference there might be between the reflection factors of the two surfaces of the photometer and any effects of stray light, a substitution method is generally employed. In this a third lamp, the comparison lamp, which need not be of known candle-power, is used with the standard and the test lamp, i.e. the lamp to be tested. Balance is first obtained between the comparison lamp and the test lamp. The distance between the comparison lamp and the photometer is then kept fixed and the standard substituted for the test lamp. By moving the standard lamp balance is again obtained. Thus the same side of the photometer is illuminated in turn by both standard and test lamps. Then if I_1 and I_2 are the candle-powers of test lamp and standard respectively and d_1 and d_2 their respective distances from the photometer when balance is obtained, equation (11.03) may be used.

11.7 PHOTOELECTRIC PHOTOMETERS

The limitations of comparison photometers are never more evident than when a simple portable instrument is required for measuring the luminous intensity of actual lamps in offices, on roadways etc., because they have reflectors which may become dirty or misaligned. A photoelectric cell (or photo-cell) offers a solution provided:

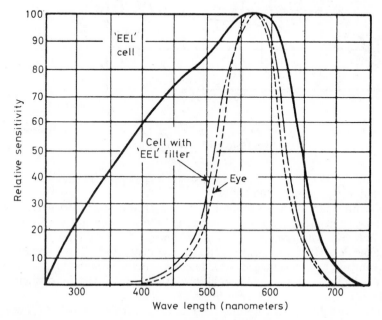

Figure 11.8 Spectral response of EEL cells (courtesy, Evans Electroselenium Ltd)

(1) It responds to light in the same way as the eye.
(2) It has an output proportional to the luminous intensity.
(3) It does not change from day to day.

Very few cells satisfy the first requirement and the nearest is the **selenium barrier-layer photo-cell**. The response of this device to light of different wavelengths is shown in *Figure 11.8* compared with the visual response. If a correction filter which absorbs ultra-violet light is placed in front of the cell the response can be made very close to that of the eye.

The cell produces a current through a resistance connected across it and the variation with light falling on it is shown in *Figure 11.9*. It can be seen that this

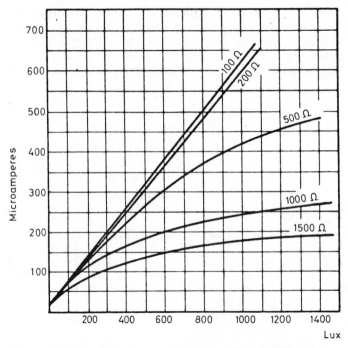

Figure 11.9 Typical output curves for 45 mm EEL cell connected to various external resistances (courtesy, Evans Electroselenium Ltd)

is linear for low values of the resistance and a micro-ammeter in the circuit could be calibrated directly in foot-candles or lux.

In manufacture pure molten selenium is poured on to a steel base plate. A very thin transparent metal layer is spluttered over the selenium to form a conducting surface. The light reaches the selenium through the metal layer and the current is generated between this layer and the base plate. Rather like standard sources of light, these cells are 'aged' after manufacture by exposing them to a very strong light under short-circuit conditions. This helps to ensure constant values to the curves of *Figure 11.9* throughout their life.

Figures 11.8 and *11.9* refer to products of Evans Electroselenium Ltd (now Diffusion Systems Ltd) and are reproduced with their permission.

11.8 PHOTOMETRY OF SCREENS

In the section on illumination the concept of the perfectly diffusing surface was discussed and defined as a surface which appears equally bright from all directions. If such a surface is used as a screen for the projection of a colour transparency or cine-film the projected image would appear equally bright in all directions. Most surfaces are not perfect and return the light in a partially specular reflection. As the audience tends to sit between the screen and the projector this can give a brighter image. Provided no objectionable high spots occur this can be a good thing.

The ratio by which a given screen appears brighter than a perfectly diffusing surface with the same illumination is often referred to as the **gain** of the screen in a particular direction. It is important to realise that there is no gain in the sense of extra light being generated. The additional brightness in a given direction is obtained at the expense of less light in another direction. The reflectivity of the screen can affect the apparent gain particularly in the case of a dirty surface.

Curves of the gain of typical screens are given in *Figure 11.10*. The curve for retroreflective screens can reach gain figures above 1000% and these have application in road signs and vehicle number plates where the driver sits within a few degrees of the projection axis of his headlamps.

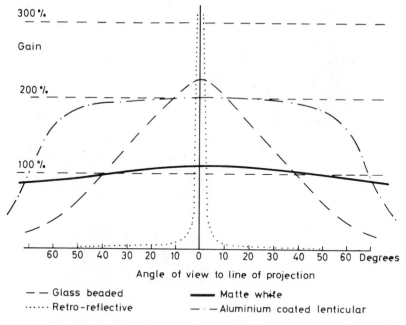

Figure 11.10 Gain of various screens

11.9 THE PHOTOMETRY OF SOURCES, INTEGRATING SPHERES

Luminous intensity or candle-power refers to the quantity of light emitted by the source in one given direction. The standard sources described have their

intensities specified in a particular direction. In ordinary commercial work, however, the given candle-power is usually an average of the intensities in all directions when it is known as the **mean spherical candle-power**.

The determination of this value for a lamp or lighting system may be done
by one of two methods. The intensity may vary in all directions although a
proportion of systems may be symmetrical about a vertical axis. In this case the
total luminous flux may be measured by supporting the lamp in its operating
position and measuring the intensity at different vertical angles as in *Figure 11.11*.

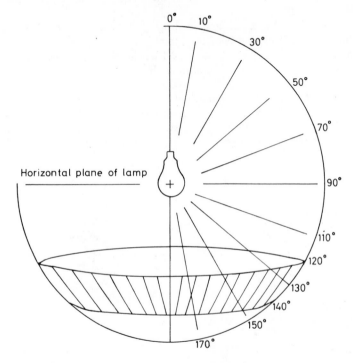

Figure 11.11 Measurement of mean candle-power

If the luminous intensity is measured at equal angles as shown, the results
cannot simply be averaged as the light emitted near the horizontal plane of the
lamp is also being emitted over a considerable solid angle all round the fitment.
Similarly the light measured directly above or below the lamp is being emitted
nowhere else.

This geometrical feature leads to the adoption of zone factors which are
proportional to the area swept out by the region around the angle chosen.
These zone factors are given by the difference in the cosines of the *bounding*
angles multiplied by 2π. For zones of $20°$ the factors are given in the table on
the following page.

The total of each luminous intensity reading multiplied by its zone factor
gives the total flux, Φ, from the lamp. When this is divided by 4π the mean
spherical candle-power is obtained in lumens per steradian or candela. If the
zone factors in the table are summed the result is 4π.

Angle of measurement	Zone	Zone factor
10°	0°−20°	0.378
30°	20°−40°	1.091
50°	40°−60°	1.671
70°	60°−80°	2.051
90°	80°−100°	2.182
110°	100°−120°	2.051
130°	120°−140°	1.671
150°	140°−160°	1.091
170°	160°−180°	0.378

As a more approximate method, a series of angles may be taken which in themselves give zones of equal solid angle. These are known as **Russell Angles** and those giving six equal zones would be −33.6°, 60°, 80.4°, 99.6°, 120°, and 140.4°. The average of luminous intensity values obtained at these angles gives the mean spherical candle-power directly.

When the source has large variations in its intensity in different *horizontal* directions then the same approach may be used but the lamp rotated completely about its vertical axis for each reading so that an average reading may be obtained. This is sometimes referred to as the **mean horizontal candle-power**. The values obtained in this manner may be used in the same way as for the symmetrical source described above.

An alternative method for total flux measurement involves the use of an **integrating sphere**. This device is a large sphere with a diffusely reflecting inner surface. The theory of its use depends on the application of the cosine law of illumination to the geometry of a sphere from which it can be shown that the surface illumination on the sphere wall *due to reflected light* is proportional to the total flux of a source inside. No direct light from the source must reach the point at which the illumination is measured.

This means that baffles must be inserted in the sphere and a hole cut in the wall to allow the illumination to be measured. These detract from the accuracy of measurement as do the presence of fittings inside the sphere. If the test lamp and standard source can be introduced successively into the same place within the sphere for measurements, these inaccuracies can be minimised.

11.10 INCANDESCENT LAMPS

Tungsten-filament lamps are so called because the source of light is a fine filament of tungsten metal through which is passed an electric current. The resistive heating of the filament raises its temperature to a maximum of about 3500°K. At this temperature the tungsten would rapidly oxidise. To avoid this the filament is enclosed in a glass envelope which is evacuated or filled with an inert gas. Even with this precaution the tungsten is slowly evaporated from the filament on to the walls of the envelope reducing its transmission and eventually ending the life of the lamp after a few hundred hours of use. Longer life can be obtained by reducing the power through the lamp (underrunning) but this reduces the colour temperature (see section 13.6) and also reduces the efficiency of the conversion of electrical power into light. At its design voltage a tungsten filament lamp will normally give between 15 to 25 lumens per watt. This is the total amount of light radiated by the lamp. The candle-power in specific directions depends very much on the shape and type of the filament. The conversion figures given above represent an efficiency of conversion of electrical energy into visible energy of between 8 and 13%.

These efficiency figures improve with higher filament temperatures but the worse evaporation of the filament material delayed the exploitation of this feature until the invention of the 'quartz-iodine' lamp (more correctly termed the tungsten–halogen lamp). In this lamp the inert gas contains the halogen, iodine or bromine which has the property of redepositing evaporated tungsten on to the filament. This action requires a high bulb temperature and as glass proved unsuitable quartz is used in a much smaller shape so that it becomes hotter, but the quartz does not play an active role.

11.11 LUMINESCENT SOURCES

Although a source of light gives out energy which it must obtain somehow, the method of heating used in incandescent sources is not the only way. Substances called **phosphors** have the property of absorbing energy from electrons, electric fields, chemical reactions and even other light, particularly ultra-violet. They then re-emit this energy in the form of light. Phosphors are frequently sulphides and oxides or silicates and phosphates of such metals as zinc, calcium, magnesium, cadmium, tungsten and zirconium. They are used in such commonplace items as fluorescent lamps and television picture tubes and more complex devices such as X-ray machines and electron microscopes. They are all luminescent sources but as they are often of large dimensions and flat the term luminescent screen is generally used.

Phosphors are used in fluorescent lamps to obtain a much broader range of colours than can be obtained with incandescent sources. They are also more efficient, with luminous efficiency values towards 90 lm/watt and less heat radiation. The light is generated by a two-stage process as the electric current is discharged through a mercury vapour which produces green and ultra-violet light. This ultra-violet light is then absorbed by the phosphor coating on the inside of the glass tube containing the gas and re-emitted as white light.

In television picture tubes the phosphor is selected not only for its colour and efficiency but also for its short response time. The light emission is stimulated by high-energy electrons striking the phosphor on the inside of the screen. As the picture changes it is important that the phosphor can start and stop emitting light very quickly as the electrons are switched on and off.

11.12 THE MEASUREMENT OF ILLUMINATION AND LUMINANCE

The photometers described in sections 11.6 and 11.7 are used to measure the candle-power of sources, although they do this by measuring the illumination due to a particular source at a given distance. If more than one source is present (other than the standard) then the illumination due to both will be measured. The comparison photometers are rather too cumbersome and the photoelectric system is commonly used for this purpose being placed on desk-tops and roadway surfaces to measure the illumination due to many sources. Indeed it is not capable of differentiating between the sources.

If the photoelectric cell is used as a camera exposure meter it measures the total light reaching the camera from the scene. No account can be taken of whether this contains a few bright sources or an even distribution and exposure errors can be caused by this. The cell essentially measures the illumination on the camera due to the scene and not the luminance of the surfaces comprising the scene.

An instrument which does the latter is the **S.E.I. Photometer** (*Figure 11.12*). The surface to be measured is viewed through a low-power telescope, T. By means of a modified Lummer-Brodun prism in the telescope a comparison spot, C, is seen superimposed on the centre of the image field. This spot subtends an angle of $^1/_2\,^{\circ}$ at the eye and is reflecting light from a diffusing screen, S, illuminated by a small electric lamp fed from a dry battery, B, through a rheostat, R. The lamp also illuminates a ring-shaped photoelectric cell, E, which is connected to a microammeter, M. By adjusting the luminous output of the lamp by means of the rheostat the needle of the microammeter can be made to coincide with a

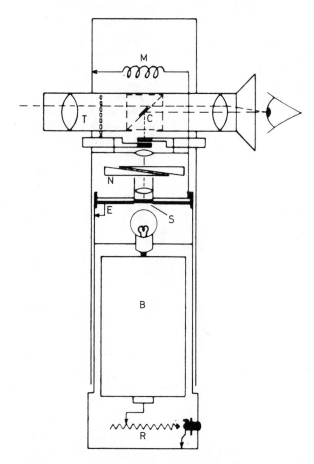

Figure 11.12 The S.E.I. photometer

standardising mark and in this way the luminance of the diffusing surface, S, can be kept at a constant value.

Between the diffusing surface and the prism are two opposed neutral wedges, N, which can be moved in opposition to one another by means of a rack and pinion mechanism operated by rotating the base of the instrument. The light reaching the comparison spot can in this way be varied over a range of 1 to 100. By interposing neutral filters in the path of the light from the surface under test the range of the instrument can be greatly increased.

The surface to be measured is viewed through the telescope with the comparison spot appearing superimposed upon it. The luminance of the spot is then adjusted to match that of the surface by rotating the base of the instrument and so moving the wedges. Colour filters below the comparison spot allow for matching surfaces illuminated by daylight or artificial light. In addition to the usual scales for obtaining photographic exposures the instrument is scaled for luminance in log. foot-lamberts.

To measure illumination a block of magnesium carbonate is placed in the required position and the luminance of the surface of the block is measured. As the surface is an almost perfect diffuser and has a reflectance factor of nearly 100%, the luminance in foot-lamberts corresponds very nearly to the illumination in lumens per square foot.

The S.E.I. photometer still depends on the eye as the final arbiter and variations between observers often occur. A more sophisticated instrument is the **Spectra-Pritchard Photometer** (*Figure 11.13*). This uses the eye solely to define the surface to be measured. The eye views the scene via a mirror having a hole in it which therefore appears as a black dot to the observer. The light from the scene is imaged on to the mirror so that the black dot is sharp and the light passing through the hole is coming only from the part of the scene covered by the black dot. This transmitted light reaches a specially calibrated photodetector via a series of filters. The mirror has holes of different sizes which can be rotated

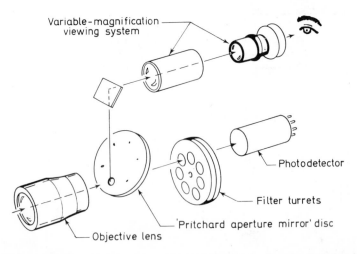

Figure 11.13 Aperture mirror disc system of the Spectra-Pritchard Photometer, model 1980 manufactured by the Photo Research Division of the Koll Morgan Corporation

into the observed scene while the filters have a considerable range of optical densities so that a very versatile broad-range instrument is made.

11.13 THE ACTION OF FILTERS

Unless the light is travelling in a complete vacuum, there is a continuous reduction in its intensity due to the media through which it passes. This reduction may be due to absorption such as occurs with dark glasses, scattering such as occurs with ground glass or reflection such as occurs with glass surfaces. In principle there is no difference between reflection and scattering but all three processes reduce the amount of light transmitted from one place to another. Even the atmosphere has finite values of absorption and scattering. These vary considerably with meteorological conditions but the transmitted light is unlikely to be better than 95% over a kilometre.

For optical and ophthalmic glass the absorption value is considerably more but a further loss occurs at the surfaces of any sample due to reflection. Thus the absorption of most optical glasses is below 2% per centimetre but curves of the transmission of 1 cm samples often show values about 90% due to the 4% reflection loss at each surface.

Transmission values may vary considerably with wavelength (Chapter 13) but filters can be obtained with very little transmission change with wavelength. Such filters are neutral to the colour of the light passing through them and are known as **neutral density filters**. If two such filters have transmission values of 0.4 and 0.6 then the transmission of both used together is found by *multiplying* these values, i.e. 24%. This fact has led to the adoption of the concept of **optical density** which is the $\log_{10}$ of the reciprocal of the transmitted fraction. Thus a transmission of 50% means that half the light is transmitted so the density is $\log_{10}(^1/0.5)$ i.e. 0.3. This is very useful when the transmission values are very low as the density values increase as in the table:

Optical Density:	0.1	0.25	0.4	0.5	0.75	1.0	2.0	3.0	4.0
Transmission %:	80	56	40	32	18	10	1	0.1	0.01

The usual advantage of a logarithmic scale is found in that when two filters of density 0.5 and 0.25 are used together they give a density of 0.75 by simple addition. In transmission terms this is: $0.32 \times 0.56 = 0.18$ or 18%. This calculation is only correct when the spectral absorptions are identical for the two filters. Thus the densities of two differently coloured pieces of glass cannot be added in this way.

The multiplying of the transmission values for filters in combination is an outcome of **Lambert's Law** which states that equal paths in the same homogeneous absorbing medium absorb equal fractions of the light entering them. Thus light entering a small section of the medium of thickness dt, will have its intensity reduced from I to I − dI so that we can say that

$$\frac{dI}{I} = -\alpha \, dt$$

where α is a constant called the **absorption coefficient**. If we integrate this we have the intensity at any value of t is given by

$$\log I_t = -\alpha t + C$$

where C is the constant of integration. When t is zero, the intensity is the incident intensity, I, so that $C = \log I$. Thus, the emergent intensity, I', from a filter of absorption coefficient α, and thickness, t is given by

$$I' = Ie^{-\alpha t} \tag{11.04}$$

Its transmission is therefore given by

$$T = \frac{I'}{I} = e^{-\alpha t} \tag{11.04a}$$

from which it may easily be shown that an increase in thickness by a factor m gives a new transmission value of T^m.

11.14 CONTRAST

The contrast pattern in the Lummer-Brodhun Photometer was used in a *null* method, that is the instrument was in correct adjustment when a pattern could not be seen. Provided the pattern subtends a reasonable size and is reasonably bright the eye can detect a contrast difference somewhat better than the 1% value of Fechner's Fraction. It is found that the ability of the eye to just detect a small step in luminance, the **threshold contrast** of the eye, is very dependent on background luminance once this is below 100 Trolands, and also varies with the type of pattern, a single edge being less easily seen than a series of edges or bars (see Chapter 20).

The word **contrast** has been used in different ways and care must be taken when comparing authors. If B_t is the luminance of the object (or target) and B_b the luminance of the background then contrast is defined as

$$C_d = \frac{B_b - B_t}{B_b} \text{ for objects darker than the background} \tag{11.05a}$$

and

$$C_1 = \frac{B_t - B_b}{B_b} \text{ for objects brighter than the background} \tag{11.05b}$$

When repetitive object patterns are used (see section 15.15) the term **contrast modulation** is more generally used which is defined as

$$C_m = \frac{|B_t - B_b|}{B_t + B_b} \tag{11.06}$$

The modulus of the difference means that only positive values of C_m occur as is the case with all the other definitions. Strictly contrast modulation should only be applied to sinusoidally varying patterns but it is generally used for any repetitive pattern. It does not matter whether the bright parts or dark parts are used for B_t.

When an object is viewed from greater and greater distances not only does its apparent size diminish and its luminance reduce due to the absorption of the atmosphere but its contrast is reduced due to scattering. Both these latter effects are very variable with weather conditions*. The **visibility** or **meteorological range**, V is a measurement of the greatest distance at which a large black object can be seen against the sky background.

The absorption coefficient and scattering coefficient of the atmosphere are often combined together in the **extinction coefficient,** σ. We can obtain an *approximation* of this for visible light from $\sigma = 3.9/V$, where σ is in kilometres^{-1} if V is in kilometres.

The contrast C_r at range r is given by

$$C_r = C_0 e^{-\sigma r}$$

where C_0 is the contrast at zero range. Both C_0 and C_r are defined as in (11.05a) and (11.05b) and the range must be in kilometres if σ is in kilometres^{-1}.

CHAPTER 11   EXERCISES

1. Define the illumination of a surface. Explain exactly what is meant by the statement that the illumination due to a source of light varies inversely as the square of the distance from the source.
If an electric lamp of 16 candle-power is set up at 40 inches from a screen, find the illumination on the screen (*a*) in lumens per sq. ft, (*b*) in lumens per sq. metre.

2. Define the terms candle-power, foot-candle and metre-candle, and find the relation between the foot-candle and the metre-candle. An illumination of 6 foot-candles is required on a study table. What candle-power lamp should be installed at a height of 4 feet above the table if 70 per cent of the light is received indirectly?

3. What is meant by the candle-power of a source of light? Describe carefully an experiment to determine this value.

4. Explain what is meant by (*a*) the illumination and (*b*) the luminance of a surface. Describe how these qualities may be measured.

5. Given that the illumination of the earth in full moonlight is 0.02 lumens per sq. ft and that the distance of the moon is 235 000 miles, find its candle-power.

6. Express lumens per sq. ft in lumens per sq. cm and lumens per sq. m. A 50 C.P. lamp is 12 feet above a surface: find the illumination in lumens per sq. ft and lumens per sq. m.

7. Define 'Candle-power,' 'Mean spherical candle-power,' 'Foot-candle' and 'Lumen.' What is the minimum illumination of the roadway in a street where the lamps are 2000 candle-power on standards 50 feet high and 60 yards apart?

8. Explain exactly what is meant by intensity of illumination and deduce the expression for the intensity on a surface placed obliquely to the light reaching it from a small source.
Describe briefly one form of standard source.

9. How does the intensity of illumination at a point on a screen vary (*a*) with the distance of the point from the source, (*b*) with the angle of incidence of the light?
A small source of 100 candle-power is suspended five feet above a horizontal table. Draw a curve showing how the intensity of illumination on a table produced by direct light varies along a straight line passing directly under the source.

10. Distinguish source intensity and intensity of illumination. How can the latter be measured and in what units is it expressed?
A surface receives light normally from a source at a distance of 2.82 m. If the source is moved closer so that the distance is only 2 m through what angle must the surface be turned to reduce the illumination to its original value?

*The book *Vision through the atmosphere* by W. E. K. Myddleton is the accepted text on this subject.

11. A point source of light of 20 candle-power is situated 50 cm from a plane mirror and on a normal to its centre. If the mirror reflects 90 per cent of the incident light, find the illumination on a screen 3 m from and parallel to the mirror.

12. A 50 C.P. lamp is suspended above a table. It is found that by lowering the lamp 2 feet the illumination is increased three times. What was the original height of the lamp?

13. A small source of light of 100 C.P. is placed 2 feet from a circular plane mirror of 8 inches diameter; find the quantity of light in lumens falling on the mirror, and find also the illumination due to the direct and reflected light on a screen 6 feet from the mirror—the mirror reflecting 90 per cent of the incident light.

14. A 100 C.P. lamp is 15 feet above the ground. Find the illumination on the ground (*a*) directly under the lamp, (*b*) 20 feet to one side.

15. A lens of 3 inches diameter is placed 10 inches from a small source of light of 20 C.P. Find the total quantity of light falling on the lens and also the illumination on a surface 4 inches from the lens if the light be brought to a focus 6 inches from the lens.

16. A small source of 30 C.P. is placed 10 cm from a + 5D lens of 10 cm diameter. Find the illumination on a screen 1 metre from the lens, neglecting losses by reflection or absorption at the lens. What change in the illumination will be produced by reducing the aperture of the lens to 5 cm diameter?

17. The illumination given out by a lamp is not necessarily the same in all directions. The candle-power of an arc lamp was measured in various directions; it was 250 C.P. directly downwards; 1070 C.P. at 30° to the vertical. If this lamp is 25 feet above the pavement, find the illumination directly under the lamp, and also on a screen at the place on the pavement corresponding to the 30° direction and placed perpendicular to the direction of the light.

18. A room 20 by 15 feet and 10 feet high is illuminated by a 50 C.P. lamp suspended from the centre of the ceiling and 8 feet above the floor. Find the illumination on the floor (*a*) directly under the lamp, (*b*) in the corner of the room.

19. A diagram is illuminated for photographic copying by two lamps of 50 C.P., 4 feet apart, and 3 feet from the plane of the diagram. If the diagram is placed centrally between the lamps, find the illumination at the centre of the diagram.

20. A sight-testing chart 22 inches by 11 inches is illuminated by a 50 C.P. lamp placed 24 inches in front of the chart and in line with the centre of the top edge; what will be the illumination of the chart at the centre and at the bottom corners?

21. Two lamps, each of 200 candle-power, are hung 5 feet apart and 6 feet from the floor above a table 2 feet 6 inches high. Find the illumination on the table directly under each lamp and at a point midway between them.

22. A 30 C.P. lamp is suspended over the middle point of a rectangular table of 5 feet by 3 feet; find the height of the lamp above the table in order that the illumination on the table directly under the lamp may be 1.5 lumens per sq. ft. What will be the illumination at the corners of the table?

23. A newspaper is held in a railway carriage in such a position that the plane of the paper and the centre of the paper are 3 feet and 7 feet respectively from the lamp. (*a*) If the candle-power of the lamp be 32, calculate the illumination on the centre of the paper. (*b*) What would the candle-power of the lamp have to be in order that the illumination on the paper might be 1.5 lumens per sq. ft.

24. A room 15 feet by 12 feet is illuminated by a 50 C.P. lamp in the centre of the room, and 12 feet above the floor. Find the illumination in one corner of the room and 4 feet above the floor.

25. What are meant by the illumination and luminance of a surface? A test chart 3 feet by 2 feet is illuminated by two 30 C.P. lamps one on each side of the chart, in line with the middle of its longer edges and 3 feet in front of it. Find the illumination at the centre of the chart.

26. (*a*) A table is situated directly under a 64 C.P. lamp which is 7 feet above the table. What will be the illumination on the centre of the table?

(*b*) If the table is moved 10 feet from its original position, calculate the candle-power of the lamp required to give the same illumination as before on the centre of the table in the new position.

27. Light from a lens 40 mm diameter is converging to a point 75 cm from the lens: find the diameter of the pencil at 15, 30 and 80 cm from the lens. Compare the illumination on a screen in these three positions.

28. A rectangular table 8 feet by 5 feet is illuminated by two lamps each of 50 candle-power suspended 3 feet above the opposite corners. Find the illumination on the table at its middle point.

29. State Lambert's law on the variation of the emission of light with direction. A perfectly diffusing incandescent surface A of area 1 sq. mm and brightness 20 candles per sq. mm is set up parallel to and 50 cm from a screen, the line AB being normal to the screen.
Find: (*a*) the luminous intensity of the surface A in candles along a direction AC inclined $30°$ to the normal AB. (*b*) the illumination in lumens per sq. metre on the screen. (*i*) at B; (*ii*) at C. C is the point where AC cuts the screen.

30. Describe an accurate form of photometer and explain the principles on which its action depends. How would you use it to determine the percentage of light reflected from a surface?

31. Describe two methods for measuring the mean spherical candle-power of a lamp in its fitting.

32. Two lamps, one of 10 candle-power, are placed at opposite ends of a 2 m bench and a screen placed between them is found to be equally illuminated on its two sides when 80 cm from the 10 C.P. lamp. Find the power of the other lamp. What form should the screen take in order that the illumination of the two sides may be easily compared?

33. In comparing two sources of light by means of a photometer, it is necessary to arrange that:

(*a*) The photometer head is on the line joining sources (axis);
(*b*) All extraneous light is excluded;
(*c*) Both sides of screen are equally clean;
(*d*) Both sides of screen are equally inclined to axis of photometer.

Explain *clearly* the need for each of these precautions.

34. State and briefly discuss Lambert's law on the emission of light. A small source of light of area 2 sq. mm is set up 20 mm from a + 60D condensing lens of diameter 40 mm. The luminous flux falling on the lens from the source is 40 lumens. Find the luminance (brightness) of the source in candelas per sq. mm and, assuming no losses, the illumination of the image of the source in lumens per sq. metre.

35. Two lamps of 30 and 20 candle-power respectively are placed 2 m apart. Find the positions on a line joining them where a screen would be equally illuminated by these two lamps. How would you determine these positions experimentally?

36. Explain the meaning of the term optical density as applied to an absorbing medium. Illustrate by finding the optical density of a filter which transmits 20% of the light flux incident upon it.
What will be the transparency (i.e. percentage of light transmitted) and optical density of a filter of the same material and twice the thickness?

37. Describe a method of measuring the absorption of a tinted glass. What are the difficulties that are met with and how may they be overcome?

38. A photometer bench 5 feet long has a lamp at each end and a balance is obtained on a grease spot disc at 3 feet 6 inches in front of the left-hand lamp. On interposing a sheet of dark glass between this lamp and the disc, balance is obtained at 2 feet from the same lamp. What percentage of light is transmitted by the glass?

39. Two lamps are mounted at opposite ends of a 2 metre bench and a photometer screen is found to balance when placed 140 cm from the brighter lamp. On placing a tinted glass in front of the brighter lamp it is found that the screen is balanced when the distances

from the two lamps are reversed. What is the percentage absorption of the glass?

40. A photometer was placed between two lamps 2 metres apart and was found to balance when 80 cm from one of the lamps. Where must it be placed to again balance if a tinted glass absorbing 55 per cent of the light be placed in front of the brighter lamp?

41. Two lamps whose candle-powers are in the ratio of 2 to 1 are placed at opposite ends of a 2 metre bench. In front of the more powerful lamp is placed a screen which absorbs 30 per cent of the incident light. What will be the position of a photometer between the lamps when equally illuminated?

42. What is an illumination photometer? Describe a simple form of such an instrument and explain how you would use it to measure the variation of illumination at selected points in a room as twilight draws on.
A lamp of 125 C.P. is suspended at a height of 8 feet above a horizontal table. At what distance from the foot of the perpendicular let fall from the lamp on to the table will the intensity of illumination of a sheet of paper laid on the table be one foot-candle?

43. State Lambert's cosine law of emission. Explain what is meant by a perfectly diffusing surface.
The luminance (brightness) of a flat perfectly diffusing surface of area 2 sq. mm is 25 candela per sq. mm. What is its luminous intensity along (a) the normal direction; (b) a direction inclined at $60°$ to the normal?
If the light leaving the surface normally falls on a circular screen 1 cm in diameter and 50 cm from the surface, what amount of luminous flux, expressed in lumens, will fall on the screen?

44. A *uniform diffuser* is a surface having the same luminance (L) in all directions. Show that, for such a source of light, the luminous intensity (I) in any direction varies as the cosine of the angle between that direction and the normal to the surface. (NOTE: 'luminance' is the new term for 'brightness').
Such a source, small in area, is suspended parallel to and 2 m above a horizontal table. The illumination on the table at a point A vertically beneath the source is 5 lumens per square metre. Find the illumination on the table at a point B distant 1 m from A.

45. (a) Light from a point source of 50 candela enters an eye of effective pupil diameter 3 mm, the source being 50 cm from the eye. What amount of luminous flux, in lumens, enters the eye?
(b) Calculate the illumination on the ground midway between two lamp-posts 100 yards apart and 16 feet high, each lamp being of 450 candle-power in the direction considered. Express your result in lumens per square foot.

46. Two small lamps A and B give equal illuminations on the two sides of a photometer head when their distances from it are in the ratio 2:5. A sheet of glass is then placed in front of lamp B and it is found that equality of illumination is again obtained when the distances A and B are in the ratio 6:5. Find the percentage of light transmitted by the glass.
From this figure calculate the optical density of the glass.

Chapter 12
The nature of light

12.1 INTRODUCTION

So far we have considered that light can be represented by straight lines drawn on paper. Indeed this is why the name **Geometrical Optics** is applied to that branch of the subject. This description is adequate where we are concerned with the way light reacts with prisms, lenses and mirrors, but is no longer sufficient when we wish to understand how light reacts with itself or when it meets optical devices with very fine structure.

In these circumstances a more complete description is required, which includes the wave motion described in Chapters 1 and 2 and incorporates the way in which this wave motion is produced and propagated. This branch of the subject is called **Physical Optics**.

12.2 THE DUAL NATURE OF LIGHT

Our understanding of the propagation of light began some centuries earlier than our understanding of how it is produced, and the same order will be followed here. Very early experiments with burning-glasses showed light to be a form of energy. In the seventeenth century **Römer's** observations on the moons of Jupiter revealed that the times between their eclipses by the planet steadily increased and decreased as the earth orbited the sun. Calculations showed that these changes could be accounted for by the different distances the light had to travel if it had a velocity of about 190 000 miles per second.

This finite velocity means that the energy has to travel in some sort of self-contained packet across the intervening distance. The possible ways in which this might occur reduce to two. In the first case energy may be propagated as moving matter as in the case of energy received when a ship is hit by a projectile; or it may be propagated as a wave motion travelling through a continuous medium which does not move as a whole, as in the case of the movement of a ship caused by water arising at a distant disturbance.

Each of these methods has formed the basis of a theory of light and indeed the modern concept of radiant energy propagation is based on a duality which utilises both ideas. A full description of these concepts is beyond the scope of this book and would be more than is needed to understand a large part of physical optics. The essential idea that light is propagated in discrete packets of energy whose actions *can be described* by a wave motion will be developed as required.

Isaac Newton considered the evidence of rectlinear propagation and the laws of reflection and refraction and championed the **corpuscular theory**, which assumed that light consisted of exceedingly minute particles shot out by the source. Refraction was explained by assuming that the particles travelled faster in the denser medium. Although this sounds inherently unlikely, a greater problem was the partial reflection and partial refraction which was found to take place on glass-air surfaces and similar boundaries.

It would seem that a particle is either reflected or refracted, and if so, why should identical particles be treated differently? Newton was forced to assume that the boundaries between media were subject to 'fits of easy reflection and easy refraction' which were in turn determined by vibrations set up by the particles.

In this last hypothesis, Newton came surprisingly close to the modern concepts of the dual nature of light. However, the assumptions regarding the speed of the particles in the denser medium were the first to be challenged. Measurements made by Foucault and Michelson in the nineteenth century gave contrary results and the **wave theory** became totally accepted until the present century.

The original wave theory was propounded somewhat vaguely by both Grimaldi and Hooke: but to **Huygens** in 1690 is accorded the honour of the first written explanation of the principle described in Chapters 1 and 2. This was not altogether accepted and its opponents maintained that waves on water and sound waves required a medium to carry them whereas light could pass through a vacuum. Huygens assumed the existence of a medium known as the *ether,* filling all space. All experiments have failed to detect the nature or even the existence of such a medium and modern theory regards it as a mathematical concept rather than an actuality.

A more pertinent objection to the wave theory was that waves could bend round objects while light gives sharp shadows and appeared to travel in straight lines. It was left to the work of Young and Fresnel to show that light did bend round objects but that its vibrations were of such short wave-length that the effect could be seen only in special experiments. This will be discussed more fully later in this and in succeeding chapters. An added virtue of the wave theory was that if a transverse vibration was assumed instead of longitudinal as in sound, the phenomenon of polarisation (Chapter 16) could be explained.

These successes ensured the acceptance of the wave theory of light throughout the whole of the last century and Maxwell was able to integrate these ideas with those of electricity and magnetism. The observed actions of reflection and refraction, interference and diffraction and polarisation could all be explained.

It was only after the discovery of the electron as the fundamental particle of electricity and the ability of light to eject these from metals (photo-electricity) that problems were seen to arise. It was found by experiment that although the numbers of electrons ejected was proportional to the intensity of the light, the energy of the individual electrons was only proportional to the frequency of vibration of the light. In other types of vibrating energy the energy flow is proportional to the amplitude and it was difficult to see why the energy of the ejected electrons did not vary with this. The theories of Planck regarding the emission and absorption of radiant energy not as a continuous action but in discrete packets or quanta and Einstein's concept of **photons** carrying the concentrated energy of the light surrounded by a wave motion of finite length were found to be viable.

Further deductions from Einstein's theories led to the idea of stimulated emission which culminated in the invention of the **laser** in 1960 as described in section 12.13. Thus the history of man's ideas on the nature of light is a story of change and modification as new discoveries were made. We have no reason to suppose that the present concepts represent an ultimate description.

12.3 WAVE MOTION

Because the original concept of a continuous wave motion goes so far in explaining the basic phenomena of interference and diffraction, it is necessary to study the fundamental characteristics of a simple wave motion. In any case such a study is a prerequisite to understanding the more complex theories to follow. If we observe the ripples on the surface of a pond, two things are at once evident; first, that any floating object executes an up-and-down motion without appreciably altering its general position, and, secondly, that crests and troughs, places of maximum displacement of the surface above and below its normal position, are repeated at regular intervals. It is obvious that there is no general forward movement of the water as the ripples or waves continue to travel out from a disturbance.

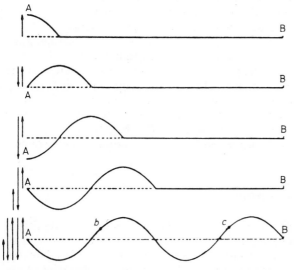

Figure 12.1 Vibrating cord

Let AB, *Figure 12.1*, represent a stretched flexible cord. If a regular up-and-down motion is given to the end A each particle will communicate its movement to the next particle in turn, and a wave will travel along the cord. The form taken up by the cord for different positions of A is shown in *Figure 12.1*. Two particles such as *b* and *c*, having the same displacement and direction of motion, are said to be in the same **phase**, and the distance between two successive particles in the same phase is a **wave-length**. If the end A moves in such a manner that it retraces its path in regular intervals of time, its motion is said to be periodic, and the time between successive passages in the same direction through any point is termed the **period** or **periodic time** of the motion. From *Figure 12.1*, it will be evident that during the time that A makes one complete up-and-down vibration, the disturbance has travelled a distance equal to one wave-length; during two complete vibrations it will have travelled a distance of two wave-lengths, and so on. Then if

> λ = **wave-length**
> V = **velocity at which disturbance travels**
> T = **period**

we have

$$\lambda = VT \tag{12.01}$$

which is the fundamental equation of wave motion.

The number of vibrations taking place per second is termed the **frequency**, N, of the vibration, and

$$\lambda = \frac{V}{N} \tag{12.01a}$$

Each vibration is commonly called a **cycle**, being that movement before the motion repeats itself. Frequency is therefore measured in cycles per second and this unit has now been given the name **Hertz** (Hz), after the German physicist who showed that light and radiated heat are electromagnetic waves.

The maximum displacement from its normal position of a particle, i.e. when it is at the crest or trough of a wave, is termed the **amplitude**, a, of the wave.

The form of the wave will depend upon the nature of the vibration of the individual particles, and each particle will be vibrating in the same way according to the periodic motion imparted by the source, such as the end of the cord in *Figure 12.1*. The motions of the individual particles in a wave motion may be very complex, but it can be shown that any complex vibration can be resolved into a number of simple to-and-fro motions of the type known as **simple harmonic motion** (S.H.M.).

12.4 SIMPLE HARMONIC MOTION

If the point P (*Figure 12.2*) moves around a circle with uniform velocity the to-and-fro motion of the projection of P on *any* diameter is termed a simple harmonic motion; thus the point R moves between XOX' and the point Q between YOY'. The maximum displacement of Q and R is equal to a, the radius of the circular path of P. This is the **amplitude**, of the wave form generated if Q is plotted against time as shown. The **phase** is the angle turned by OP from some arbitrary fixed point such as X. If y is the displacement of Q from O a time t after P was at X then

$$y = a \sin POX$$

or

$$y = a \sin \omega t = a \sin \frac{2\pi t}{T}$$

where $\omega/2\pi$ is the number of cycles P completes in a second or where T is the time for P to complete a cycle. Each revolution of P completes one cycle of simple harmonic motion. If t is measured from some other moment such as P at Y then we would have

$$y = a \cos \omega t = a \sin (\omega t + 90^\circ)$$

or, more generally,

$$y = a \sin (\omega t + \delta) \tag{12.02}$$

where δ is the initial phase between the start location of P and some arbitrary location.

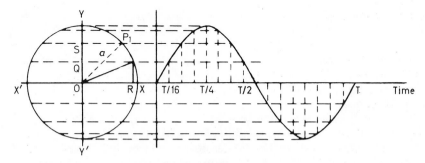

Figure 12.2 Graphical representation of simple harmonic motion

A simple harmonic motion is characteristic of the vibration in a straight line of a particle acted on by a restoring force linearly proportional to the displacement of the particle from the centre of attraction. An example is seen in the motion of a weight suspended from the end of a spring. Many of the phenomena of light, especially those of interference, diffraction and polarization, may be explained by considering the resultant effect as if a particle were being acted on simultaneously by two or more simple harmonic motions. The effects due to each motion at successive values of t are calculated and the resultant at each value of t is the algebraic sum of the separate effects at the corresponding times. This is the **principle of superposition** and has been shown to be true where the effect is small compared to the energies and separation distances of the particles affected. This proviso ensures the linearity of the restoring force described above. Different effects can occur with high-powered lasers where the energies are large.

The composition of a number of motions acting in the same straight line may be solved graphically by plotting, in the way shown in *Figure 12.2*, each motion to the same scale and with the same co-ordinates. The curve of the resultant displacement is then the sum of the separate displacement curves. An example is shown in *Figure 12.3* where the broken line is the curve of the resultant displacement due to two motions,

$$y_1 = a \sin \frac{2\pi t}{4} \text{ and } y_2 = b \sin \frac{2\pi t}{8} \text{ executed in the same vertical line.}$$

At any moment in time the displacement of the resultant can be found by drawing the lines OA and OB at the correct phase angles and constructing, by the

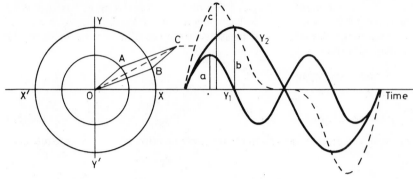

Figure 12.3 Composition of two simple harmonic motions

method of the parallelogram of forces, the vector addition to obtain the resultant OC. As the vectors OA and OB rotate at different frequencies in this example,
OC has a varying length which when projected on to YY′ produces the broken
line wave motion.

If the two motions are of the same frequency then the phase difference between them must remain constant. In *Figure 12.4a*, the two motions have different amplitudes, *a* and *b*, but the resultant found by the parallelogram of
forces has a constant amplitude A because the angle between *a* and *b* remains
the same. Because *a, b* and A all rotate together it is easy to see that the frequency of the resultant is the same as that of the two composing motions but
it has its own phase angle.

From the triangle the resultant amplitude A is given by:

$$A^2 = a^2 + b^2 + 2ab \cos \epsilon \tag{12.03}$$

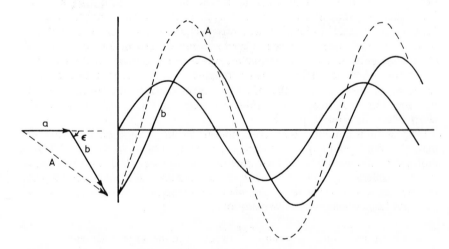

Figure 12.4a Composition of two simple harmonic motions

where ϵ is the phase difference between a and b. Notice that A can be zero if a equals b and ϵ equals $180°$. It often occurs in optics that there are two or more motions to be summed of equal or nearly equal amplitudes. Furthermore the phase differences between them are often equal. *Figure 12.4b* shows four such motions and the broken line representing the resultant emphasises the large amplitude this can achieve. It may be shown however that the resultant is zero if the phase differences add up to $360°$.

Although *Figure 12.4* shows only two and four motions of the same frequency it is a general rule that any number of motions of the same frequency will yield a resultant of that frequency.

Figure 12.4a shows the case when the two motions are of equal frequency but differ in phase. We have

$$y_1 + y_2 = a \sin (\omega t + \gamma) + b \sin (\omega t + \delta) \tag{12.03a}$$

where a and b are the amplitudes of the constituent motions and γ and δ are their phases. Expanding (12.03a) we have*

$$a \sin \omega t \cos \gamma + a \cos \omega t \sin \gamma + b \sin \omega t \cos \delta + b \cos \omega t \sin \delta$$

$$= \sin \omega t \, (a \cos \gamma + b \cos \delta) + \cos \omega t \, (a \sin \gamma + b \sin \delta)$$

which may be set equal to

$$\sin \omega t \, (A \cos \epsilon) + \cos \omega t \, (A \sin \epsilon) = A \sin (\omega t + \epsilon) \tag{12.03b}$$

which is a motion of the same frequency but different amplitude and phase.
We find A from

$$A^2 = (A \cos \epsilon)^2 + (A \sin \epsilon)^2$$

$$= (a \cos \gamma + b \cos \delta)^2 + (a \sin \gamma + b \sin \delta)^2$$

$$= a^2 + b^2 + 2ab \cos (\gamma - \delta) \tag{12.04}$$

We find ϵ from

$$\tan \epsilon = \frac{a \sin \gamma + b \sin \delta}{a \cos \gamma + b \cos \delta} \tag{12.05}$$

When more than two motions are summed of equal amplitudes but consistently differing phases as often occurs in interference we have, for example:

$$a \sin (\omega t + \epsilon) + a \sin (\omega t + 2\epsilon) + a \sin (\omega t + 3\epsilon) + a \sin (\omega t + 4\epsilon)$$

$$= a(2 + 2 \cos \epsilon)^{1/2} \sin \left(\omega t + \frac{3\epsilon}{2} \right) + a (2 + 2 \cos \epsilon)^{1/2} \sin \left(\omega t + \frac{7\epsilon}{2} \right)$$

$$= a(2 + 2 \cos \epsilon)^{1/2} (2 + 2 \cos 2\epsilon)^{1/2} \sin \left(\omega t + \frac{5\epsilon}{2} \right) \tag{12.06}$$

This resultant is a motion of the same frequency as the component motions but whose amplitude and phase depends on ϵ. The amplitude is 4 when ϵ is zero and zero when ϵ is a multiple of $\pi/4$. The resultant for ϵ equal to $\pi/8$ is shown in *Figure 12.4b*. If the value of ϵ is varied, the amplitude of the resultant varies between $4a$ and zero.

The resultant waveform of *Figure 12.3a* occurs when OA and OB produce the simple harmonic motion by projection on to the same vertical line YY'. If the projections on to XX' were used a similar waveform would result. If, however, we combine OA projected on to YY' and OB projected on to XX' a more complex result occurs. If the analysis is restricted to cases where the frequency of

*From $\sin(A + B) = \sin A \cos B + \cos A \sin B$ used twice.

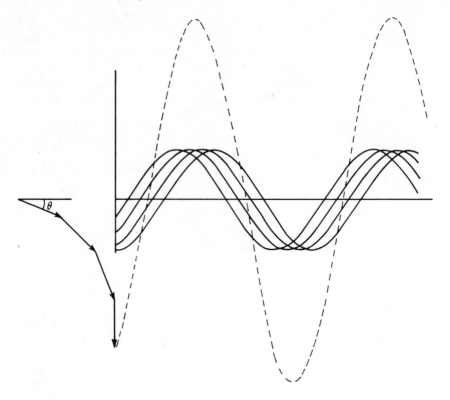

Figure 12.4b Composition of four simple harmonic motions of equal frequency and amplitude but progressively out of phase

the two contributing motions is the same a result is obtained of great importance in the study of polarized light.

Suppose the two contributing wave forms are represented by

$$x = 11 \sin \left(\frac{2\pi t}{T} + \frac{\pi}{4}\right) \quad \text{and } y = 7 \sin \left(\frac{2\pi t}{T}\right)$$

which are executed along XOX′ and YOY′ respectively of *Figure 12.5a*. If two circles are drawn with radii proportional to 11 and 7 they can be divided into equal parts representing equal time intervals. As the two motions are of the same frequency the number of divisions is the same but the difference in phase is represented by the different starting points. Remembering that the resultant is derived from the components of the rotating vectors on *different* axes, these must be taken *before* the parallelogram of forces is drawn. The result of these at any instant is found to lie on the ellipse shown. In fact the result of any two orthogonal S.H.Ms of equal frequencies is an ellipse except for the special case where the phase difference is zero or π when the result is a straight line. A further special case occurs when the phase difference is an odd multiple of $\pi/2$ and the amplitudes are also equal. These conditions result in a circular motion.

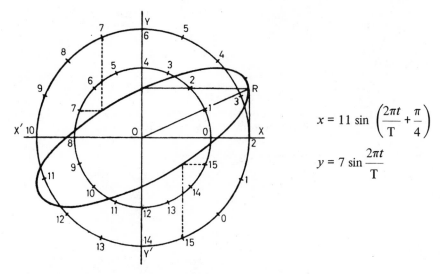

Figure 12.5a Two simple harmonic motions executed in directions at right angles

$$x = 11 \sin\left(\frac{2\pi t}{T} + \frac{\pi}{4}\right)$$

$$y = 7 \sin \frac{2\pi t}{T}$$

In more general terms we have

$$y = b \sin \omega t$$

$$x = a \sin (\omega t + \delta)$$

$$= a \sin \omega t \cos \delta + a \cos \omega t \sin \delta$$

$$\frac{x}{a} = \frac{y}{b} \cos \delta + \left(1 - \frac{y^2}{b^2}\right)^{1/2} \sin \delta$$

$$\left[\frac{x}{a} - \frac{y}{b} \cos \delta\right]^2 = \left(1 - \frac{y^2}{b^2}\right) \sin^2 \delta$$

$$= \frac{x^2}{a^2} - \frac{2xy}{ab} \cos \delta + \frac{y^2}{b^2} \cos^2 \delta = \sin^2 \delta - \frac{y^2}{b^2} \sin^2 \delta$$

Therefore

$$\frac{x^2}{a^2} + \frac{y^2}{b^2} - \frac{2xy}{ab} \cos \delta = \sin^2 \delta \qquad (12.07)$$

This is the equation of an ellipse the principal axes of which, like that in *Figure 12.5a*, are not aligned with the x and y axes, but contained in the rectangle of sides $2a$ and $2b$ as shown in *Figure 12.5b*. If δ is made equal to $\pi/2$ the principal axes do align and the familiar equation is obtained.

$$\frac{x^2}{a^2} + \frac{y^2}{b^2} = 1 \qquad (12.08)$$

This becomes a circle when $a = b$, while (12.07) does not, remaining an ellipse with its principle axis at 45° to the x and y axes. Thus for circular motion we need equal amplitudes, $a = b$, *and* $\delta = \pi/2$ or some multiple of this.

On the other hand when $\delta = 0$ or some multiple of π we have from (12.07)

$$\frac{x^2}{a^2} + \frac{y^2}{b^2} - \frac{2xy}{ab} = 0$$

that is,

$$\left(\frac{x}{a} - \frac{y}{b}\right)^2 = 0 \text{ or } y = \frac{b}{a}x$$

which is the equation of a straight line for all values of a and b. The resultants for equal a and b but different phase differences are shown in *Figure 12.5b*.

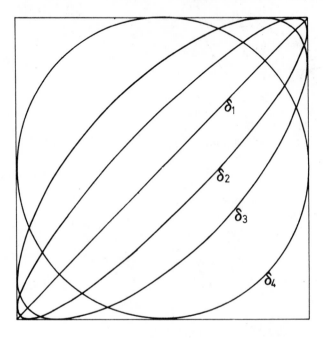

$$\delta_1 = 0 \text{ (or } 180°, 360° \text{ etc.)}$$
$$\delta_2 = 22.5° \text{ (or } 157.5°, 202.5° \text{ etc.)}$$
$$\delta_3 = 45° \text{ (or } 135°, 225° \text{ etc.)}$$
$$\delta_4 = 90° \text{ (or } 270°, 450° \text{ etc.)}$$

(b)

Figure 12.5b Composition at right angles of equal-amplitude, equal-frequency, simple harmonic motions with phase difference

12.5 WAVE TRAINS AND THEIR SUPERPOSITION

It was seen in section 12.3 that the form of a wave will depend upon the motion imparted to the particles by the source. If this motion is simple harmonic, the simultaneous displacement of all the particles may be found using equation

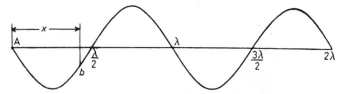

Figure 12.6 The form of a wave motion. A wave train

(12.02). This applies in the case where particles are in actual motion and in the case where 'motion' is really that of a varying electromagnetic field.

Let us suppose that a source A, *Figure 12.6*, is executing a simple harmonic motion given by the expression

$$y = a \sin \frac{2\pi t}{T}$$

and that this is transmitted from one location to the next with a velocity V. Then a location b at a distance x from A commences its motion x/V seconds after that at A and its motion is therefore given by,

$$y = a \sin \frac{2\pi}{T} \left(t - \frac{x}{V} \right) \tag{12.09}$$

As the wavelength λ is given by $\lambda = VT$ we have

$$y = a \sin 2\pi \left(\frac{t}{T} - \frac{x}{\lambda} \right) \tag{12.10}$$

This equation gives the complete form of any simple wave motion,* for by giving a constant value to x we have the equation of the motion at this point, and by giving any constant value to T the curve obtained gives the form of the train of waves at that instant. The curve, *Figure 12.6*, so obtained should not be confused with that in *Figure 12.2* which shows the displacement at a *single* location for various values of T.

When two or more waves are superposed, the resultant displacement at any point may be found from the separate displacements by the methods of section 12.4. As the resultant displacements will still be periodic, a new wave is formed. If the waves are travelling with equal velocities in the same direction and the vibrations are taking place in the same plane, the resultant waveform may be found graphically. *Figure 12.7* shows the resultant waveform (full line) when two waves (broken lines) are superposed. The two waves, of the same amplitude, are travelling in the same direction with equal velocities and one has twice the wavelength of the other. Any point on the resultant wave has a displacement OR equal to the algebraic sum of the separate displacements.

This process may be repeated for any number of simple waves and the most complex waveforms may be built up. By the same reasoning, the most complex waveform can be expressed as a sum of a number of simple waves of simple

*Provided $y = 0$ when x and t are zero.

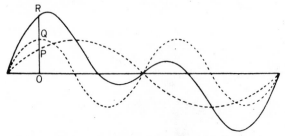

Figure 12.7 Superposition of waves

harmonic form. The mathematics of this was developed by Fourier (1768–1830) and is beyond the scope of this book. However, the principle is readily understood and is applied to optical phenomena in Chapters 15 and 18.

The difference in phase at any instant between two locations in a simple waveform may be found from (12.10). Thus, if x_1 and x_2 be distances of the two locations from the source the displacement of each is

$$y_1 = a \sin 2\pi \left(\frac{t}{T} - \frac{x_1}{\lambda} \right)$$

and

$$y_2 = a \sin 2\pi \left(\frac{t}{T} - \frac{x_2}{\lambda} \right)$$

Therefore the difference in phase at any instant

$$\epsilon = 2\pi \left(\frac{t}{T} - \frac{x_1}{\lambda} \right) - 2\pi \left(\frac{t}{T} - \frac{x_2}{\lambda} \right)$$

$$= \frac{2\pi}{\lambda} (x_2 - x_1)$$

$$= \frac{2\pi}{\lambda} \text{(path difference)} \tag{12.11}$$

This will also express the difference in phase between two disturbances arriving at a single location from two sources at distances x_1 and x_2 providing the two sources are producing exactly similar motions. Expression (12.11) is of great importance in the phenomena of interference (Chapter 14).

By considering *Figure 12.2* again, it can be seen that a particle in simple harmonic motion has, on passing through its mean position, its maximum veloctiy, $2\pi a/T$ which is the velocity of the point P. Its velocity then diminishes as it recedes from its mean position becoming zero as it reaches its maximum displacement. The energy of the particle, being all kinetic at the moment of maximum velocity, is given by $\frac{1}{2}mv^2$ where m is the mass and v the velocity. Thus energy is proportional to $(2\pi a/T)^2$ or, more importantly, to the amplitude squared, a^2. This again applies where the 'motion' is really that of a varying electromagnetic field. As such, it is found that I, the intensity of the light, represented by the wave motion, is proportional to the amplitude squared. This result is important in interference and diffraction.

12.6 WAVEFRONTS

The two previous sections have analysed the wave motion as a one-dimensional vibration. In Chapter 1 we loosely considered a wavefront moving out from a source and saw its propagation in terms of wavelets generated by each point on the wave. The wave itself had travelled from the source in a given time and this time is the same for all parts of the wavefront. If we isolate different points on the wavefront they will all be in the same phase and the wavelets generated will be in phase. Thus if exactly similar vibrations be travelling out uniformly in all directions in a homogeneous medium from the point source B (*Figure 1.1*) any sphere DAE having its centre at B, will be a surface of equal phase generating wavelets of equal phase. The only way to obtain wavelets of different phase is to take them from points at different distances from B.

12.7 THE ELECTRO-MAGNETIC THEORY

For many years after the wave theory of light had become generally accepted it was supposed that the vibrating atoms of a luminous body set up waves in an elastic-solid medium which filled all space, and most of the more important phenomena of light have been satisfactorily explained on this theory.

In 1873, however, Clerk Maxwell (1831–1879) put forward a new form of the wave theory in which light is considered to consist of alternating electro-magnetic waves. This **electro-magnetic theory**, which conceives an alternating condition of the medium in place of an oscillation of position of its particles as in the elastic-solid theory, satisfactorily explains most of the known facts about light and overcomes the difficulties met with in the older theory. Maxwell showed mathematically that the electro-magnetic vibration would be transverse to its direction of travel and would travel in free space with a velocity of 3×10^8 metres per sec, which, as we have seen, agrees with the measured velocity of light.

About 1886 Hertz (1857–1894) succeeded in producing electro-magnetic waves by purely electrical means, and this was a most important discovery in confirmation of Maxwell's theory. These waves, which are those now used for radio, have many of the properties of light waves, but their wave-lengths extend to many thousand times those of light. The discovery in 1895 of the Röntgen or X-rays by Röntgen (1845–1923) and later of the gamma rays emitted by radio-active-substances, gave us another series of electro-magnetic waves with the same characteristics as light waves, but of very much shorter wave-length.

Later researches have succeeded in filling up the gaps that existed between the various groups of radiation and we now have a complete range of radiations, differing only in frequency and therefore in wave-length, extending from the longest electrical waves of some thousands of metres wave-length, as used in radio, through the infra-red or heat radiations to the extremely small portion of the complete range, the visible spectrum, which we know as light, where the wave-lengths vary from 7×10^{-4} mm in the red to 4×10^{-4} mm in the violet. Beyond the violet we have the still shorter wave-lengths, the ultra-violet radiations, with wave-lengths extending from that of the extreme violet down to about 14×10^{-6} mm, and beyond these again the X-rays, the wave-lengths of which extend down to 6×10^{-9} mm. In the gamma rays the shortest wave-length found is about 6×10^{-10} mm. More recent work has shown the existence of radiations of very great penetrating power and having wave-lengths as short as 8×10^{-12} mm. To these radiations the name cosmic rays has been given.

The various regions of this enormous range of radiations have been shown to possess many properties in common; they all travel at the same speed, and all give rise to certain phenomena, such as interference and diffraction. The different regions are, of course, produced by different sources and detected in different ways. The known range of electro-magnetic waves is shown diagrammatically in *Figure 12.8*.

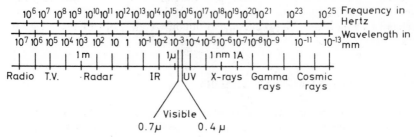

Figure 12.8 The electromagnetic spectrum

The wave-lengths of light and other short wave-length radiations are usually expressed in terms of one of the following units:

(1) The **angstrom** (Å). This is equal to 10^{-10} metre or a ten-millionth part of a millimetre.
(2) The **millimicron** (mμ) or **nanometre** (nm), which is equal to 10Å, 10^{-9} metre or one-millionth part of a millimetre. The nanometre is now the preferred unit.
(3) The **micron** (μ), or micrometre (μm) which is equal to one-thousandth part of a millimetre, 10^{-6} metre.

Thus the wave-length of the yellow light of the sodium flame is 5890 Å = 589 nm = 0.589 μm = 0.000589 mm. We shall adopt the nanometre in referring to wave-lengths. Although the nanometre is the internationally recommended term, the millimicron is still commonly used and some exercises will use the old unit.

12.8 INCANDESCENCE AND CHARACTERISTIC COLOURS

When the heated poker was discussed in Chapter 1, the change in colour with temperature was noticed. As the poker becomes hotter it gives out more light and appears brighter. At the same time the colour of the light emitted changes from red only towards blue. In other words the poker finds it more easy to emit blue light the hotter it gets. This is the first indication that different colours are associated with different amounts of energy.

If instead of a solid poker a pure element such as sodium is heated in a flame, the light it gives out is an intense orange-yellow. The same colour appears when an electric discharge is passed through a sodium vapour as in a common type of street light. If the sodium vapour is replaced by mercury vapour blue and green colours are generated to give a bluish-white appearance. Thus it is seen that particular colours are associated with particular elements.

12.9 THE QUANTUM THEORY AND ATOMIC STRUCTURE

The Huygens conception of light travelling out from a luminous point as a *continuous* wave surface is equally applicable to both the elastic-solid and the electro-magnetic forms of the wave-theory, and this conception can satisfactorily account for the observed facts of reflection, refraction, interference, diffraction and polarisation. There are, however, certain phenomena, notably those concerned with an interchange of energy between radiation and matter, absorption, and the photo-electric effect, which cannot be explained in terms of the radiation being transmitted as a continuous wave surface.

At the beginning of the present century Planck successfully explained certain facts regarding the emission and absorption of energy on the hypothesis that energy is not emitted or absorbed continuously, but in definite minute units or **quanta**. Later Einstein extended the quantum theory to light, and considered that light travelled, not as a wave surface over which the energy was equally distributed, but in minute bundles of energy—**light quanta** or **photons**—the energy of which remains concentrated as they travel through space.

When the light emitted by energised gases is examined by the spectroscopic methods described in Chapter 13 a bewildering array of spectral colours is obtained. The simplest arrangement is found when the gas is hydrogen, *Figure 12.9*. This shows a series of lines which become closer together with decreasing wavelength in a very regular manner. It was these lines which led to the establishment of the whole subject of Atomic Physics when Niels Bohr postulated a structure of matter in which light-weight electrons orbit a heavy nucleus. The electron orbits are restricted in number and each orbit requires a specific amount (or level) of energy in the electron which occupies it. Bohr postulated that light is emitted when an electron moves from a high energy level orbit to one having a lower

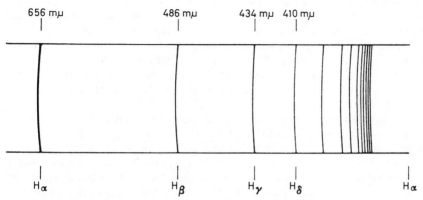

Figure 12.9 Emission spectrum of the hydrogen atom

energy level. The difference in energy is carried away by the emission of a single photon of light, the wave-length of which is defined by the equation

$E = h\upsilon$
where E is the energy difference
        h is Planck's constant
        υ is the frequency of the light

This equation is of fundamental importance in physics and the constant h (6.63×10^{-13} joule. second) is a fundamental constant of nature. Thus the energy radiated by the heated poker increases with increasing temperature as more atoms are emitting energy more frequently but the energy of each photon of light emitted depends solely on the change in energy of the electron actually emitting the light.

The particular photon and the electron energy levels are very closely related. This is evidenced by the fact that if the photon passes another electron in another atom in the lower of the energy levels there is a high probability that the photon will be absorbed by the electron which jumps to the higher energy level. This is further discussed in sections 12.10 and 13.7.

12.10 STIMULATED EMISSION AND OPTICAL PUMPING

When an incandescent filament emits light we can imagine many atoms having their electrons raised to high energy levels by absorbing energy from the free electrons constituting the electric current through the filament. These electrons subsequently fall to lower levels and emit light. No specific colour is generated as many different electrons and energy levels are involved. The actual moment when a particular electron emits light is not predictable as it does so on a *spontaneous* basis although the time spent in the higher (or excited) level may be very short. However, for transparent sources such as gas discharge lamps a second process exists.

It is now known that if a photon of the same wave-length as that about to be emitted encounters an electron in the excited state a high probability of *stimulated emission* occurs. In this case the emitted photon is naturally of the same wave-length as the stimulating photon but more importantly it is emitted in the same direction and in phase. However, this is not very usual as there are many electrons in the lower state ready to absorb the stimulating photon and the excited electron may also emit spontaneously before the stimulating photon arrives. Thus the processes of stimulated emission, spontaneous emission and absorption can be seen as competing actions within the luminous volume. Very careful arrangements are needed if the stimulated emissions are to predominate over the normal luminescence of the material.

On approach is to ensure an over-abundance of electrons in the excited state. This can be achieved without upsetting the stimulation process by indirect *optical pumping*. With this method intense light is focused into the material so that its particular wave-length is absorbed by the electrons which enter a higher energy level than that required. A proportion of these will then emit light spontaneously and descend to the required high energy level. From here they may spontaneously or by stimulation descend to the lower state from which they can be again raised by the incoming light beam. If this light is sufficiently intense it is possible to maintain more electrons in the high level than in the lower. This *inverted population* of electronic energy levels is essential in the design of **lasers.**

12.11 COHERENCE AND RESONANT CAVITIES

It is seen from the preceding sections that light is normally emitted when an electron in a higher energy level *spontaneously* jumps to a lower level and emits a photon of light. This hardly relates to the analogy of ripples on the surface of water used in Chapter 1.

If, however, the ripples are caused by a single stone dropped into the water then they are seen to have a definite duration. A circle of a finite number of ripples moves out from the point of impact. When these have passed no other ripples occur until another stone has been thrown in. If the second stone entered the water before the first set of ripples had died away it would be possible, by carefully judging the correct moment, to make the second set of ripples in phase with the first. Indeed a whole series of stones could be dropped at specific times to generate a continuous ripple system moving out from the point of impact. If the stones were dropped on a random basis over a small area then the ripples system would be a complex superposition of several ripple trains even though they would have the same wave-length and speed of propagation.

The difference in these two cases is similar to that between **coherent** and **incoherent light**. The latter case is typified by the incandescent filament where the atoms are spontaneously emitting at random times. In this example many different frequencies are also involved. The former case requires stimulated emission by a controlling wave. This wave can be generated only in a **resonant cavity**. Resonance occurs in sound when a travelling wave is reflected at one end of an enclosed space and returns to the other end for a second reflection. If it now finds itself in phase with the remainder of the wave not yet reflected the two superpose to give a greater amplitude. If they are out of phase zero amplitude will result.

It is easy to show that the length of the cavity is the important dimension. For a fixed length l, any wave for which $n\lambda = 2l$ will resonate, n being an integer. If the material with an inverted population (section 12.11) is contained in an optical resonant cavity a controlling light wave reflected at each end will stimulate emission from the material to build up its amplitude as it travels backwards and forwards.

12.12 GAS LASERS

The first gas laser was built in 1961 by Javan, Bennett and Herriott at Bell Telephone Laboratories. It consisted of a tube, 100 cm long, filled with a mixture of helium and neon. At both ends of the tube reflectors were fitted to make the resonant cavity. These were flat mirrors having a flexible coupling to the tube so that they could be adjusted parallel to each other to within a few seconds of arc. In the original model the discharge in the gas was generated by radio-frequency radiation from electrodes wrapped round the tube. Electrodes

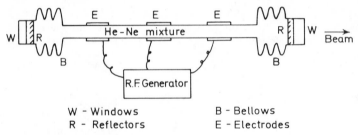

W – Windows   B – Bellows
R – Reflectors   E – Electrodes

Figure 12.10 Layout of the first gas laser

can also be used inside the tube between which a current can be passed through the gas. *Figure 12.10* shows the general arrangement.

The stimulated emission occurs in the neon but the helium helps to absorb the electrical energy and generate the inverted population in the neon. When spontaneous emission occurs one of these photons is reflected normally by an end mirror and proceeds down the tube stimulating other excited electrons to emit. On reflection the same action occurs on the return journey and again after a further reflection.

When the stimulating wave passes over an excited electron the stimulated emission may not start until most of the wave has passed and need not occur at all. This means that after two reflections of the stimulating wave the stimulated action may still be in process and so the need to remain in phase is paramount. Energy is continually lost in the absorption of the stimulating wave by electrons in the lower energy levels and in the less than perfect reflection at the mirrors. A lasing action will only occur and be maintained if these losses are smaller than the gains due to stimulated emission. Once this has been achieved one of the mirrors may be made partially transparent so that the beam of coherent light emerges from the cavity. This also represents loss to the system.

In general terms the efficiency of the system is of the order of 0.1% and the energy radiated a few milliwatts—far less than a simple flash lamp. On the other hand the beam is extremely well collimated so that its luminous intensity is very high, making it a safety hazard (see section 12.15).

Since the first helium-neon laser many other gases have been found to lase although this original combination has proved to be the most tractable. Low powered types can now be purchased for a few tens of pounds while specially stable types are used in measuring systems (section 14.14). The light emitted is red, 632.8 nm but other electronic levels will also lase giving better efficiencies although the light emitted is infra-red. For visual work a green colour is preferred and after some delay the argon ion laser was invented. Now a large number of gases are known to lase giving a large choice of wave-lengths.

In all cases the light emitted has an extremely narrow spread of wave-lengths. The need to remain in phase after a reflection at each of the end mirrors means that the cavity length must contain an integral number of wave-lengths. Whereas photons spontaneously emitted may have a small range of wave-lengths as the various atoms are influenced by collisions with other atoms which give rise to slight differences in the energy levels; photons in the light beam from a laser being related to the resonant cavity length are the most monochromatic known to man.

12.13 PULSED LASERS

Although the He–Ne laser described in the previous section was the world's first *continuous* source of stimulated radiation it was preceded by some months by the pulsed ruby laser of **Maiman** at Hughes Research Laboratories. The major problem to be overcome was the too great a loss of energy by spontaneous emission, non-optical losses and absorption in the end mirrors. Previous exploratory work by **Townes, Basov and Prokhorov** for which they received the Nobel Prize had shown the extent of these other means of energy dissipation compared with the required method.

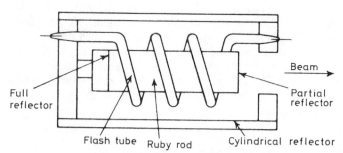

Figure 12.11 Layout of the first ruby laser

Maiman's solution (*Figure 12.11*) was to wrap a xenon flash tube around the working element which was a cylinder of pink ruby containing 0.05% chromium. The end faces of the cylinder were polished flat and parallel and coated to provide the resonant cavity, although one face is only partially reflecting to allow the emergence of the laser beam. The flash lamp surrounding the ruby is operated by discharging a bank of capacitors across it. A large proportion of this energy is dissipated as heat but a fraction is emitted as blue-green light which is absorbed by the ruby. For a short time the ruby has an inverted population in its chromium atoms. With high-speed detectors it is possible to observe that spontaneous incoherent emission from the ruby starts almost immediately radiating in all directions but after about 0.5 milliseconds the coherent radiation emerges from the partially reflecting face of the ruby rod. This laser radiation may last for only a few milliseconds but peak powers in the order of tens of kilowatts have been recorded.

Once again the precise wave-length emitted depends on the length of the rod, the resonant cavity. During the discharge this heats up and expands and it is possible to observe the change in wave-length due to this. The wave-length is about 694.3 nm, which is governed by the energy levels in the atoms of chromium known as the *active* element or *working* element. Again other wave-lengths can be found although the red line first discovered is the most efficient.

An alternative lasing material is neodymium which can be contained in a glass rod or a crystal. This gives out light at 1.09 nm in the infra-red part of the spectrum. Solid lasing materials such as these can now be made to lase continuously as gas lasers but any defects in the material absorb energy and can heat up and fracture. Later research turned to liquid lasers where any local hot-spot can be healed by the movement of the liquid.

12.14 SEMI-CONDUCTOR LASERS

The population inversion so necessary to coherent light generation can be obtained in the junctions of semi-conductor devices. Instead of inverting the population by putting energy into the working volume from other lamps, the direct passage of a current across the junction produces this effect. The excited electrons may emit light spontaneously and such devices are often called **light emitting diodes** or **LEDs**. If the ends of the junction region are plane and well polished a resonant cavity is formed as the refractive index of the material is so high that substantial reflectivities occur without any reflective coatings.

The most common material is gallium arsenide and this emits in the infra-red part of the spectrum. LEDs are available which emit red and green light. The efficiency is about 50% but large powers are difficult to obtain. For the infra-red emitters a number of suitable detectors are available and such devices have been used on eye-movement experiments where the infra-red beam is reflected off the sclera but not from the cornea. There is very little heat generated and at the low power levels which suffice, emitters and detectors can be mounted close to the eye on a spectacle frame without danger and without the subject being able to see the light as it is infra-red.

12.15 PHOTOMETRY OF LASERS—VISION HAZARDS

If plane mirrors define the resonant cavity of a laser then the emitted laser beam will be plane as far as diffraction effects allow (see Chapter 15). Plane mirrors can be difficult to adjust and often concave mirrors are used but with a further curved surface to give a parallel output beam having a divergence of about $^1/_3$ minute of arc. This means that the total solid angle containing the energy is about 10^{-8} steradians while more common sources of light radiate into 4π steradians.

Because of the large loss of energy by other means which occurs in laser systems the efficiency is usually about 0.1%. Thus a watt of energy must be supplied to a He–Ne laser to obtain a milliwatt of light. Most small gas lasers produce between one and 30 milliwatts of red light. In photometric terms this is equivalent to little more than 5 lumens. However, candle-power is measured in lumens per steradian and as these few lumens are concentrated into such a narrow beam the intensity of the source seen looking along the beam is 500 million candela! Because this is in a parallel beam the eye focuses this source on to the retina. In all but the very weakest laser such an intensity will cause burn damage on the retina before the muscles of the eyelid can close to protect the eye. It is therefore very important that operating lasers are not directed towards observers or on to mirrors which could reflect the beam into unprotected eyes.

With pulsed lasers the hazard is even greater as the chance reflection of the pulse from a metal surface such as an optical bench or equipment case could direct sufficient energy into an eye to cause blindness. Workers in laser laboratories are normally supplied with protective goggles containing glass which absorbs those wave-lengths at which the laser is operating. Other parts of the spectrum are transmitted so that the experimentor can see what he is doing.

The use of lasers for interference and holography (Chapters 14 and 15) normally requires the laser beam to be divergent and a mirror or lens is used for this purpose. This gives the beam a solid angle of about one-tenth a steradian and the apparent intensity of the source is now only 500 candela–similar to tungsten filament lamps. Nevertheless all lasers should be treated with the utmost respect and any examination of the beam is best done by allowing it to fall on a piece of white card or paper even if the result looks quite dim.

12.16 APPLICATION OF LASERS

The potential of lasers was recognised from their beginning. Many uses were proposed but as these did not come to fruition within a few years some cynicism has been generated culminating in the tag that lasers were 'a solution looking for a problem'. This is far from the truth. Two major uses of lasers occur in Optical Metrology and in Holography which are discussed in Chapters 14 and 15 respectively. Other uses involve one or more of the three most obvious attributes of this unique source:

(1) Narrow wave-length range
(2) High collimation
(3) Large radiant energy density (high intensity)

Referring to the last attribute first, the use of pulsed lasers for photo-coagulation in cases of retinal detachment is a case of a hazard being put to controlled use. Other energy based applications include the cutting of metals, glass, cloth and even the erasing of typewritten letters by the vaporisation of material. The absorption of the laser energy can be so immediate that no burning is caused although in the case of metal cutting some systems incorporate a jet of oxygen to cause combustion.

The high degree of collimation available particularly from gas lasers allows aerial straight-edges to be created. A laser beam shone into a tunnelling or road grading machine can be used to ensure a straight line of travel. In setting up optical bench experiments a laser can be very useful but stray reflections must be carefully avoided.

The narrow wave-length range permits lasers to be used as sources in various types of spectrophotometer (Chapter 13); of great value in the chemical and allied industries. Various laser systems have been proposed for monitoring pollution.

The major attribute of lasers is, of course, their coherence as described in section 12.11. The applications which use this are Optical Metrology and Holography, which, as mentioned above, are dealt with in subsequent chapters; Optical Communications and Optical Processing are largely beyond the scope of this book.

12.17 THE MEASUREMENT OF THE VELOCITY OF LIGHT–HISTORICAL METHODS

The deduction by Römer (section 12.1) in 1675 that light travelled at a finite velocity was based on his observations of Jupiter's moons when the earth was at opposite locations in its orbit so that the light was travelling over different distances. A second astronomical method was used by Bradley (1728) on stars located perpendicular to the earth's orbit. When these were viewed at six-monthly intervals the orbital velocity of the earth is in opposite directions so that the vector addition of the earth's velocity and the light velocity results in a slight angular displacement of a telescope aligned with the star. This was called the **aberration of light** although the word is now applied primarily to image defects.

Galileo had already tried a century earlier to measure the velocity by terrestrial observations. Two observers A and B, with lanterns, were stationed on hills some miles apart. A uncovered his lantern and B uncovered his as soon as he saw the light from A. A noted the time which elapsed between showing his light and seeing that of B, as this would be the time taken for the light to travel from A to B and back to A. Obviously such a crude method would not give a sensible result when the transit velocity is as fast as that of light. The same general principle, however, was used by Fizeau in 1849 in the arrangement shown in *Figure 12.12*.

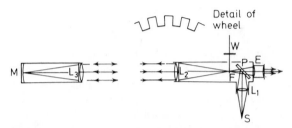

Figure 12.12 Measurement of the velocity of light. Fizeau's method

Light from the source S passes through a lens L_1, is reflected at a plane glass plate P and brought to a focus at F. It is then made parallel by the lens L_2 and travels a great distance to the reflector L_3M. This reflector consists of a concave mirror placed at the principle focus of the lens L_3 and having its centre of curvature at the optical centre of the lens. This arrangement ensures that the principle ray of any pencil reaching the lens is reflected back along its original path*. The reflected light again passes through L_2 and some of it passes through the plate P and is received by an eyepiece E and the eye. In the plane of the focus F is placed a toothed wheel W which can be rotated at a high speed, and is in such a position that light from S alternately passes through the space between two teeth or is cut off by a tooth as the wheel rotates.

Light is sent out when the space is coincident with the image F. Depending on the speed of the wheel the time taken for the light to go out to the reflector and back will be such that a tooth or a space is in the position F at the moment of its return. In Fizeau's original experiment the distance MF was 5.36 miles (8.63 kilometres). By steadily increasing the speed of the wheel which had 720 teeth the image seen in the eyepiece grew alternately brighter and darker and by measuring the number of revolutions of the wheel at the brightest and darkest conditions the velocity can be calculated.

*A similar arrangement is used in road signs and markings to reflect the light from vehicle lamps back towards the driver.

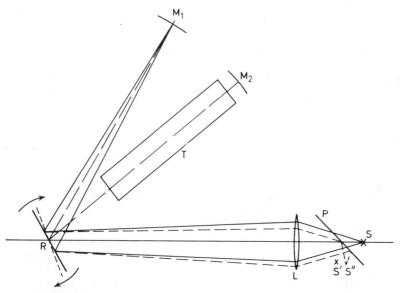

Figure 12.13 Measurement of the velocity of light: Foucault

Foucault, one time collaborator of Fizeau, published a different method in 1850 using a rotating mirror. The principle of this is shown in *Figure 12.13*.

In this method the light is sent out via a rotating mirror R. During the transit time of the light between RM, and back the mirror rotates so that the image of S is no longer at S' but shifted to S''. In Foucault's experiment RM was 20 m and the displacement S'S'' was 0.7 mm; the velocity was found to be 298 000 000 m/s. This apparatus was also used by Foucault to make the crucial determination as to whether light travelled faster or slower in a denser medium. For this the second mirror M2 (equidistant from R) and the tube T were added and two displaced images could be seen when the tube was filled with water. The image from the water beam was displaced more than the other showing that the rotating mirrors had had time to rotate further. Thus the velocity of light in water is less than that in air.

A more accurate experiment was done by Michelson in 1885 using white light when he found that the ratio of the velocity in air to that in water was 1.330. A further experiment using carbon disulphide gave a short spectrum at S'' showing that the light had different velocities for different wave-lengths. The ratio of velocities for yellow light was 1.75 while the refractive index of carbon disulphide is only 1.64. This discrepancy is dealt with in section 12.18.

12.18 THE MEASUREMENT OF THE VELOCITY OF LIGHT—MODERN METHODS

The measurements made with liquids give a refractive index different from that calculated by refraction measurements on prisms, etc. This is because in any medium except a vacuum light has two velocities. Returning to section 12.10 it was shown that most light is emitted in rather short trains of waves when an electron changes its energy. Light emitted in this way, although monochromatic, is far from the very narrow range of frequencies found when a laser is used to generate a continuous beam from a resonant cavity using simulated emission. It was shown by Fourier in the early part of the last century that the length of the wave train and the range of frequencies within it are mathematically linked together. A short wave train is formed by using a wide range of frequencies as discussed in section 15.12.

If the different frequencies have different velocities it can be shown that the velocity of the wave train or **group** of waves is considerably slower than the wave or **phase** velocity of the component wave-lengths. Only in a vacuum does the group velocity equal the phase velocity. In modern determinations of the velocity of light it is important to distinguish which velocity is being measured, although most measure the group velocity. When this is measured in air a correction must be made for the refractive index of the medium. Michelson devised a method using a more sophisticated rotating mirror system for which the light path was an evacuated pipe one mile long. By successive reflections at mirrors the total distance traversed amounted to 10 miles. The pipe could be evacuated to a pressure of 0.5 mm of mercury and values at various pressures were obtained. The rotating mirror had 32 sides and roated at such a speed that the subsequent face of the mirror replaced the first face during the transit time of the light. The mean of almost three thousand measurements was found to be 299 774 km/s ± 11 km/s.

Other methods involving Kerr-cells and ultrasonic waves have been used as well as radio waves. These latter have been used in radar systems and also in cavity resonators where the frequency as well as the wave-length can be measured.

CHAPTER 12   *EXERCISES*

1. What is meant by the wave-length of light, and how is it related to its frequency of vibration and velocity of propagation?

2. State the fundamental equation of wave-motion. Given the velocity of light in air as 3×10^8 metres per second, determine the wave-lengths for red, yellow and blue light, the frequencies being

  Red 395×10^{12} per second
  Yellow 509×10^{12} per second
  Blue 617×10^{12} per second

Give the wave-lengths in Angstrom Units, nanometres, microns and inches.

3. A series of waves of wave-length 100 cm is travelling across a pond on which three corks A, B and C are floating at a distance of 1.5, 2.25 and 3.8 metres respectively from a fixed post. In which direction will each of the corks be moving when a trough of a wave is passing the fixed post? State also whether each cork is above or below its normal position.

4. What is meant by a simple harmonic motion? Define the terms, period, amplitude and phase.

5. Plot the graph of the S.H.M. represented by

$$y = 3 \sin \frac{2\pi t}{8}$$

6. A particle is acted on simultaneously in the same straight line by two S.H.M. re-presented by

$$y = 3 \sin \frac{2\pi t}{6} \text{ and } y = 5 \sin \left(\frac{2\pi t}{8} - \frac{\pi}{2} \right)$$

Plot a graph showing the resultant motion.

7. Find graphically the resultant motion of a particle acted upon by two perpendicular S.H.M. of equal period and amplitude, and differing in phase by

$$(a)\ 0,\ (b)\ \frac{\pi}{4},\ (c)\ \frac{\pi}{2},\ (d)\ \frac{3\pi}{4},\ (e)\ \pi,\ (f)\ \frac{3\pi}{2}$$

8. Repeat the graphical constructions of question 7 for the case when one S.H.M. has twice the amplitude of the other.

9. A particle B executing a simple harmonic motion given by the expression $y = 8 \sin 6\pi t$ is sending out waves in a continuous medium travelling at 200 cm per second. Find the resultant displacement of a particle 150 cm from B, one second after the commencement of the vibration of B.

10. Explain the significance of the work of Römer, Foucault, Maxwell and Planck in the establishment of the theory of light.

11. List and describe the prime requirements for a system to emit coherent light.

12. Describe a laser and one application of it.

13. In using Fizeau's method of determining the velocity of light the disc has 200 teeth and revolves 2800 turns per minute to cut off the light when the mirror is 5 miles distant; find the velocity of light.

14. In measuring the velocity of light by Fizeau's methods the distance from the toothed wheel to the distant mirror is 6 kilometres. If the wheel has 750 teeth, what must be the least number of revolutions per second in order that the light returning from the mirror shall be cut off?

15. The maximum displacement of a very distant star from its mean position is found to be 20.44 seconds. If the orbital velocity of the earth be 30 557 metres per second, what is the velocity of light?

16. In astronomy stellar distances are measured in 'light years,' the distance which light travels in a year. The star Sirius is 8.55 light years away, what is its distance in miles?

17. Write a short account of the complete radiation spectrum extending from the long electrical waves at one end to the short gamma and cosmic rays at the other end. Indicate by name the various regions of radiation in their order on a diagram, inserting figures giving their approximate wave-lengths.

Why is the spectrum sometimes call the electro-magnetic spectrum?

Make a brief mention in your essay of the work of Hertz and Röntgen in developing our knowledge of radiation.

Chapter 13

Dispersion and colour
— optical materials

13.1 DISPERSION

From the observations described in the last chapter it is known that light of all colours travels with the same velocity *in vacuo*. As was stated in Chapter 2, this is no longer the case when light enters a denser medium. The light of each different colour or frequency then has a different velocity and in most cases the red light travels the fastest and the violet the slowest. This effect is always evident when the light passes obliquely through a refracting surface; the different colours are refracted different amounts according to their velocities, and are therefore separated or dispersed.

13.2 THE PRODUCTION OF A PURE SPECTRUM

Newton, in his original experiments on the spectrum, allowed light from the sun to pass through a small circular aperture into a darkened room, and received the small circular patch of light on a white screen. On interposing a prism this circular patch was broadened out into a band of colour, each of the colours, which together constitute white light, being deviated a different amount by the prism. Such a spectrum is, of course, very far from pure, since, as can be seen from *Figure 13.1*, the different coloured patches overlap considerably. If the screen is removed and the eye placed in the light leaving the prism, the aperture is seen as a spectrum in the position V'G'R'. This spectrum, while being reasonably pure if the aperture is quite small, will have very little width. Substituting a narrow slit parallel to the refracting edge of the prism for the circular aperture, a

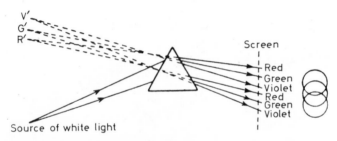

Figure 13.1 Dispersion by a prism

226

wide fairly pure spectrum will be seen on looking through the prism, the spectrum consisting of virtual images of the slit in each of the constituent colours.

Usually we shall require a spectrum that can be received on a screen or examined with an eyepiece, and in order that the spectrum shall be as pure as possible, that is, that there shall be little overlapping of the various colours, certain conditions are necessary. These are as follows:

(1) The light must pass through a narrow slit with its length parallel to the refracting edge of the prism.
(2) A well-defined real image of this slit must be formed by means of a positive lens.
(3) Each pencil of light must be parallel on passing through the prism, otherwise various parts of a pencil will be deviated by different amounts and a sharp image cannot be formed.

The arrangement shown in *Figure 13.2* satisfies these conditions. A narrow slit S, usually adjustable in width, is illuminated by light from the source of which the spectrum is required. The slit is situated at the first principal focus of a positive lens A_1 and slit and lens then form a collimator from which emerge

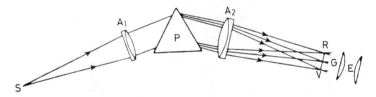

Figure 13.2 Optical arrangement of spectroscope

parallel pencils from each point on the slit. The light passes through the prism P where it is dispersed, each different colour being refracted in a different direction, and is then brought to a focus by a second achromatic lens A_2. A well defined spectrum is therefore formed in the second focal plane of the lens A_2, each different colour being a fine line, an image of the slit S. When the source is an incandescent solid, the spectrum will be a continuous band of colour consisting of an infinite number of these slit images. The image in the focal plane of A_2 can best be examined by means of an eyepiece E, and it will be seen that the lens A_2 and the eyepiece form an astronomical telescope. The complete instrument constitutes a **spectroscope**.

13.3 THE SPECTROMETER

We frequently require to measure the angle through which light of various colours is deviated by a prism, as, for example, in the measurement of refractive index (Chapter 4), and the instrument used for this purpose is the **spectrometer** (*Figure 13.3*). The arrangement of the instrument is shown in plan in *Figure 13.4*. Its essential parts are a collimator, an astronomical telescope, a table for carrying a prism and a divided circle. The collimator, with an adjustable vertical slit, is mounted with its axis directed towards the axis of the divided circle and

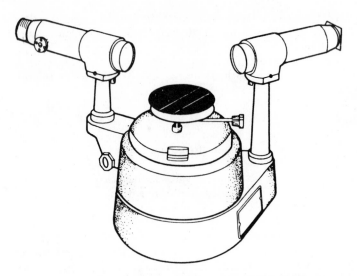

Figure 13.3 Modern spectrometer (courtesy P.T.I. Co. Ltd.)

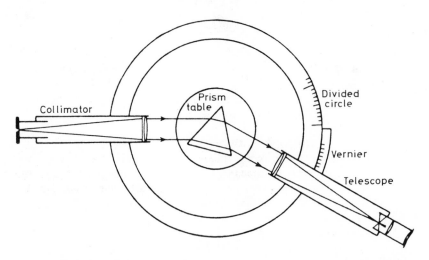

Figure 13.4 Plan of spectrometer

is usually fixed. The telescope, which has a Ramsden eyepiece with cross-lines at its focus, rotates in a horizontal plane about the axis of the circle and its position is read on the circle by means of a vernier. The prism table, which is provided with levelling screws, also rotates about the axis of the circle, and in some instruments its position can likewise be read on the circle by means of a vernier. In adjusting the spectrometer the telescope eyepiece is first focused on the cross-lines and the telescope then focused on a distant object. The slit of the collimator is illuminated and the telescope having been brought into line with the collimator the latter is adjusted by moving the slit in or out until a sharp image of the slit is formed in the plane of the cross-lines of the eyepiece.

13.4 MEASUREMENT OF THE REFRACTIVE INDEX WITH THE SPECTRO-METER

Measurement of the angle of a prism

The angle of a prism may be measured on the spectrometer by either of the following methods:

(1) The prism is placed on the table of the spectrometer with its refracting faces vertical and the angle to be measured facing the collimator (*Figure 13.5a*). A certain amount of the light is reflected from the two faces of the prism, and the angle between the two reflected beams is measured by reading the position of the telescope when an image of the slit is received on the cross-lines, first by reflection at the face AC and secondly at the face AE. It can easily be proved that the angle between these two reflected beams is equal to twice the angle of the prism.

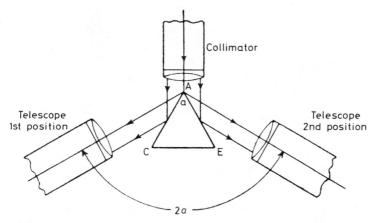

Figure 13.5a Measurement of angle of prism

(2) The telescope is fixed at an angle of roughly 90° with the collimator (*Figure 13.5b*). The prism is rotated until an image of the slit formed by reflection at the face AE is received on the cross-lines, and the position of the prism table is read on the circle. The prism table is then rotated until the image of

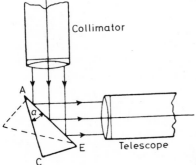

Figure 13.5b Measurement of angle of prism

the slit reflected from the face AC falls on the cross-lines, and the reading of the table again taken. The angle through which the table has turned will then be equal to $180° - a$, where a is the angle of the prism.

Measurement of minimum deviation

To determine the refractive index of the prism it will be necessary to measure the angle of minimum deviation; this may be done as follows: The collimator slit is illuminated by a source giving the light for which the refractive index is wanted and the telescope is brought into line with the collimator, so that an image of the slit falls on the cross-lines. The reading of the telescope in this position is taken. The prism, which will usually have a refracting angle of about $60°$, is then placed on the table with one refracting face making an angle of about $45°$ with the beam from the collimator, *Figure 13.6*. The telescope is turned until an image of the slit is seen. To find the position of minimum devia-

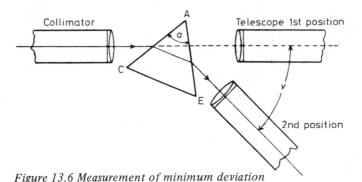

Figure 13.6 Measurement of minimum deviation

tion the prism is slowly rotated in the direction that causes the image to move towards the first position of the telescope, i.e. towards the undeviated direction. Following the image with the telescope a position is found where the image commences to move in the opposite direction and this is therefore the direction of minimum deviation. The image of the slit in this position is brought on to the cross-lines, and the reading of the telescope again taken. The difference in the readings gives the angle of minimum deviation v, and the refractive index is found from the expression (4.06a)

$$n = \frac{\sin\left(\dfrac{a+v}{2}\right)}{\sin\left(\dfrac{a}{2}\right)}$$

13.5 REFRACTOMETRY

Although the refractive index of a material may be found by grinding and polishing it into a prism shape and using it on a spectrometer to give an angle of minimum deviation the method is time consuming and expensive as a high degree of flatness is required on the incident and emergent faces for good accuracy. The methods described in section 4.14 are insufficiently accurate when the precise value of index is required at different wave-lengths so that achromatic lenses (section 13.11) may be designed and manufactured. For this purpose the Hilger-Chance V-block Refractometer was developed and a side view of this instrument is shown in *Figure 13.7*.

The V-shaped block B is made as indicated of two prisms of medium-index glass. The angle between the faces of the V is exactly 90 ° and the sample of material is ground and polished to about 90 °. The quality of the polish and the accuracy of the angle is not of great importance as a film of liquid is placed between the sample and the block to ensure optical continuity. Monochromatic light from the source S is collimated by the lens L and passes through the loaded V-block which forms an Amici prism system (section 13.14). The deviation of the emergent beam is measured by the horizontal telescope T which can be rotated as shown. This deviation depends on the refractive index of the sample and may be in either direction. In use the instrument is usually calibrated with samples of known materials.

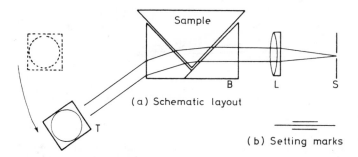

Figure 13.7 Hilger-Chance V-block Refractometer

The setting system received special attention during the design and utilises two short parallel lines in the eyepiece of the telescope. These can be set equidistant about the image of the slit seen in the telescope as in *Figure 13.7b*. The eye is very sensitive to any non-symmetry and maximum accuracy can be achieved. This instrument is commonly used for quality checking by optical and ophthalmic glass manufacturers.

13.6 WAVE-LENGTH, COLOUR AND TEMPERATURE

By methods described in Chapters 14 and 15, it is possible to measure experimentally the wave-length and hence, as the velocity is known, the frequency of any light. It is found that when light is dispersed to form a spectrum, each portion of the spectrum consists of light of a particular wave-length and therefore frequency. Thus in the case of a continuous spectrum of white light, there is a continuous range of wave-lengths extending, in the portion visible to the eye, from about 390 nm at the violet end to 760 nm at the red end. We may therefore specify any *spectrum colour* or any position in a spectrum in terms of the wave-length or frequency of the light dispersed to that position (Plate 6a).

The continuous spectrum of *white* light is usually obtained from an incandescent solid or liquid. The word incandescent means glowing with heat and the hotter the source the more it glows. As well as becoming more intense the source also changes colour from red through white to blue-white. When the intensity for each wave-length is plotted for a perfect (or black-body) source the curve varies with temperature as in *Figure 13.8.*

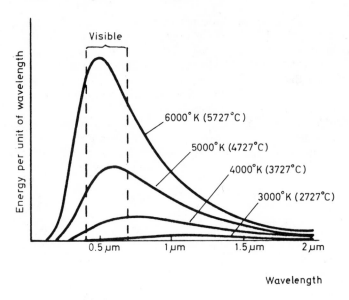

Figure 13.8 Black-body radiation

It is therefore possible to describe the red-white or blue-white colour seen in terms of the temperature and this is known as the colour-temperature usually given in degrees Kelvin. This need not be the same as the actual temperature for if a blue filter is placed in front of a tungsten lamp the red parts of its spectrum will be absorbed more than the blue and through the filter the colour-temperature will appear to be higher even though the lamp itself remains the same. Due to atmospheric absorption the colour-temperature of the sun is about 6000 °K even though its actual temperature is higher than this. The colour temperature of most lamps ranges from 2000 °K to 4000 °K. Standard lamps are manufactured to specific colour temperatures (Plate 6c).

Each colour temperature can be stated in terms of its 'mired' value (micro-reciprocal degree) which is given by 10^6 divided by the colour temperature in degrees Kelvin. The reciprocal nature of this unit allows the filters described above to be assigned a mired shift value, which may be positive or negative. Lamps of different colour temperatures will be 'shifted' by the same number of mireds by a given filter. This analysis is used in photography but care must be exercised as it assumes that the lamps are nearly black-body radiators. Some fluorescent lamps depart considerably and photographic results show these differences.

13.7 SPECTRA AND SPECTRAL ANALYSIS

There are two classes of spectra: *emission spectra* and *absorption spectra.* Within these classes there are three types: *continuous spectra, line spectra* and *band spectra.* Emission spectra are obtained when light coming from a source is examined by a dispersive instrument such as a spectroscope. Absorption spectra occur when light from a source having a continuous spectrum is passed through an absorbing material before entering the spectroscope. The electrons in the absorbing material absorb specific wave-lengths and then re-emit these by spontaneous emission. This occurs in all directions and so the beam passing through the material is depleted of these wave-lengths and dark areas or lines appear on the spectrum seen. Absorption spectra carry as much information about the structure of materials as emission spectra and as the sample does not have to be heated it is usually preferable to use this class of spectra in absorption spectrophotometers (section 13.8).

Continuous spectra

Solids and liquids tend to have continuous spectra in which every possible wave-length is emitted within a broad range without any sharp edges or details. Absorption spectra of liquids especially solutions are also broad but often show some structure particularly in the infra-red part of the spectrum. The absorption spectrum of an alcoholic solution of chlorophyll, the green colouring matter of plants, for example, shows a fairly narrow absorption band in the orange-red and another broad band extending from the blue to the violet end of the spectrum.

Line spectra

When a gas is made incandescent, the spectrum consists of bright lines on a dark background. These lines are, of course, images of the slit. Any gas, therefore, when made incandescent, emits radiations of certain definite wave-lengths, and each gas gives its own characteristic spectrum, although this may differ somewhat according to the means used to render the gas incandescent. Thus the spectrum of incandescent sodium vapour, which may be obtained by placing a small quantity of common salt in the flame of a bunsen burner, will be found to contain two bright yellow lines. These two lines are very close together, that is, they differ little in wave-length, and with a small spectroscope will appear as one.

The usual method of rendering a gas luminous, in order to examine its spectrum, is by means of a discharge tube. A glass or quartz tube is filled with the gas or vapour at low pressure. Two metal electrodes pass through the walls of the tube so that an electric current may be passed through the gas. This usually requires a high voltage to start with until the gas ionises. If the gas is hydrogen the tube will glow with a purplish coloured light and on passing this through a spectroscope the visible spectrum will be found to consist of four bright lines, the first red, the second greenish-blue, the third deep blue, and the fourth violet. These are shown in *Figure 12.9.* The other lines are outside the visible region.

The spectra of other gases and some of the more volatile metals, such as mercury and cadmium, can be obtained in the same way. Sometimes the electrodes used to pass the current through the gas can give rise to impurities and extra spectral lines. It is possible to excite the atoms of the gas by an electromagnetic field at radio frequencies. The gas is then sealed in a complete glass envelope. These are known as electrodeless lamps and are being increasingly used in spectroscopy.

It should be understood that with a given spectroscope the line corresponding to any wave-length will always occupy the same position; thus the yellow line of the sodium spectrum will fall in the same position as the corresponding yellow in the continuous spectrum. The same thing applies if the lower temperature discharge tube is inserted into a beam of white light falling on to a spectroscope slit when absorption lines are found in exactly the same position as the emission lines due to the absorption and re-emission of particular wave-lengths. A special case of absorption lines is considered in section 13.9.

Band or fluted spectra

In certain cases a spectrum will consist of broad luminous bands, which are sharply defined at one edge, and gradually fade away at the other giving a fluted appearance. When examined with a spectroscope giving great resolving power, it is found that these bands are composed of a great number of lines closely packed together towards the bright edge, and becoming more and more openly spaced towards the other. The band spectrum is characteristic of molecules and hence is produced by chemical compounds in the gaseous state. Line spectra are associated with atoms.

The line spectrum of hydrogen (*Figure 12.9*) can be seen to be a band spectrum except that it is spread out over a very wide range of wave-lengths. The progression shown is due to the very regular construction of the hydrogen atom.

13.8 SPECTROPHOTOMETERS

Because of the very great importance of spectral analysis to modern science, the design and manufacture of instruments to give accurate and repeatable results has also become important. These measure the *intensity* of each spectral line or band as well as its wave-length so the term **spectrophotometer** is used. Although many of these use a prism to disperse the light a larger number use a diffraction grating (Chapter 15) which has a more regular dispersing action. Interest in spectral lines extends well outside the visible spectrum and it becomes impossible to find materials for prisms and lenses. It is then necessary to construct a system using a reflecting diffraction grating and concave mirrors to collimate the light. In the extreme UV it is necessary to evacuate the instrument so that the atmosphere does not absorb the radiation.

Careful mechanical design is needed to rotate the prism or grating so that different wave-lengths are imaged on to an exit slit just in front of a photo-detector. It is commonly arranged that the same mechanical system drives a pen over a chart so that the x-axis of the pen position is related to wave-length.

The output of the photo-detector is used to drive the pen in the y-axis so that a graphical plot of the emission or absorption spectrum is recorded. Photo-detectors vary considerably in their response to different wave-lengths. This response may have to be carefully calibrated so that the pen records the true values. Alternatively, absorption instruments may be designed with two alternate beams, one of which passes through the sample while the other is direct. If the detector is used to measure intensities of these two beams alternately any differences due to detector response can be cancelled out. It is often necessary to switch in different detectors for different regions of the spectrum as some are satisfactory only in the UV to visible while others respond in the IR.

13.9 FRAUNHOFER LINES

The sun and most of the stars provide important examples of absorption spectra. The spectrum of sunlight is seen to be crossed by a great number of fine dark lines distributed throughout the length of the spectrum. These lines were first observed by Wollaston (1766–1828) in 1802 and were more fully studied by Fraunhofer (1787–1826); they are known as the **Fraunhofer lines**. They occupy definite constant positions in the spectrum and indicate that certain wave-lengths are absent in sunlight when it reaches the earth. Fraunhofer observed about 600 of these lines in the solar spectrum, and denoted the most prominent ones by letters from A in the extreme red to H in the violet; the wave-lengths corresponding to them are given in Appendix 2. Many more Fraunhofer lines have since been discovered not only in the visible, but also in the ultra-violet and infra-red portions of the solar spectrum.

The explanation of these lines was given by Kirchhoff who assumed the sun to consist of an incandescent solid or liquid core giving a continuous spectrum, surrounded by an envelope, the *reversal layer,* at a lower temperature, in which the various elements are in the form of gases. Most of the lines in the solar spectrum have been identified as due to elements also found on earth, while a few of the lines are due to absorption in the earth's atmosphere. The existence of the element helium was discovered in this way in the sun, before it was found on the earth.

The Fraunhofer lines are extremely useful in enabling us to specify particular colours or regions of the spectrum, for, as has been seen, each one occupies a perfectly definite and constant position corresponding to a particular colour or wave-length.

13.10 DISPERSIVE POWER

It has been shown in Chapter 2 that, for practically all transparent substances, the dispersion is such that the refractive index is greater the smaller the wave-length. The change in refractive index with change in wave-length will, however, differ with different media, and it is necessary that the dispersion of the various glasses and other media, of which lenses and prisms are to be made, shall be accurately known. The refractive indices of a glass are usually determined for the following colours:

The yellow line given by the sodium lamp, corresponding to the Fraunhofer line D.

The red and green-blue lines given by hydrogen, coinciding with the Fraunhofer lines C and F respectively.

In addition refractive indices are given by the glass manufacturer for the violet, G', line of the hydrogen spectrum and a number of other lines in the spectra of helium and mercury. These are the red, b, and the yellow, d, lines of the helium spectrum and the green, e, and the violet g and h lines of the mercury spectrum. The various indices will be denoted by n_D, n_F, n_C, n_d, etc.

The refractive index of a medium for light corresponding to the D line is known as the **mean refractive index** and is the value understood by the term unless otherwise stated. The value $n_F - n_C$ is termed the **mean dispersion** and the **partial dispersions** are given by $n_C - n_b$, $n_d - n_C$, $n_e - n_d$, etc.

The sodium D line has a broad structure making its use in measurements difficult. Optical glass companies adopted the helium d line but lately values have also been quoted for the mercury e line (546 nm) as being nearer the peak of the visual luminosity curve.

The mean dispersion of a glass is usually greater for glasses of high mean refractive index, but the increase in mean dispersion is not proportional to the increase in mean refractive index. This is illustrated by the following figures for the deviation of the C, D and F lines produced by two prisms, one of crown glass and the other of flint glass, the refracting angles being such that the minimum deviation for the D line is equal in each case.

	n_C	n_D	n_F
Hard Crown Glass	1.5150	1.5175	1.5235
Dense Flint Glass	1.6176	1.6225	1.6349

	v_C	v_D	v_F	$v_F - v_C$
Crown Prism: $a = 60°$	$38°\,31'$	$38°\,42'$	$39°\,14'$	$0°\,43'$
Flint Prism: $a = 52°\,6'$	$38°\,27'$	$38°\,42'$	$39°\,42'$	$1°\,15'$

It will be seen that while the mean deviation, v_D, produced by the two prisms is equal, the dispersion between the F and C lines, $v_F - v_C$, for the flint prism is almost double that for the crown. The effect is shown diagrammatically in *Figure 13.9,* the dispersion being exaggerated.

From expression (4.08) the deviation of a prism of small refracting angle for the D line is given by

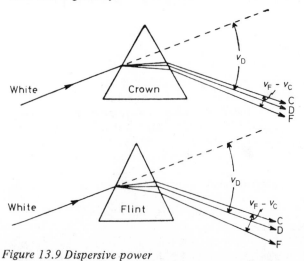

Figure 13.9 Dispersive power

$$v_D = (n_D - 1)a$$

Similarly

$$v_F = (n_F - 1)a$$

$$v_C = (n_C - 1)a$$

The dispersion between the F and C lines is therefore

$$v_F - v_C = (n_F - n_C)a = \frac{n_F - n_C}{n_D - 1} \cdot v_D$$

This ratio of the dispersion of the C and F lines to the mean deviation

$$\frac{v_F - v_C}{v_D} = \omega = \frac{n_F - n_C}{n_D - 1} \tag{13.01}$$

is called the **dispersive power** or **relative dispersion** of the medium.

In the calculation of achromatic lenses and prisms the reciprocal of dispersive power is frequently useful and is generally denoted by the symbol V and called the V-**value** or **constringence**.

$$V = \frac{n_D - 1}{n_F - n_C} \tag{13.02}$$

The **relative partial dispersions** give the ratios of the several partial dispersions to the mean dispersion, e.g. $(n_C - n_b)/(n_F - n_C)$, etc.

Lists are published by glass manufacturers giving these values for each type of glass manufactured, and as each melting will vary slightly from the type, exact figures for the actual melting are usually also supplied.

The two common types of optical glass are the crown glasses which will have values for n_D of a little over 1.5 and for V of about 60, and the flint glasses having n_D values of about 1.6 and V values of about 36. In addition to these many other types of glass are now made having special properties with respect to their dispersive powers and relative partial dispersions. A selection of varieties of optical glass is given in Appendix 3 (see also section 13.17).

13.11 THE ACHROMATIC LENS

We have seen in Chapter 6 that any thin lens will have chromatic aberration—light of different colours focusing in different places—and such a lens will not, therefore, produce a sharp image when light of different wave-lengths is coming from the object. It will be shown below that by combining two lenses made of different glasses it is possible to produce a combination that will focus light of two different colours at the same place; such a combination is said to be **achromatised** for these two colours. When the lens is to be used in conjunction with the eye, as for example the objective of a telescope or microscope, it will be achromatised for light corresponding to the C and F lines, but for photographic purposes it will usually be achromatised for light corresponding to the D and G′ lines; since the ordinary photographic plate is chiefly sensitive to the blue and violet end of the spectrum.

For a thin lens we have

$$F_D = (n_D - 1)(R_1 - R_2)$$

$$F_C = (n_C - 1)(R_1 - R_2)$$
$$F_F = (n_F - 1)(R_1 - R_2)$$

Therefore

$$F_F - F_C = (n_F - n_C)(R_1 - R_2) = \frac{n_F - n_C}{n_D - 1} \cdot F_D$$

or representing the mean power F_D by F,

$$F_F - F_C = \omega F = \frac{F}{V} \tag{13.03}$$

In an achromatic combination it will be necessary that the difference in power for the two colours in the one lens shall be neutralised by that in the other and as the power of two thin lenses *in contact* is equal to the sum of the separate powers, the condition to be satisfied is that

$$\frac{F_1}{V_1} = -\frac{F_2}{V_2} \tag{13.04}$$

From (13.04) it is obvious that the two lenses must be made from glasses having different V values as otherwise F_1 will equal $- F_2$ and the combination will have no power. Also it will be seen that for a positive combination the glass from which the positive lens is made must have the higher V value, and the positive lens will usually be of crown glass and the negative lens of flint.

Combining the expressions

$$F_1 + F_2 = F$$

and

$$\frac{F_1}{V_1} = -\frac{F_2}{V_2}$$

we have

$$F_1 = \frac{V_1}{V_1 - V_2} \cdot F \tag{13.05a}$$

and

$$F_2 = -\frac{V_2}{V_1 - V_2} \cdot F \tag{13.05b}$$

Having found the powers of the two lenses that when combined will give an achromatic combination, the curves can be chosen to correct or reduce other aberrations, especially the spherical aberration. It is usual, where possible, with lenses up to two or three inches diameter to arrange that the adjacent surfaces of the two lenses shall be of equal curvature, and the lenses are cemented together with optical cement. *Figures 13.10a* and *13.10b* show two usual forms of small telescope objectives; (*b*) is frequently used as the objective of the prism binocular, and when made of suitably chosen glasses, will give correction for oblique aberration over a small angle.

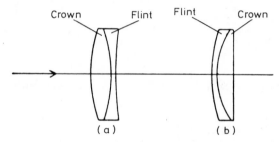

Figure 13.10 Forms of achromatic lenses

Example:
An achromatic lens of + 3D power is to be made from the glasses of which the constants are:

	n_D	$n_D - n_C$	V
Crown	1.5175	0.0085	60.9
Flint	1.6225	0.0173	36.0

From (13.03)

for crown lens

$$F_F - F_C = \frac{F_1}{V_1} = \frac{F_1}{60.9}$$

for flint lens

$$F_F - F_C = \frac{F_2}{V_2} = \frac{F_2}{36}$$

In order that the two lenses shall achromatise when in contact

$$\frac{F_1}{60.9} = -\frac{F_2}{36}$$

or

$$F_2 = -\frac{36}{60.9} F_1$$

Substituting in

$$F_1 + F_2 = F$$

we have

$$F_1 - \frac{36}{60.9} F_1 = 3$$

Therefore

$$F_1 = +7.34 \text{ D}$$

$$F_2 = -4.34 \text{ D}$$

The crown lens is to be equi-convex and cemented to the flint lens, then $R_1 = -R_2 = -R_3$

$$R_1 - R_2 = \frac{F_1}{n_{D_1} - 1} = \frac{7.34}{0.5175} = +14.18 \text{ D}$$

$$R_3 - R_4 = \frac{F_2}{n_{D_2} - 1} = \frac{-4.34}{0.6225} = -6.97 \text{ D}$$

Thence we have

$R_1 = +7.09\text{D}$	$r_1 = +\ 14.1$ mm
$R_2 = -7.09\text{D}$	$r_2 = -\ 14.1$ mm
$R_2 = -7.09\text{D}$	$r_2 = -\ 14\ 1$ mm
$R_2 = -0.12\text{D}$	$r_2 = -833\ 3$ mm

Newton maintained that the dispersion of any medium was proportional to its refractivity, that is, that the dispersive powers of all media were equal and therefore the construction of an achromatic lens was impossible. For this reason he and his followers turned their attention to perfecting the reflecting telescope. Later it was argued, notably by Euler (1707–1783) in 1747, that the achromatic lens was a possibility since the eye was achromatic. That this is not the case can be very easily proved by looking at a bright object whilst half covering the pupil of the eye with a card, and it is now known that the eye has considerable chromatic aberration. In 1733 Chester Moor-Hall (1703–1771) had succeeded in producing two or three telescope objectives which gave images fairly free from colour defects, but it was not until 1757 that, through the work of John Dollond (1706–1761), the regular construction of these achromatic lenses was commenced.

13.12 CHROMATIC DIFFERENCE OF MAGNIFICATION, LATERAL COLOUR

Unless the achromatic combination is quite thin, as was assumed above, it is not possible to get both the principal foci and the principal points respectively coinciding for the different colours. In the case of the telescope objective it is necessary that the foci for different colours shall coincide, that is, the back focusing distances shall be equal. In eyepieces, however, it is necessary that the images in the different colours shall be the same size and hence the equivalent focal lengths must be equal. If this be not so, images seen through the eyepiece will be fringed with colour, the defect increasing towards the edge of the field. When this second condition is fulfilled the system is said to be corrected for **chromatic difference of magnification** or **lateral chromatic aberration**.

The Huygens eyepiece is an example of a system where this condition is satisfied by the use of two lenses of the same glass, the blue and red rays from any point on the image emerging parallel, and the images in these two colours therefore subtending the same angle.

For two thin separated lenses in air we have

(1) $F = F_1 + F_2 - dF_1F_2$

putting

$F_F - F_C = \delta F$

(2) $F + \delta F = (F_1 + \delta F_1) + (F_2 + \delta F_2) - d(F_1 + \delta F_1)(F_2 + \delta F_2)$

where d is the distance separating the lenses.

Subtracting equation (1) from equation (2) and neglecting the term involving the product of the small quantities δF_1 and δF_2 we have

$\delta F = \delta F_1 + \delta F_2 - (F_2 \delta F_1 + F_1 \delta F_2)d$

In order that

$\delta F = 0$

$$d = \frac{\delta F_1 + \delta F_2}{F_2 \delta F_1 + F_1 \delta F_2}$$

now

$$\delta F_1 = \frac{F_1}{V_1}, \quad \delta F_2 = \frac{F_2}{V_2}$$

$$d = \frac{V_2 F_1 + V_1 F_2}{(V_1 + V_2)F_1 F_2} = \frac{V_1 f'_1 + V_2 f'_2}{V_1 + V_2}$$

If the two lenses are made of the same glass, $V_1 = V_2$. In order that there shall be no difference between F_C and F_F we have

$$d = \frac{f'_1 + f'_2}{2}$$

as the condition for achromatism of magnification.

13.13 THE ACHROMATIC PRISM

For certain purposes it is sometimes required to use a combination of prisms to produce a deviation of the light without, as far as possible, any dispersion. Such an **achromatic prism** may be formed by combining two prisms giving equal dispersions with the base of one against the apex of the other. Rays of light of the two colours, for which the prism is achromatised, will then emerge in parallel directions as shown in *Figure 13.11*.

If we consider prisms of small angle for which the approximate expression (4.08) may be used, we have

$v_D = (n_D - 1)a$

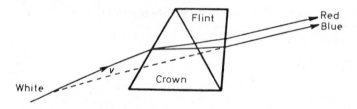

Figure 13.11 Achromatic prism

and the dispersion

$$v_F - v_C = (n_F - n_C)a$$

In order that two prisms when combined shall achromatise, $v_F - v_C$ for the two prisms must be equal.

Therefore

$$(n_{F1} - n_{C1})a_1 = (n_{F2} - n_{C2})a_2 \qquad (13.06)$$

If the two prisms be made of different glasses, the mean deviations v_D will be unequal when expression (13.06) is satisfied and their difference is the resultant mean deviation.

Example:

A prism of $15°$ refracting angle is made of the crown glass, particulars of which are given in section 13.11. What prism of the flint glass will be required to give achromatism for the C and F lines, and what will be the resulting mean deviation?

For the crown prism the dispersion

$$v_{F1} - v_{C1} = (n_{F1} - n_{C1})a_1 = 0.0085 \times 15°$$

For the flint prism the dispersion

$$v_{F2} - v_{C2} = (n_{F2} - n_{C2})a_2 = 0.0173\, a_2$$

and this must be made equal to $v_{F1} - v_{C1}$

Therefore

$$0.0173a_2 = 0.0085 \times 15°$$

$$a_2 = 7.37°$$

Then

$$v_{D1} = 0.5175 \times 15 = 7.77°$$

and

$$v_{D2} = 0.6225 \times 7.37 = 4.59°$$

Therefore, resultant mean deviation, $v_D = 3.18°$.

13.14 THE DIRECT VISION PRISM—AMICI PRISM

If the refracting angles of two prisms made of different glasses be such that the mean deviations they produce are equal, the two prisms when placed base to apex will produce no deviation on light corresponding to the D line, but the other colours will be dispersed. Thus, with the two prisms for which the deviations are given in section 13.10 combined in this way, the D line will be undeviated, while the C line is deviated 4' to one side, and the F line 28' to the other side of the D line. Such a combination forms a **direct vision prism**, and is frequently used in the spectroscope, particularly in the small pocket instrument.

To obtain sufficient dispersion the direct vision prism is usually composed of three or five prisms, the usual form being that shown in *Figure 13.12*. This was first made by Amici (1786-1863) in 1860, and consists of a prism of very dense flint glass cemented between two prisms of crown glass. The glasses should be chosen to have the greatest possible difference in their V values. In the construction of some D.V. prisms liquids of high dispersive power, such as carbon disulphide and cinnamic ether, are used.

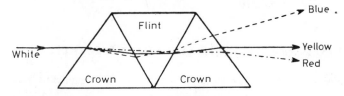

Figure 13.12 Amici direct vision prism

13.15 THE CONSTANT DEVIATION PRISM

A very convenient form of spectroscope can be constructed by making use of a **constant deviation prism**. This may take the form shown in *Figure 13.13* and may be considered as being composed of two 30° prisms ABE and BCD and a right-angled reflecting prism AED, although in practice it will be made in one piece. For a certain angle of incidence i_1 the ray refracted at AB is normal to AE and is totally reflected at the surface AD undergoing a deviation of 90°. As the angle ABC = 90° the angle of incidence at BC is equal to the angle of refraction at AB. Therefore $i'_3 = i_1$ and the emergent ray is at right angles to the incident ray. The prisms ABE and BCD may be considered as two halves of a 60° prism and when the angle of incidence i_1 at the surface AB is equal to the angle of refraction i'_3 at the surface BC the deviation is a minimum. Thus any ray that is deviated through 90° by the constant deviation prism is passing through at minimum deviation. The collimator and telescope of the spectroscope are permanently fixed at right angles to one another and the prism is rotated about a vertical axis. As the angle of incidence of the light from the collimator changes the spectrum moves across the field of the telescope and light of any wavelength is at minimum deviation when it falls in the centre of the telescope field. The prism can be rotated by means of a screw attached to which is a divided drum, which gives the wave-length for the particular portion of the spectrum falling on the cross-lines.

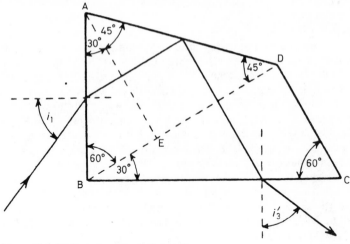

Figure 13.13 Constant deviation prism

13.16 IRRATIONALITY OF DISPERSION. SECONDARY SPECTRUM

An achromatic lens composed of two different glasses can be designed to focus any *two* colours in the same place, but all other colours will focus in slightly different positions. The reason for this small residual colour defect, which is known as the **secondary spectrum**, will be seen if we examine the dispersions

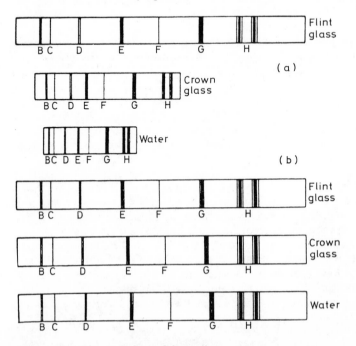

Figure 13.14 Irrationality of dispersion

of various media between different parts of the spectrum. Thus, while the mean dispersion is higher in flint glass than in crown, the dispersion towards the red end of the spectrum is relatively higher in crown glass than in flint and relatively lower at the blue end as will be seen from the figures for the relative partial dispersions, Appendix 3. These relative differences in the dispersions in different parts of the spectrum are still more marked if the dispersion of water is compared with that of flint glass.

Figure 13.14a shows the relative lengths of the spectra formed at the same distance from prisms of equal refracting angle made of water, crown glass and flint glass. In *Figure 13.14b* the spectra of the crown glass and water prisms have been made the same length as that of the flint glass prism between the B and H lines; this could be done either by suitably choosing the refracting angles, or by receiving the spectra at different distances. It will be seen that the other lines do not coincide. This effect is known as **irrationality of dispersion**.

The amount by which the power or focal length for any colour differs from the combined power or focal length for the C and F lines for which the lens has been achromatised may be found as follows:

For the D and F lines we have

$$F_D = (n_{D1} - 1)(R_1 - R_2) + (n_{D2} - 1)(R_3 - R_4)$$

$$F_F = (n_{F1} - 1)(R_1 - R_2) + (n_{F2} - 1)(R_3 - R_4)$$

$$F_F - F_D = (n_{F1} - n_{D1})(R_1 - R_2) + (n_{F2} - n_{D2})(R_3 - R_4)$$

$$= \frac{(n_{F1} - n_{D1})F_{D1}}{n_{D1} - 1} + \frac{(n_{F2} - n_{D2})F_{D2}}{n_{D2} - 1}$$

From (13.05a and b)

$$F_F - F_D = \left(\frac{n_{F1} - n_{D1}}{n_{D1} - 1} \cdot \frac{V_1}{V_1 - V_2} \cdot F_D \right) - \left(\frac{n_{F2} - n_{D2}}{n_{D2} - 1} \cdot \frac{V_2}{V_1 - V_2} \cdot F_D \right)$$

$$\frac{F_F - F_D}{F_D} = \left(\frac{n_{F1} - n_{D1}}{n_{D1} - 1} \cdot \frac{V_1}{V_1 - V_2} \right) - \left(\frac{n_{F2} - n_{D2}}{n_{D2} - 1} \cdot \frac{V_2}{V_1 - V_2} \right)$$

$$= \left(\frac{n_{F1} - n_{D1}}{n_{F1} - n_{C1}} \cdot \frac{1}{V_1 - V_2} \right) - \left(\frac{n_{F2} - n_{D2}}{n_{F2} - n_{C2}} \cdot \frac{1}{V_1 - V_2} \right)$$

Calling the relative partial dispersions

$$\frac{n_D - n_C}{n_F - n_C}, \frac{n_F - n_D}{n_F - n_C}$$

and

$$\frac{n_{G}' - n_F}{n_F - n_C}$$

a, b and *c* respectively, we have

$$\frac{F_F - F_D}{F_D} = \frac{f'_D - f'_F}{f'_F} = \frac{b_1 - b_2}{V_1 - V_2}$$

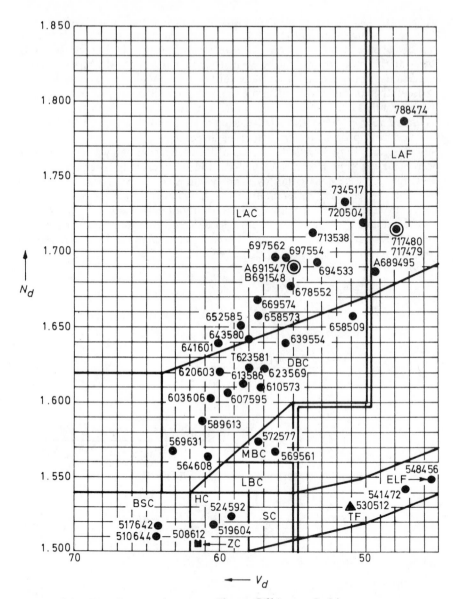

Figure 13.15 Glass Diagram (courtesy Chance-Pilkington Ltd.)

Similar expressions may be obtained for the other spectrum lines. Thus it is seen to reduce the secondary spectrum glasses should be chosen having their respective relative partial dispersions as nearly equal as possible or having a large difference in their v values.

Lenses in which the secondary spectrum has been practically eliminated are called **apochromatic** and form the highest class of microscope and telescope objectives.

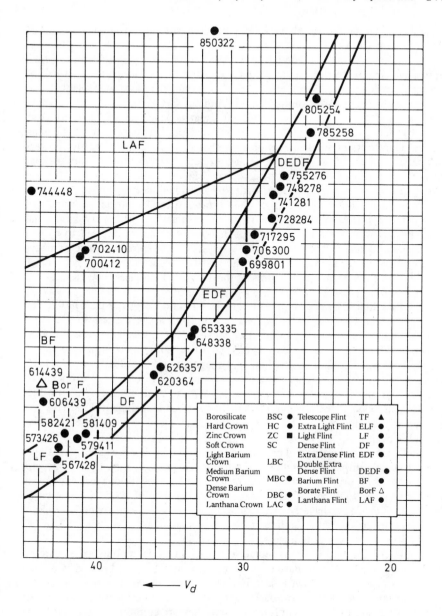

13.17 TYPES OF OPTICAL GLASS

In designing a lens system to correct as far as possible the various aberrations considered in Chapter 18, advantage has to be taken of all the possible variables. These are the curvatures of the surfaces, the thicknesses and separations of the component lenses and the refractive indices and dispersive values of the various media. It is therefore obviously an advantage to have available a variety of glasses with different characteristics. The elimination of the secondary spectrum is an example of the need for glasses with characteristics differing from those of ordinary crown and flint.

At the time of the introduction of the achromatic lens, the only optical glasses available were the ordinary crown and flint types. During the first half of the nineteenth century systematic efforts to extend the range of available types of glass were made by Fraunhofer and Guinand in Germany and by Faraday, Harcourt and Stokes in England, but the great advance in this direction was made by Schott and Abbe, who in 1885 issued the first catalogue of the Jena glasses. This contained particulars of a number of glasses, the dispersive powers and partial dispersions of which differed considerably from those of earlier types of crown and flint. Other types of glass have since been produced and all these glasses and many others are now made by any manufacturer of optical glass.

Figure 13.15 shows a diagram of glass types plotted by refractive index and reciprocal dispersion (V-value). The double line somewhat arbitrarily divides the types into crowns and flints. Each area shown is commonly referred to by the descriptions indicated but for specific glass types six-figure reference numbers are also used. These are derived from the first three significant figures of n_d-1 and the first three significant figures of the V-value. Thus glass type HC 524592 has a refractive index of 1.52400 and a V-value of 59.21.

This glass, like all glasses, requires a glass former and a series of glass modifying materials which control the characteristics of the glass. Hard crown, 524592, may be taken as a starting point. For most of the optical diagram the glass former is silica and in the case of hard crown the modifiers comprise oxides of calcium, sodium, potassium, barium and magnesium. In order to move to the lower left part of the diagram, part of the silica is replaced by phosphorus or fluorine. To move to the right, that is from crown to flint, requires the introduction of lead oxide at the expense of some of the silica and alkali oxides. This was a very early change as it gave a glass composition which would melt more easily, the index change being a by-product.

Lead is a heavy element and increasing amounts of lead oxide gave increasingly heavier glasses with higher indices. The terms, dense flint, extra dense flint, etc, mean what they say. With about 50% or more of the constituent materials being lead oxide the EDF and DEDF types have densities of 4 to 5 grams cm^{-3}. Recently flint glasses have been introduced in which the lead oxide is replaced by titanium oxide with significant reductions in density to below 3 grams cm^{-3}—little more than crown glass.

If some of the lead oxide is replaced by barium oxide, high index is achieved without higher dispersion. Barium oxide can also be introduced into crown (lead free) glass types to give high index. For even further increase in index, elements from the rare earth series have been added to the barium crowns and flints. As the principal rare earth element is lanthanum, they are commonly called lanthana crowns and lanthana flints. Such a two-dimensional plot of glass types does not cover all the features. For instance, the addition of boron to flint glass gives a higher index but with a relative partial dispersion increased in the red. A table of glass types is given in Appendix 3.

13.18 OTHER OPTICAL MATERIALS

As an alternative to glass, organic polymers (plastics) have advantages and disadvantages. Although they can be formed into lenses by moulding, which is cheaper than grinding and polishing, the accuracy of curve is not as good.

Although having about half the weight of the equivalent glass lens, their resistance to scratching is usually lower. The homogeneity attainable with plastics is considerably poorer than that with optical glass and the choice of index and V-value is restricted to two locations for readily obtainable plastics:

Polycarbonate	1.58	29.5
Polystyrene	1.59	30.5
Polymethyl methacrylate	1.49	57.5 (Perspex)

Apart from spectacle lenses, optical plastics are used in cheap cameras and binoculars and also in some display systems where weight is important. The limitation to two main types is a very serious restriction on lens design and the extra elements required to give equal performance often offset the weight and cost advantage.

Most glasses have very high transmission between 450 nm and 1000 nm although some of the special types tend to absorb in the blue. For work outside these wave-lengths, in the ultra-violet and infra-red, other materials have been used—crystals of calcium fluoride, lithium fluoride and quartz are commonly used in the ultra-violet. While for the infra-red oxide glasses such as calcium aluminate are used for wave-lengths up to 5 μm. In the so-called thermal region (8–13 μm) germanium and silicon are used as well as the polycrystalline materials in the Kodak 'Irtran' range.

13.19 ANOMALOUS DISPERSION

For most substances the refractive index increases as the wave-length decreases, the increase becoming more rapid with the shorter wave-lengths. In some substances, such as cyanine and fuchsine which are strongly coloured, the refractive index for some of the longer wave-lengths is greater than that for some of the shorter. If a prism is made by pressing some fused cyanine between two glasses inclined at a very small angle, the spectrum of white light seen through it will be found to have the colours in the order green, blue, red and orange, the green being the least deviated. The yellow is completely absorbed. This effect is known as **anomalous dispersion**. As cyanine absorbs a very great deal of the light, only a small amount of light will be transmitted, even at the edge of a prism of very small angle.

Substances giving this so-called anomalous dispersion have a strong absorption band in the visible spectrum. On the blue side of this absorption band there is an abnormal drop in the refractive index which gradually rises again towards the blue end of the spectrum. On the red side of the absorption band there is an abnormal increase of refractive index on approaching the band. Thus with cyanine, which gives an absorption band in the yellow, the refractive indices will have increasing values in the following orders of colours, green, blue, red and orange, as is seen in the spectrum with the cyanine prism. As with ordinary transparent substances such as glass there is a well marked absorption band in the ultra-violet and another in the infra-red, it would appear that the dispersion of such substances is of the same general form as that of substances having an absorption band in the visible spectrum. That is, the refractive index rapidly decreases for wave-lengths slightly less than those for the absorption band in the infra-red and rapidly increases for wave-lengths slightly greater than those for the absorption band in the ultra-violet. It is therefore supposed that there is no real anomaly in the so-called anomalous dispersion, and the general nature of dispersion has been explained by considering the interaction of the particles of the medium and the light vibrations passing through it.

13.20 PHOTOCHROMIC GLASS

This material has the property of darkening to a sunglass tint in strong sunlight and fading to almost clear in dusk and dark conditions, rather like a reversible photograph. The photographic process works because crystals of silver halide are soluble in developers while silver is not. The action of exposing the photosensitive material is that of breaking the molecular bond between the silver and the chlorine, bromine or iodine. The latter diffuse away leaving the silver to record the image. Silver halides are transparent substances while silver is not.

The same principle is used in photochromic glass except that the silver and halide when separated by the activating light only move apart by a very short distance and recombine shortly afterwards. The time of recombination depends on temperature—at low temperatures there is little thermal motion to bring the ions into contact. At too great a temperature (above 400 °C) the ions diffuse into the glass to give permanent darkening. At normal temperatures a state of dynamic equilibrium is achieved when the rate of separation is equal to the rate of recombination.

The activating light is more effective when in the UV and blue end of spectrum. Usually, a mixture of halides is used—bromide and iodide being sensitive to yellow light as well as blue and UV.

Although this UV and blue light needs to be absorbed to drive the reaction, the absorption spectra of photochromic glass, that is, its appearance as a filter or tint is a different spectral curve usually with a peak at 500 nm giving a grey to brown appearance. This is affected by the size of the silver halide crystals used. If these are too small, below 5 nm, light of 500 nm wavelength is not affected. If they are above about 500 nm the crystalites scatter the light and the glass becomes opalescent. Choice of crystal types and size and concentration, glass type and sensitizing dopants such as copper are all factors influencing the final product.

13.21 PHOTOCHROMIC SPECTACLES

The most important parameters are the working colour, density and speed of response both to darkening and to fading. *Figure 13.16a* shows the transmission curves for the Reactolite Rapide* when clear and when exposed to bright sunlight as found at latitudes of about 60°. This particular material has a very fast reaction time and *Figure 13.16b* shows the changes in transmittance at 550nm when the activating light is suddenly increased from zero to full sunlight and then back to zero.

As indicated on these graphs, these figures apply to a specific thickness and temperature. If the thickness is increased the change in transmittance in the fully exposed condition does not follow the density rule (section 11.13) as the dark-

*Reactolite Rapide is a Trade Mark of Pilkington Bros. Ltd.

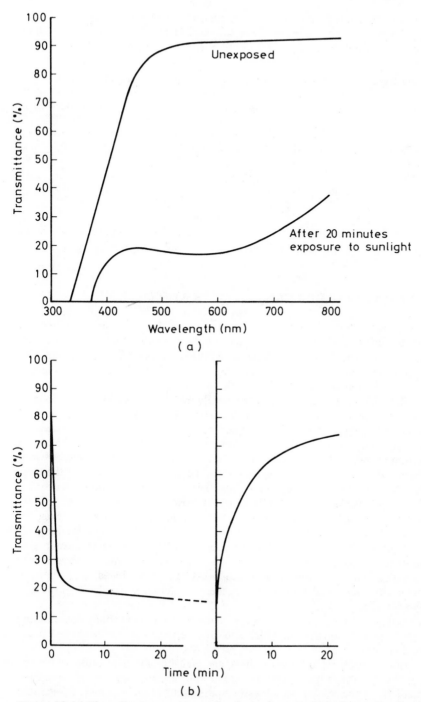

Figure 13.16 The salient parameters of a photochromic glass–Reactolite Rapide. These curves relate to glass 2 mm thick at 25 °C (a) Spectral transmittance (b) Darkening and fading

ening near the front of the lens reduces the activating light intensity—and hence the darkening—in the rear portions of the lens. The table and the graphs allow 8% loss in transmittance due to reflections at the surfaces.

Transmittance values for various thicknesses of Reactolite Rapide at 25 °C

Thickness	Fully faded	Fully darkened
1 mm	90%	28%
2 mm	88%	16%
3 mm	86%	10.5%
4 mm	84%	8%
5 mm	83.5%	6%

The effect of temperature was indicated in section 13.20. In the table above the fully exposed transmittance of 2 mm material would change to 11% at 15 °C and 20% at 30 °C.

13.22 THE ULTRA-VIOLET AND ITS APPLICATIONS

Vision is dependent only on radiations having wave-lengths from about 760 nm to about 390 nm. Most sources of light, however, emit a much larger range of wave-lengths than this, comprising in addition to the visible spectrum, the shorter wave-length ultra-violet radiations at one end and the longer wave-length infra-red radiations at the other end of the visible spectrum.

Ultra-violet light may be detected by fluorescent materials and by photography. The latter was used by Becquerel (1842) to show the existence of Fraunhofer lines in the ultra-violet region of the solar spectrum. Fluorescence is still used today. Phosphors such as sodium salicylate re-radiate in the visible spectrum when excited by ultra-violet light and in some cases by x-rays.

Nowadays a large range of photo-electric detectors are available in the form of photo-multipliers or image tubes. As the energy of ultra-violet photons is higher than visible photons it is generally easier to eject electrons (photo-emission) from materials than in the visible region. Photo-detectors exist for wave-lengths down to 50 nm (500 Å). The shortest wave-length reaching the earth from the sun through a clear atmosphere is about 300 nm as wave-lengths shorter than this are absorbed by the air. For experiments at wave-lengths shorter than about 200 nm the apparatus must be evacuated leading to the term, **vacuum ultra-violet** which distinguishes these shorter wave-lengths from the **near ultra-violet**.

Ordinary glass is opaque to wave-lengths shorter than about 330 nm and materials such as quartz, sapphire and, for very short wave-lengths, lithium fluoride must be used.

Various sources of ultra-violet light are available although ordinary electric lamps emit ultra-violet radiations to some extent. For more intensity, particularly at the shorter wave-lengths, special sources are required. These radiations are chiefly emitted by sources in which the temperature is extremely high or in which the atoms are subject to intense electrical excitation. Xenon and mercury arcs are commonly used.

For a material to be affected in any way by radiations falling on it, these radiations must be absorbed by the material; radiations reflected or transmitted by the material will produce no effect.

Ultra-violet radiations, particularly those of shorter wave-length, are completely absorbed by living tissues, even in extremely thin layers, and photochemical or *abiotic* changes are produced. The extent of these changes increases as the wave-length decreases and is also dependent on the intensity of the radiation and the time of exposure.

The abiotic effects caused by the absorption of ultra-violet may be sufficient to destroy minute organisms, such as bacteria; and water, milk, etc., may be sterilised by exposing them to these radiations. In the case of large animals the effects will be confined to the surface owing to the rapid absorption of the ultra-violet by the tissues, and these effects may be beneficial or harmful according to the extent of change produced.

The health-giving properties of sunlight are well known, and the outward signs of sunburn and bronzing of the skin have been found to be caused almost entirely under natural conditions by the longer wave ultra-violet between 320 nm and 290 nm, the limit of the solar radiation reaching the earth. Extensive use is being made of the shorter wave ultra-violet produced by artificial sources, in the treatment of disease conditions, both general and local. In the latter the action is such as to cause slight damage to the tissues sufficient to set into being or to stimulate the processes of repair.

Much has been written regarding the action of ultra-violet on the eye. It can generally be accepted, that no deleterious effects are caused on the eye by exposure to the radiations present in sunlight to which the eye is adapted. In special circumstances, e.g., on snowfields particularly at high altitudes, radiations of shorter wave-length and of greater intensity than those to which the eye is normally accustomed may be present, and the wearing of protective glasses which absorb these radiations will be necessary. The shorter wave radiations produced by artificial sources, such as the arcs previously mentioned, are, however, readily absorbed by the eye tissues, and damage to these tissues, usually of a temporary nature, is produced.

Various chemical effects, for example the bleaching of dyes and the hardening of paints and varnishes, which are dependent on light action, have been found to be chiefly affected by the ultra-violet portion of the solar spectrum, and the commercial testing of these articles is considerably hastened by exposing them to a source rich in ultra-violet radiations. Another commercial application that has become of considerable importance during recent years is the exposure of certain foodstuffs to ultra-violet sources with the object of increasing their vitamin content.

Ultra-violet has been successfully applied to the microscope, the short wavelength in this case leading to considerable increase in resolving power. (See section 15.11.)

13.23 THE INFRA-RED AND ITS APPLICATIONS

The infra-red radiations are characterised by their heating effect, and their presence may therefore be detected by any sensitive heat measuring device. It is obvious that heat is radiated from any source of light, but the discovery that these heat radiations form a continuation of the visible spectrum was first made by Sir William Herschel (1738–1822) in 1800. Using a delicate thermometer with blackened bulb in order to find the heating effect of different colours of the spectrum, he found that while the temperature rose towards the red end of the spectrum, the maximum was reached in the region beyond the visible red end. In 1840 Sir John Herschel (1792–1871) succeeded in showing the existence of Fraunhofer lines in the infra-red region of the spectrum.

The detection method used was the drying of lampblack wet with alcohol. This measures the heating effect of the radiation and the **bolometer** uses the same effect to alter the resistance of platinum wire. If two wires of different materials are joined together a voltage difference can be generated by heating the junction and this forms the basis of the **thermopile**. When a gas is heated its volume or pressure must change and this is the principle of the **Golay cell**, which uses optical interference methods to achieve high sensitivity. All these detectors are known as **thermal detectors** as they depend on the heating effect of the absorbed radiation. They have a generally slow response of the order of milliseconds and a wide spectral range.

The transmission of the infra-red part of the electro-magnetic spectrum by the atmosphere shows considerable variation with wave-length. The region from visible red to 5 μm is generally transparent but a region three micrometres wide is heavily absorbed before reasonable transmission is again obtained between 8 μm and 14 μm. These transparent regions are known as windows and are commonly referred to as the near infra-red, and the far infra-red although the latter can be used for wave-lengths up to one millimetre.

Because the energy of individual photons decreases with increasing wave-length the ability to eject electrons (photo-emission) reduces and photo-emissive materials are difficult to obtain for wave-lengths below about 1.2 μm. However, semi-conductor materials do have the ability to absorb photons with a resultant change in conductivity (photo-conductivity) when their wave-length is as low as 20–30 μm. These **photo** or **quantum detectors** have a much faster response (typically microseconds) but they also generate a lot of electrical noise due to random changes in the electron energies. In practice they are often cooled by liquid nitrogen or helium to reduce this noise.

Ordinary incandescent lamps radiate far more in the infra-red than in the visible regions. At normal temperatures every object has appreciable radiation in the thermal infra-red region. The human body, car exhausts and tyres can be easily detected as can any object at a different temperature (or contrast) with its surroundings.

Absorption of infra-red radiation produces a rise in temperature of the absorbing material; the radiant heat or infra-red from the sun and other heat sources is utilised in this way.

Excessive exposure of living tissue to infra-red radiations will cause damage to the tissue in the form of burns or *thermal lesions,* and this damage will often be more deep-seated than the abiotic changes produced by ultra-violet, because of the greater penetrability of the longer wave-length. Radiant heat is used in the treatment of disease conditions existing in positions too deeply seated to be reached with ultra-violet.

It has been found that serious and permanent damage may be caused to the eye by excessive exposure to the short-wave infra-red, such as may occur in certain industrial processes (glass and furnace work, welding, etc.) and the eyes of workers in these processes must be adequately protected from these radiations. Glass containing iron in the ferrous state very largely absorbs the short-wave infra-red.

On the other hand, body tumours have a slightly higher temperature than normal cells and so **thermography** can be used, under controlled conditions, to detect their existence.

13.24 COLOUR

It is convenient to refer to light of any particular frequency or wave-length in terms of the colour sensation to which it gives rise when received by the normal eye, but it should be understood that colour is a purely visual sensation. Light, which we call white, consists of a continuous band of frequencies, as is seen from its spectrum, and this complex vibration gives rise to the sensation of white when received by the eye. Portions of this band of frequencies or a predominance of any portion of it give rise to the sensation of colour as distinct from white. Thus light of wave-lengths from about 760 nm to about 620 nm gives the sensation of red, from about 580 nm to about 510 nm the sensation of green, and so on.

That the type of sensation is not specifically determined by the wave-length of the radiation can be shown in a number of ways. Thus the sensation of yellow as produced by a radiation of wave-length 589 nm may also be evoked by presenting to the eye a suitable mixture of pure red and pure green, that is, by light containing no true yellow radiation. Also the sensation of white will result from the mixture in suitable proportions of certain pairs of pure colours known as complementaries.

Any colour can be specified in terms of three variables, its hue, its saturation or **purity** and its **luminosity**. The hue is the property which depends on the frequency of the light, the saturation depends on the amount of white present in addition to the light giving the hue, the less white light the more saturated the colour is said to be. Thus for the same hue we have the intense saturated yellow of the spectrum, the paler yellows when this hue is mixed with white and the various browns as the luminosity is reduced.

In the spectrum there are five very distinct hues, red, yellow, green, blue and violet. There is also purple, not found in the spectrum, but formed by a mixture of red and blue. The following divisions of the spectrum with the corresponding wave-lengths were given by Abney (1844-1920):

Violet	446 nm to End
Ultramarine	464 nm to 446 nm
Blue	500 nm to 464 nm
Blue-green	513 nm to 500 nm
Green	578 nm to 513 nm
Yellow	592 nm to 578 nm
Orange	620 nm to 592 nm
Red	End to 620 nm

More recently, H.B. Tilton has suggested the adoption of equal hue bands each containing 10 just-noticeable differences in wave-length for the colour-normal observer except for the Violet (8 just-noticeable differences) and the red (15). The suggested names with their wave-length limits are:

Violet	388 nm to 429 nm
Indigo	429 nm to 458 nm
Blue	458 nm to 481 nm
Cyan	481 nm to 499 nm
Turquoise	499 nm to 513 nm
Green	513 nm to 528 nm
Emerald	528 nm to 546 nm
Chartreuse	546 nm to 561 nm
Yellow	561 nm to 575 nm
Amber	575 nm to 587 nm
Ochre	587 nm to 599 nm
Orange	599 nm to 610 nm
Tangerine	610 nm to 622 nm
Scarlet	622 nm to 636 nm
Red	636 nm to 782 nm

13.25 COLOUR MIXING AND MEASUREMENT*

When a mixture of light of two or more colours reaches the eye, new colour sensations are produced, and the eye is incapable of recognising the simple colours contained in the mixture. Light of different colours may be mixed in a number of simple ways. Two or more projection lanterns with coloured filters may be arranged to project overlapping patches of coloured light on a white screen. A disc may be coloured with sectors of the colours to be mixed, and on quickly rotating the disc the colours blend and produce the sensation due to the mixture. An improvement on this method is the Maxwell's colour disc. Three discs coloured say red, green and blue-violet, and having a radial slot cut in each, as in *Figure 13.17a*, may be fitted together to form a single disc, with three sectors, the angles of which can be varied by sliding one disc over the others.

The results of adding colours may be stated in a general way as follows:

Red + Green + Blue-violet = White
Red + Green = Yellow
Green + Blue-violet = Green-blue (Peacock Blue)
Red + Blue-violet = Purple

Therefore

Yellow + Blue-violet = White
Green + Purple = White

By altering the proportions of the colours one to another, all the hues are produced; thus starting with red and adding increasing quantities of green, the hue changes from red to orange-red, orange and yellow and then through the greenish-yellows to green. Adding increasing amounts of blue-violet to green the

*For a full account of colour and colour measurement see *Visual Optics* Vol. II by H. H. Emsley, Hatton Press, 1953, and *The Measurement of Colour* by W. D. Wright, Hilger and Watts Ltd. 1958.

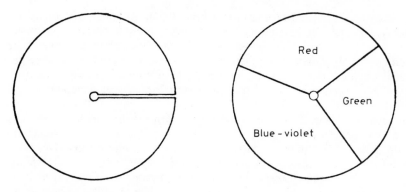

Figure 13.17a Maxwell's discs

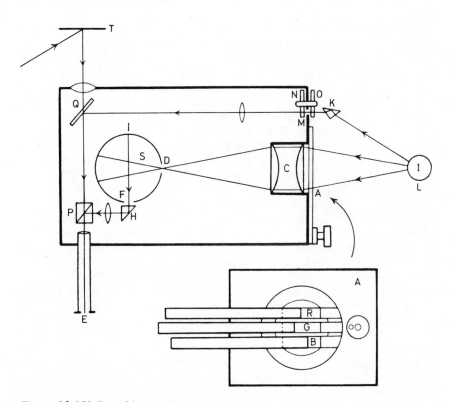

Figure 13.17b Donaldson colorimeter

hue changes from green through the various hues of green-blue and blue to blue-violet. In the same way by adding the blue-violet to the red we get magenta, the various purples and violet.

It is generally more useful, however, to study the effects of mixing purer colours than can be used in the above methods. Three suitably chosen spectrum colours, usually red, green and blue, when mixed in certain proportions produce the *sensation* of white and any colour sensation can be matched in *hue* by varying proportions of the three primaries or **matching stimuli**. Although any colour can be matched in hue a complete match is not always possible because the colour produced by the mixture may be less saturated than the colour to be matched, the *test colour*. To obtain a complete match the test colour will have to be desaturated by the required amount by adding to it either a known amount of white or, as can be shown to produce the same effect, a certain amount of one of the matching stimuli (Plate 6c).

Any colour can therefore be expressed in terms of the three matching stimuli, R, G, and B and this is the basis of **colorimetry**, a subject of considerable importance in industry. A number of instruments, known as **colorimeters**, have been devised for colour measurement and these differ mainly in the methods by which the matching stimuli are produced, mixed and varied in intensity. Coloured light, desaturated if necessary, from the specimen under test fills one-half of the observation field of the instrument and the mixture of the matching stimuli the other half.

In some instruments, such as the Wright Colorimeter, three narrow selected bands of the actual spectrum are mixed together in varying proportions, but in others the matching stimuli are obtained by the use of three colour filters giving highly saturated colours. This enables a simpler instrument, more adapted for industrial use to be produced. An example of an instrument using colour filters is the Donaldson Colorimeter (*Figure 13.17b*). The light source L is a 250 watt projection lamp; this is run at 90% its rated voltage, which gives a longer life and more uniform light output. The light is received on the plate A in which are three rectangular apertures R, G and B, the dimensions of which can be independently varied by shutters sliding in grooves. Each shutter is operated by a separate rack and pinion and the extent of the opening is read on a scale on the shutter. Behind the apertures are mounted the three colour filters.

An image of the lamp filament in each of the three colours is formed by means of the lens C on the aperture D in the integrating sphere S. This sphere has a diameter of 6 inches and its interior surface is silver plated and coated with magnesium oxide. By repeated reflections in the sphere the light from the three filters is completely mixed and this mixed light from the sphere wall at I passes through a second aperture F in the sphere. The light from F after reflection by the right angle prism H is reflected by the Photometer prism P to the eye at E. The eye sees one-half of the field evenly illuminated by light which is a mixture of the light passing through the three filters and the proportion of the three component colours is adjusted by means of the shutters in A. Light from the test surface T fills the other half of the field.

When it is necessary to desaturate the test colour light from the source is deflected by the prism K through a small aperture M fitted with a diffusing screen. A disc O carries three filters similar to the three main filters which can be brought in turn in line with the aperture and N is a circular neutral wedge which can be rotated over the aperture to regulate the amount of desaturating stimulus. The desaturating light is then reflected into the field of the test colour by the plane parallel glass plate Q.

When a given colour has been matched by the mixture of the three matching stimuli its colour can be expressed in the form of an equation, as was first suggested by Clerk-Maxwell in 1856.

Thus $c = r(R) + g(G) + b(B)$, where c is the colour and the numerical coefficients *r*, *g* and *b* represent the amounts of the red, green and blue stimuli required for the match. The units in which the matching stimuli are measured are arbitrarily chosen so that an equal number of each stimulus will match a standard white when added together. These units are called **trichromatic units** and are units of *colour* as distinct from units of *light*. In the **unit trichromatic equation** the **trichromatic coefficients** *r*, *g* and *b* will always add up to unity, thus for standard white we have

$$1W \equiv \tfrac{1}{3}(R) + \tfrac{1}{3}(G) + \tfrac{1}{3}(B)$$

For some colour, say a yellowish-green, the equation might be

$$C^* \equiv 0.3(R) + 0.6(G) + 0.1(B)$$

or in the case of a colour, such as a pure spectrum green which must be desaturated before a match can be obtained, the equation might read

$$C \equiv 0.480(R) + 0.543(G) - 0.023(B)$$

or

$$C + 0.023(B) \equiv 0.480(R) + 0.543(G)$$

that is when the given amount of blue is added to the colour a match can be obtained by the mixture of appropriate amounts of red and green.

The fact that most colour sensations could be produced by the mixture of three suitably chosen primaries led to the formulation of the **trichromatic** theory of colour vision. This theory, first put forward by Young in 1802 and later elaborated by Maxwell and Helmholtz, is usually known as the Young-Helmholtz theory, and supposes the existence in the retina of three types of receptors, which when stimulated give rise to the sensations of red, green and blue respectively.

Plate 6c shows a colour reproduction of the colour triangle. The edge colours of this shape represent the spectral hues while the central 0.33, 0.33, 0.33 point is pure white. A straight line from this point to the edge represents the locus of saturations from 0 to 100% at that hue. It is seen that the values for which r = 1; or g = 1; or b = 1 are three different colours with saturations above 100%. They are therefore not real colours and cannot be duplicated by any dyes. Essentially they are theoretical points which allow the whole of the observable colour diagram to lie within the positive axes of the graph.

A plot of colour temperatures may also be done on this diagram and this is shown on Plate 6c (see section 13.6).

Any two colours which when added together give the *sensation* of white are known as **complementary colours**; thus, as was seen above, yellow and blue-violet are complementaries, as are also green and purple. In many cases pairs of pure spectral hues, that is monochromatic light, having appropriate relative luminosities may be complementaries. Some of these complementary pairs of wave-lengths with their required relative luminosities as determined by Sinden (1923) are shown in the table on the next page.

---

*The coefficient 1.0 for the test colour in unit trichromatic equations is usually omitted.

Complementaries (nm)

λ_1	650	609	586	578.5	574	573	570.5
λ_2	496	493.5	487.5	480.5	472	466.5	443

Relative luminosities

L_1	42.1	52.4	73.8	85.5	92.2	94.2	97.5
L_2	57.9	47.6	26.2	14.5	8.0	5.8	2.5

13.26 COLOURED OBJECTS

The colour of most objects is due to selective absorption. When white light falls on a coloured opaque substance, some of it passes into the substance, and certain frequencies are diffusely reflected close to the surface. A red object, for example, illuminated with white light, is reflecting some of the white light from the surface, and for a short distance inside the surface is diffusely reflecting the red and possibly also the orange and yellow, while the green, blue and violet are absorbed. The same object illuminated with green or blue light will appear almost black, as except for a small amount from the surface it is reflecting no light of this colour. Coloured transparent substances, coloured glasses, dyes, etc., owe their colour to a similar cause, a red glass transmitting only red light and absorbing light of other colours.

The colour of an object will obviously depend on the colour of the illuminating light, and in speaking of the colour of any object we are referring to its colour when illuminated with white light.

The colours obtained by reflection and transmission are usually far from pure, and examined with a spectroscope will be seen to contain a band of frequencies on either side of the predominant hue. There is also usually a considerable amount of white light reflected from the surface, and if this is diffusely reflected and mixes with the coloured light from below the surface, the colour will be less saturated than when the surface is polished, and the white light is reflected in a definite direction. This effect is seen in the difference in colour in polished and unpolished wood and marble.

The effects of mixing coloured substances, such as pigments, or of combining colour filters—dyes, coloured glasses, etc.—are quite different from those obtained in adding coloured lights. It is well known, for example, that the mixture of blue and yellow water-colours gives green, whereas we have seen that the addition of blue and yellow light gives the sensation of white. The explanation is simple: in the case of the pigments, the blue particles, when illuminated with white light, are absorbing the longer wave-lengths and reflecting the violet, blue and green; the particles of yellow pigment are absorbing the violet and blue and reflecting the green, yellow and red. A mixture of the two pigments will therefore absorb or subtract everything from the white light but the green which is reflected. If we place a red filter in one lantern and a green filter in another and project two overlapping patches on a white screen, the additive mixture produces the sensation of yellow, but if the two filters are placed together in one lantern, no light passes through the combination, as the light transmitted by one filter is absorbed by the other.

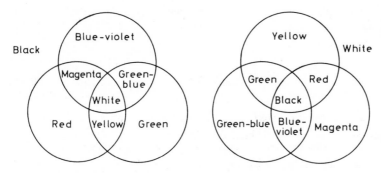

Figure 13.18 Additive and subtractive primaries

A great range of colours can be obtained by the mixture of pigments or dyes of three carefully chosen colours; these are known as the **subtractive primaries**, as distinct from the **additive primaries** dealt with earlier. The subtractive primaries are magenta, yellow and green-blue. These colours are in themselves mixtures, being composed of broad bands of the spectrum. They are such that, of white light, the magenta reflects all but green, the yellow all but blue and violet, and the green-blue all but red and yellow. The general results of additive and subtractive mixtures are shown diagrammatically in *Figure 13.18*.

The subtractive primaries are the colours used in three-colour printing. Three photographs are taken, one through a red, one through a green and one through a blue filter. Prints are made from the three negatives in colours complementary to those of the filters, and the three prints are superposed. As the red filter transmits only red light the dark portions of a print from the negative taken through the red filter represent absence of red, and a print from this negative is therefore made in white minus red, i.e., green-blue or cyan. Similarly the print from the negative taken through the green filter must be made in magenta, and the print from the negative taken through the blue filter must be made in yellow. If the colours of the filters and inks are carefully adjusted the superposed prints can give excellent reproduction of the colours of the original object. Both the additive and subtractive systems of colour mixing have been used in colour photography but all the present day processes are subtractive.

13.27 COLOUR DUE TO THE SCATTERING OF LIGHT

When light passes through a space filled with minute particles, some of it is scattered, chiefly due to diffraction. The percentage of light so scattered is found to be inversely proportional to the fourth power of its wave-length, so that the scattered light will consist mainly of the shorter wave-lengths, the blue and the violet. This is known as the **Tyndall effect** after Tyndall (1820–1893) who investigated the effect experimentally.

The blue of the sky is due to this selective scattering. A certain amount of the shorter wave-lengths of the sun's light is scattered by the particles of dust and water and the molecules of gas in the earth's atmosphere, while the longer wave-lengths are transmitted. The increasing redness of the sun, as it gets nearer the horizon, is due to the same cause. As the light is then travelling through a great thickness of atmosphere more of the shorter wave-lengths are scattered, and the remaining transmitted light is therefore redder. The light scattered in this way is also polarised (section 16.4).

CHAPTER 13   *EXERCISES*

1. Describe fully a method of obtaining a pure spectrum. What difference will be seen between the spectra of:
   (a) the sun, (b) the electric arc, (c) a tube containing incandescent hydrogen.

2. Explain clearly how you would use a spectrometer to determine the refractive index of the glass of a prism. Give the necessary theory and describe the experimental details.

3. What is meant by the dispersive power of glass? Describe carefully a practical method of measuring this value.

4. In measuring a prism with refracting angle of $60°\,4'$ on the spectrometer the following values were obtained for the minimum deviation for different colours:

Red	(C Line)	$38°\,23'$
Yellow	(D Line)	$38°\,37'$
Blue	(F Line)	$39°\,8'$

Find the dispersive power of the glass. What kind of glass is this?

5. Manufacturers' lists of optical glasses give particulars of refractive index, dispersive power and partial dispersions. Briefly explain these quantities.

6. Explain clearly the terms:
   Dispersion
   Mean dispersion from C to F
   Partial dispersions
   Chromatic aberration.

7. Explain carefully how to determine the mean refractive index n_D and the dispersive power of a sample of glass in the form of a $60°$ prism. Describe the adjustments of the instrument used and show the method of calculation.

8. A $60°$ prism is made of flint glass having the following refractive indices:

n_C	n_D	n_F
1.615	1.620	1.632

Find the angle between the emergent light corresponding to the C and F lines when a beam of white light is incident on the prism in the direction of minimum deviation for the D line.

9. Explain the construction of an achromatic lens, showing exactly why it is necessary to use two different glasses. Why does a lens of this type not focus all colours in the same position?

10. Given that for a certain flint glass $(n_D - 1)/(n_F - n_C) = 36$ and $n_D = 1.6$, calculate the total curvature of a concave lens made from the glass in order when placed in contact with a convex lens having a difference of power for the F and C lines of 0.2167 D, the combination may be achromatic.

11. A positive lens is made of crown glass the refractive indices of which are as follows: red 1.5150, yellow 1.5175, blue 1.5235, the mean power of the lens being 5D; what will be the difference in power for the red and blue? What must be the mean power of a negative lens of flint glass, the refractive indices of which are red 1.6175, yellow 1.6225, blue 1.6348, in order that when combined with the positive lens the red and blue focus at the same point? What is such a combination called?

12. Given the following glasses, find the curves of an achromatic lens of 25 cm focal length, the crown lens to be equi-convex and to be in exact contact with the flint.

Crown	$n_D = 1.517$	V = 60.5
Flint	$n_D = 1.612$	V = 37.0

13. An achromatic telescope objective is to be constructed of the following glasses:

	n_D	$n_F - n_C$	V
Crown	1.5188	0.0086	60.3
Flint	1.6214	0.0172	36.1

The radii of the crown lens are to one another as 2 to −3, while the flint curve next to the crown is at the same time represented by −2.815. Determine the radii of all four surfaces on the assumption that the lens is thin and is to have a focal length of + 25 cm.

14. Explain why two different glasses must be used in the construction of an achromatic lens. A positive achromatic lens of 30 cm focal length is to be constructed from the following glasses:

	n_D	$n_F - n_C$
Crown	1.519	0.0086
Flint	1.614	0.0166

The flint lens is to be cemented to the crown and its second surface is to be plane; find the radii of curvature of the surfaces.

15. Assuming that the optical system of the eye is roughly equivalent in its action to a spherical surface of water of radius 5.1 mm and given the following refractive indices of water: for red light = 1.332, for blue-violet = 1.340, calculate the chromatic aberration as expressed by the interval between the corresponding foci. Assuming the eye-pupil to be 4 mm in diameter, calculate the diameter of the least patch of confusion between these foci (where the retina may be assumed to be positioned) and comment on the result, remembering that the diameter of a foveal cone is about 0.0025 mm.

16. A thick lens is made of crown glass which has the following constants:
$n_D = 1.52300, n_F = 1.52924, n_C = 1.52036$
The lens is of meniscus form, its radii of curvature being + 55.0 mm and + 87.16 mm, thickness 20 mm. For each of the three colours find:
(*a*) Power and focal length of the lens.
(*b*) Position of principal points.
(*c*) Size of image of distant object subtending 10°.

17. Show by diagrams how two prisms of crown and flint glass respectively may be combined to obtain:
(*a*) dispersion without deviation, and
(*b*) deviation without dispersion.

18. Explain the principles underlying the construction of (*a*) an achromatic prism, (*b*) a direct vision prism.

19. A thin prism with a refracting angle of 8° is made of crown glass having the following refractive indices: for red 1.527, for yellow 1.530 and for blue 1.536; a thin prism of flint glass is to be combined with the crown prism in order to neutralise the dispersion, the refractive indices of the flint glass being for red 1.630, for yellow 1.635 and for blue 1.648. What must be the angle of the flint prism and what will be the total mean deviation produced by the combination?

20. A direct vision prism is to be made of crown and flint glass having the following constants:

	n_C	n_D	n_F
Crown	1.527	1.530	1.536
Flint	1.630	1.635	1.648

The D ray is to be undeviated; what will be the resultant angle of dispersion between the C and F rays if the angle of the flint prism is 5°?

21. It is required to produce an achromatic prism combination to give a deviation of the mean ray of 1 in 70. The two component prisms are to be made of crown and flint respectively, which have the following constants:

Crown	$n_D = 1.5178$,	V = 60.2
Flint	$n_D = 1.6190$,	V = 36.2

Find the angles of the component prisms.

22. Two thin prisms are to be combined to form a thin achromatic prism combination which is required to produce a deviation of the mean ray of 1 in 100. What are the angles of the component prisms? The first component prism is made of a crown and the second of a flint, the constants for these glasses being:

Crown	$n_F - n_C = 0.0086$,	V = 60.2
Flint	$n_F - n_C = 0.0171$,	V = 36.2

23. An achromatic prism is to be constructed of glasses having the following values:

	n_C	n_D	n_F
Crown	1.527	1.530	1.536
Flint	1.630	1.635	1.648

The crown prism is to have an angle of $10°$, find the angle of the flint prism and the deviation produced.

24. What is meant by the term achromatism?
Prove that a system of two lenses of the same material will be approximately achromatic if the lenses are separated by a distance equal to one-half of the sum of their focal lengths. State whether the system will be converging or diverging in character.

25. Two pieces of glass, one blue and the other yellow, are superposed. What colour is seen? If the two pieces are placed side by side in a lantern so as to illuminate a screen, will the effect be the same? Give reasons in each case.

26. Discuss the occurrence of the colour in the case of:
    1. The blue sky.
    2. A piece of blue glass.
    3. A piece of blue paper.

27. How would you specify a given colour in order that it may be reproduced from a knowledge of the specification?

28. A constant deviation prism of the form shown in *Figure 13.13* is made of dense flint glass having the following refractive indices: red 1.630, yellow 1.635, blue 1.648. A narrow beam of white light is incident on the face AB in such a direction that the angle of refraction for the yellow at this surface is $30°$. Find the angles between the red, yellow and blue light emerging from the surface BC.

29. Explain why it is possible with two pure spectrum colours suitably chosen (say a red and a green) to match daylight containing the whole range of spectrum colours. What does *match* mean here? Could such a light be used by which to paint a picture successfully?

30. The letters on a poster printed in red ink disappear altogether when looked at through a red glass. So does the lettering on a poster printed in yellow ink when seen by gas light. Explain these results.

31. A better photographic rendering of objects, particularly of blue sky and clouds, is obtained when a yellow filter is used on the photographic objective; why is this? Does it also apply for colour photography?

32. Write a short essay on infra-red and ultra-violet radiation.

33.

	C	D	F	G
Crown	1.506	1.509	1.514	1.523
Flint	1.749	1.757	1.776	1.812

The table gives the refractive indices of two glasses for the stated Fraunhofer lines. Explain the meaning of the term irrational dispersion, using the quoted figures to illustrate your answer. Find the dispersive powers and the relative partial dispersions of the two glasses.

34. Explain each of the following:
    (a) Yellow and blue paints mixed together produce green paint, but yellow and blue lights projected on to a screen produce white light.
    (b) Smoke from the end of a burning cigarette appears blue whereas smoke from the mouth of the smoker appears grey.
    (c) Materials with bright saturated colours can be obtained in the red-orange-yellow range, but in the green-blue range the colours may be saturated but not bright.

35. Using appropriate diagrams, explain the following terms: (a) Dispersive power; (b) Secondary spectrum; (c) Anomalous dispersion.

36. The face AC of a prism ABC is silvered. A ray from an object is incident on the face AB and after two reflections, first at the silvered face AC and then at the face AB, it emerges from the base BC. Determine what relation must exist between the prism angles at A and B in order that there shall be no chromatism.

37. Describe the advantages and disadvantages of using plastics instead of glass in an optical system.

38. Show that a pair of thin lenses in contact gives an achromatic combination if $\omega_A P_A + \omega_B P_B = 0$ explaining symbols used.
    An achromatic cemented pair with a nett focal length of 82½cm has for the properties of its components glasses
    (a) R.I. = 1½ $\omega = 0.015$
    (b) R.I. = $1^5/_8$ $\omega = 0.045$
and the concave external surface of the negative component has radius of curvature 25 cm. Find the radii of the surfaces of the positive component.

39. Sketch and explain the spectral radiation curves for a black body at various temperatures.
    With reference to the curves describe the change in appearance of a black body which would take place if its temperature were raised from cold to, say, seven thousand degrees Kelvin.
    Explain what is meant by 'colour temperature' and 'mired value'. What does the term 'mired' signify when applied to a filter?

40. Explain what is meant by the dispersive power and constringence of a medium.
    An achromatic combination of overall mean power +5D is required from crown and flint glasses of constringence values 60.5 and 37.0 and n_D values of 1.517 and 1.612 respectively. The flint lens is to be cemented by its curved surface to the crown lens with its second surface plane. Calculate the radii of curvature of all the surfaces.

Chapter 14
Interference and optical films

14.1 INTRODUCTION

The phenomenon of interference arises directly out of the wave theory of light. According to the principle of superposition of simple harmonic motions developed in Chapter 12 interference should occur whenever two or more waves meet. That this is not so in practice is due mainly to the departure of ordinary light from a *continuous* wave-motion. To get steady interference effects with ordinary light sources we have to satisfy rather stringent conditions. On the other hand the light from a laser source approximates much more closely to a continuous wave motion and workers with these light sources will testify that interference phenomena occur so easily that they can become a nuisance.

If, however, the correct effects can be isolated they form a very precise method of measurement being able to assess accurately changes in dimension which are less than one thousandth of the wave-length of the light used. For green light this is 5 Ångstrom Units or 2×10^{-8} inches. This branch of optics is now known as **optical metrology** and an outline of this technology will be given in this chapter.

On the other hand if optical components can be made with thicknesses less than the wave-length of light, the interference phenomena obtained can be used to modify reflection coefficients and obtain colour filters, very high transmission elements and polarising effects. This branch of optics is known as **optical thin films** or **optical coatings** and this chapter will conclude with a review of the theory and applications of this subject.

14.2 CONDITIONS FOR THE INTERFERENCE OF LIGHT

Two sets of waves of equal amplitude and frequency spreading out from point sources will neutralise the effects of one another in certain places and will give increased disturbance in others. This can be shown visually by generating two sets of ripples on the surface of mercury. If two small points are attached to a tuning fork and arranged so that they just dip into the mercury, two exactly similar sets of ripples will be created when the tuning fork is set into vibration. Each ripple system will consist of a series of concentric circles. The appearance will be as in *Figure 14.1* where B_1 and B_2 are the sources and the crests of the waves are represented by solid lines and the troughs by broken lines.

Along the lines marked with crosses the crest of one wave coincides with the

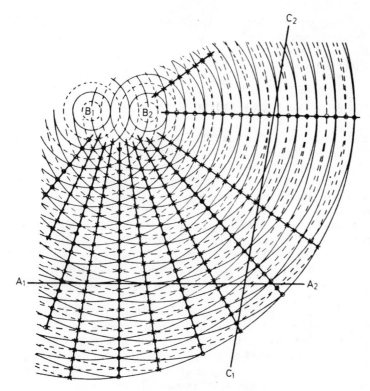

Figure 14.1 Interference between two sets of waves

trough of another and since the amplitudes are equal the displacements will be neutralised in accordance with the principle of superposition. Along the lines marked with circles the crests coincide and the troughs coincide so that the resultant disturbance has twice the amplitude. As the energy of the wave-motion is proportional to the square of the amplitude there will be four times the energy of a single source along these lines and thus there is no loss of total energy from the two sources due to interference, but only a redistribution.

If a barrier were erected at A_1-A_2, it would receive no waves where the crossed lines meet it and double amplitude waves at the places where the circled lines meet it. In moving along the barrier from A_1 to A_2 one receives doubled and zero effects in a cyclic fashion. Similarly a barrier along C_1-C_2 would show the same effect but with the different regions of energy more spread out. Their location is crucially dependent on the distance between B_1 and B_2 *measured in terms of the wave-length of the ripples.*

As has been indicated earlier, two ordinary sources of light placed in positions such as B_1 and B_2 do not produce visible interference effects. If they did we should expect to find on a screen at A_1-A_2 bright patches where the crests coincided and darkness, because of zero amplitude, in the regions where a crest always coincides with a trough. These patches would stay much the same on the screen above and below the plane of the page and would be seen as alternately bright and dark **fringes**. If either of the two sources varies in phase with respect to the other source these lines would move. With any two ordinary sources of light, the relative phase between them is constantly changing in a very fast random manner. The fringes formed are therefore blurred out to a uniform illumination. When the source has a finite size there is an equally random relationship in the phase between one part and another.

14.3 COHERENCE, INCOHERENCE AND MUTUAL COHERENCE

In section 12.10 and following the concept of coherent sources was developed and the various forms of laser described. However, interference phenomena do not require such complete coherence and indeed fringe patterns due to interference were being studied long before the laser was invented. This was achieved with largely incoherent sources because the presence of stationary fringes on screen A_1-A_2 of *Figure 14.1* depends only on obtaining a constant phase difference between sources. In other words it does not matter if the phase of B_1 changes as long as that of B_2 changes in an identical manner. In terms of superposition of waves at A_1 the wave-motion from B_1 and B_2 can vary from moment to moment provided that a crest from B_1 always coincides with a trough from B_2 and vice versa. Such a relationship between B_1 and B_2 is termed **mutual coherence**.

In all aspects of coherence it should be remembered that a continuous single-frequency simple harmonic motion cannot be perfectly achieved, nor can complete identity between two sources. It is proper therefore to speak of **degrees of coherence** which are less than 100%. In general terms the visibility of the fringe patterns described will vary with the degree of coherence obtained. However, some fading of the pattern can usually be tolerated.

In mathematical terms it may be stated that for two coherent sources the intensity on the screen is found by adding the respective amplitudes and squaring the result; whereas for incoherent sources the final intensity is found by adding the respective intensities. The latter method can never give darkness as the intensities are always positive being the square of the amplitudes shown in section 12.5.

14.4 INTERFERENCE WITH INCOHERENT SOURCES

The methods of interference with incoherent sources are aimed at obtaining a sufficient degree of mutual coherence between two sources. This is done by making the two sources both dependent on the same source. For instance the reflection of light in a mirror provides an additional source apparently behind the mirror surface. It is possible to obtain interference between an incoherent source and its reflection provided there is a region where the two sets of waves overlap and provided the distance between the source and its reflection is not greater than a few wave-lengths. These conditions are easier to achieve if two reflected sources are used and Newton's rings (section 14.10) is a particular case which constitutes the earliest study of interference phenomena.

The next example was devised by Young in the early nineteenth century. He illuminated a small aperture so that it became affectively a single small source. A short distance from this he placed a screen containing two small apertures which were illuminated only by the first aperture. These two apertures now constitute mutually coherent sources and provided that the illuminating light is reasonably monochromatic, interference fringes may be observed on a screen in front of the two apertures as $A_1 - A_2$ in *Figure 14.1*. The fringes are very faint as nearly all the original light has been obscured. Better results are obtained if narrow parallel slits are used in place of the three apertures.

Much better use of the available light was made by **Fresnel** in his **double-**

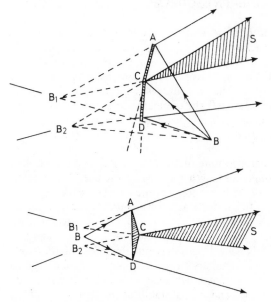

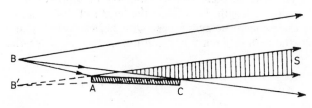

Figure 14.2 Fresnel's mirrors and bi-prism

Figure 14.3 Lloyd's mirror

mirrors, and **bi-prism** methods. These are most easily explained in *Figure 14.2*. The single source B is in each case converted to two sources B_1 and B_2, by reflection or refraction at AC and CD. So that the apparent sources are close together the angle ACD of the bi-prism and between the mirrors must be in the region of $179°$. The interference takes place in the regions S where the beams overlap.

Considerably brighter fringes may be obtained if B is an illuminated narrow slit parallel to the join of the mirrors or refracting edge of the prism. This gives clear parallel fringes in the region S with a regular spacing which may be measured experimentally using a travelling microscope. Fringes may be found anywhere in the shaded region S and are said to be non-localised.

Lloyd's single mirror method is different from the above as it uses only one reflected source. In *Figure 14.3* the illuminated slit B is placed parallel and close to the surface of a plane mirror (or black glass) and the reflected source B_1 interferes with the original light waves in the shaded region S.

14.5 INTERFERENCE WITH COHERENT SOURCES

The mutual coherence of the interfering sources described in the last section arises from the dependence of each source on a single original source. Although the precise phase and structure of the wavefront from the real source may be changing many times a second, the dependent sources are changing also. When we have a phase difference between the two sources it means that the precise parts of the waves which are interfering left the original source at different times. But the original source is changing many times a second and so if the time difference becomes too large we find that the two waves are no longer mutually coherent as the source changed between the times of their emission. The allowed time difference is called the **coherence time** of the source and when multiplied by the velocity of light gives a **coherence length** which is a more useful parameter.

It is found that this is related to how monochromatic the source is. If the source contains wave-lengths λ_1 to λ_2 then as k of equation (14.01) (section 14.6) increases: x can equal $k\lambda_1$ and $(k + \frac{1}{2})\lambda_2$ and the fringe pattern fades away, with λ_2 filling in the dark bands of the λ_1 pattern. With an incandescent source followed by a filter the coherence length may only be a few micrometers. With a spectral line source such as a mercury lamp a coherence length of 200 to 300 mm may be achieved. With continuous wave lasers source coherence over many metres may be readily obtained and special stabilising methods can extend this to many kilometres.

The above discussion takes no account of the finite size of sources which effectively reduces the coherence length. With lasers the length of the resonant cavity has an effect on their coherence and fringes with highest visibility are generally obtained when the path difference is a multiple of the cavity length.

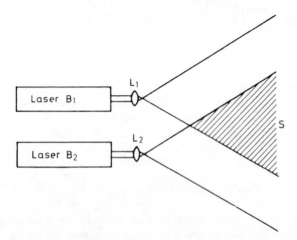

Figure 14.4 Interference with lasers—no fringes are seen

In *Figure 14.4*, two similar lasers are arranged as interference sources B_1 and B_2. Their normally collimated output is caused to diverge by lenses L_1 and L_2, so that an interference pattern is formed at S on the screen. With ordinary laboratory lasers no fringes will be seen. This is because the mutual coherence between the two lasers is limited by the random phase changes which limit their individual coherence length to a few kilometres at best. Although this is a considerable distance, the velocity of light is such that the *coherence time* is only a few microseconds. Thus the interference pattern is only stationary over this period of time and moves in a random manner. It is therefore not possible to see fringes in this way although special methods have shown they exist.

However, most laser interferometer systems use single lasers and the beam-splitting methods of section 14.4 to achieve measurements with long path differences or a multiplicity of interfering sources to give very fine fringes for high accuracy. These will be discussed in sections 14.7 to 14.11.

14.6 THE FORM OF INTERFERENCE FRINGES

Referring again to *Figure 14.1* the presence of a bright or dark fringe at a given point on $A_1 - A_2$ depends on whether the waves arrive in or out of phase. If in *Figure 14.5* B_1 and B_2 represent two mutually coherent line sources emitting waves of equal amplitude, frequency and phase then the waves will arrive in phase at point S on the screen as the distances B_1S and B_2S are equal if CS perpendicularly bisects $B_1 B_2$. S is therefore a bright fringe.

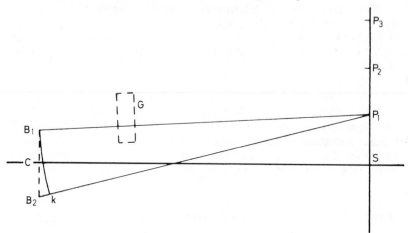

Figure 14.5 Width of interference fringes

For a point P_1 to one side of S the paths B_1P_1 and B_2P_1 will be unequal and the waves will arrive out of phase. If the phase difference is equivalent to one-half wave-length a dark fringe will result as crests and troughs are arriving together. A bright fringe will result at P_2 if the phase difference by then is equivalent to one whole wave-length. At P_3 the difference may have increased to one and a half wave-lengths and so a dark fringe results. Therefore on either side of the bright fringe at S there will be a series of bright and dark fringes as in Plate 1. The bright fringes occur when the path difference is $k\lambda$ and the dark fringes for $(k + {}^1\!/_2)\lambda$, where k is any integer.

The path difference for P_1 is found by constructing a circle centre P_1 through B_1. Then as $B_1 P_1$ and KP_1 are equal, the inequality is due to $B_2 K$. Normally $B_1 B_2$ is small compared with the distance to the screen if there is to be any appreciable separation of the fringes. Thus, $B_1 K$ approximates to a straight line perpendicular to $P_1 C$ and $B_2 K$. Representing $B_1 B_2$ by b, SP_1 by x and CS, which is approximately equal to CP_1, by d, we have by similar triangles; $B_1 B_2 K$ and $B_1 P_1 B_2$

for the centre of a dark fringe

$$x = \frac{d}{b}(k + {}^1\!/_2)\lambda \tag{14.01}$$

for the centre of a bright fringe

$$x = \frac{d}{b}(k)\lambda \tag{14.01a}$$

and the distances between centres of consecutive bright or dark bands will be

$$y = \frac{d}{b}\lambda \tag{14.02}$$

It is thus seen that the spacing depends on the wave-length of the light used. If a source of more than one wave-length is used a number of interference bands will be produced which will mix in and out of step because of their different fringe spacings. The central band at S will remain white as all wave-lengths have zero phase difference.

If in *Figure 14.5*, a block of glass is inserted into one beam only, a displacement of the fringe pattern will result. It is assumed that the frequency of any light vibration remains constant whatever medium it is travelling in. As the velocity changes, by the definition of refractive index, (see section 2.8) the wave-length inside a medium of index n will be $\lambda_n = \lambda/n$ where λ is the wave-length *in vacuo* or in air. The calculation of phase difference due to a path difference in glass must take account of the different wave-length. Thus equation (12.11) becomes

$$\text{Phase difference } \epsilon = \frac{2\pi}{\lambda_n} \times \text{Path difference}$$

$$= \frac{2\pi n}{\lambda} \times \text{Path difference} \tag{14.03}$$

Referring again to *Figure 14.5* the effect of a block of glass of index n and thickness t is to remove a path difference due to air and insert one due to glass. The resulting change in the phase difference is

$$\frac{2\pi n t}{\lambda} - \frac{2\pi t}{\lambda} = (n - 1)\frac{2\pi t}{\lambda}$$

If the displacement of the central fringe due to the insertion of the glass is m fringes this represents a phase change of $2\pi m$ and so:

$$(n - 1)t = m\lambda \tag{14.04}$$

The product nt is referred to as the **optical thickness*** of the glass block.

In the case of Lloyd's mirror (section 14.4) it is found by experiment that the fringes produced are not as predicted by equations (14.01) and (14.01a).

The fringe pattern is displaced by one-half fringe and this is due to the reflected source B_1 being π out of phase with the original source. This occurs due to a phase change of π occurring on reflection at the mirror and is considered further in section 14.9.

From expressions (14.01) and (14.02) the wave-length of the light can be determined by measuring the distances x or y, b and d in any of the methods described in section 14.4. Conversely, if the wave-length is known variations in b and d will show in the spacing of the fringes.

The interference fringes will not be sharply defined, because the resultant amplitude at different points on the screen gradually decreases as the waves get more and more out of step until a minimum is reached and then gradually increases to a maximum as the waves again come into step. The variation in intensity at different points on the screen may be found by applying the graphical method of section 12.4. If a_1 and a_2 (*Figure 14.6*) are the amplitudes of the wave motion from b_1 and b_2 respectively arriving at a general point P with a phase difference of ϵ, the resultant amplitude A at the screen is the vector addition of a_1 and a_2.

$$\text{Analytically: } A \sin\left(\frac{2\pi t}{T} + \beta\right) = a_1 \sin\frac{2\pi t}{T} + a_2 \sin\left(\frac{2\pi t}{T} + \epsilon\right) \tag{14.05}$$

where β is the phase of the resultant wave motion.

By vectors (*Figure 14.6*) for the case where $a_1 = a_2 = a$:

$$A^2 = 2a^2 (1 + \cos \epsilon) = 4a^2 \cos^2 \frac{\epsilon}{2} \tag{14.05a}$$

$$\text{or: } A^2 = 4a^2 \cos^2 \left(\frac{2\pi \text{ path difference}}{\lambda}\right) \tag{14.05b}$$

It should be clear from equation (14.05) and *Figure 14.6* that A will be a maximum when the phase difference, ϵ, is 0, 2π, 4π etc. and a minimum when ϵ is π, 3π, 5π etc. The intensity of the fringes varies as A^2 and so is proportional to the cosine squared of the path difference. These fringes are usually produced when two sources are used and are known as *cosine squared* fringes and have an intensity profile as in *Figure 14.7*.

The central fringes in Plate 9b are cosine squared fringes although in this case they are a little over exposed so that the peripheral fringes can be seen. If the

---

Optical Thickness, nt relates to the difference in phase caused by an optical material (section 2.8).

Apparent thickness, t/n relates to imaging through an optical material (section 4.6).

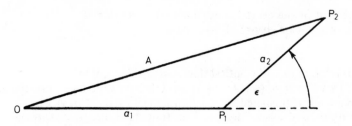

Figure 14.6 Resultant amplitude

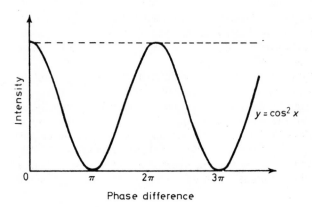

Figure 14.7 Intensity profile of two-beam fringes

number of interfering beams is increased to four, the fringes become sharper and extra fringes appear between them as in Plate 9c. The intensity profile of these is the square of the amplitude value calculated in section 12.4 for four equal interacting motions (12.06).

From that expression $(2 + 2 \cos \epsilon)(2 + 2 \cos 2\epsilon)$ may be evaluated for a range of ϵ (assuming a equal to unity). The profile obtained will show the subsidiary fringes of about $1/16$ the intensity of the main fringes. Outside the central three peaks of Plate 9c diffraction effects become evident.

14.7 MULTIPLE-BEAM INTERFERENCE

The cosine squared fringes of Plate 1 are the result of two interfering beams. When measuring their spacing it is difficult to locate accurately the centre of the fringe whether it is bright or dark. Using photo-electric methods it is possible to locate fringes to 0.01 times their spacing but better accuracies can be obtained if many more interfering beams are used. Although multiple-beam methods were employed in 1897 for accurate wave-length measurement the advent of the laser has brought them into common use.

In these methods the effective sources produce waves which are out of phase by multiples of the phase difference between the first two beams. This means that equation (14.05) becomes:

$$A \sin \left(\frac{2\pi t}{T} + \beta \right) = a_1 \sin \left(\frac{2\pi t}{T} \right) + a_2 \sin \left(\frac{2\pi t}{T} + \epsilon \right) + a_3 \sin \left(\frac{2\pi t}{T} + 2\epsilon \right)$$

$$+ a_4 \sin \left(\frac{2\pi t}{T} + 3\epsilon \right) + a_5 \sin \left(\frac{2\pi t}{T} + 4\epsilon \right) \text{ etc.}$$

The amplitudes a_1, a_2, a_3 etc. are generally reducing and the graphical calculation of the resultant appears as *Figure 14.8*, where the result of a small change in phase difference can be seen.

The analytical result gives the intensity of the fringes as proportional to:

$$\frac{E \sin^2 \frac{\epsilon}{2}}{1 + E \sin^2 \frac{\epsilon}{2}}$$

where ϵ is the phase difference and E is a factor depending on the number of interfering beams and their amplitudes. *Figure 14.9* shows the profile of the fringes for different values of E.

If the interfering beams have the same amplitude then the fringe profile is still sharpened up but other fringes of lower intensity appear between them. (See section 15.13.) The most common method of producing multiple-beam fringes

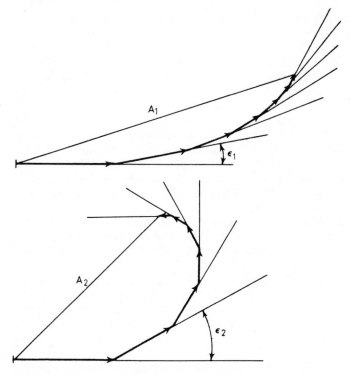

Figure 14.8 Vector addition for seven beams

276 *Interference and Optical Films*

uses partial reflection (section 14.9) and this gives the fringe profile of *Figure 14.9* in the transmitted beam. The reflected beam normally shows the inverse of this, that is, sharp dark fringes on a bright background. A photograph of this type is shown in Plate 10a and b.

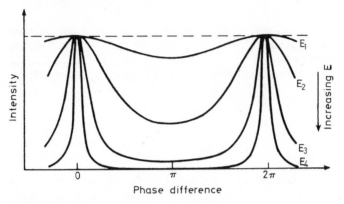

Figure 14.9 Intensity profile of multiple-beam fringes

14.8 INTERFERENCE BY PARTIAL REFLECTION

In *Figure 14.1* the interference effects of two sources were shown as generating bright and dark regions on screens at A_1-A_2 and C_1-C_2. The methods of interference considered so far have formed apparent sources side-by-side to the screen as indicated by A_1-A_2 of *Figure 14.1*. However, interference effects can also occur when the sources are effectively one behind the other as for screen C_1-C_2. Apparent sources displaced along the line of viewing are readily generated by partial reflection*.

The best known phenomenon due to the interference of light is the brilliant colouring produced when white light is reflected from a thin film of a transparent substance. Examples of this are seen in the soap bubble, a patch of oil on a wet road, and a very thin film of air between two pieces of glass. Unlike the effects of interference previously described, which require the use of a small source, the interference colours of a thin film are best seen when the film is illuminated by an extended source such as the sky.

This action may be set up in the laboratory without the two surfaces being in close proximity. The Michelson interferometer is shown in *Figure 14.10*. This uses the partially reflecting element A to divide the light from source S into two mutually coherent beams which are reflected by M_1 and M_2.

The partially reflecting element is often called a beam-splitter from this first action but it is also able to recombine the beams reflected by M_1 and M_2. Although some light is lost back to the source the appearance to the eye at E is that the two reflecting surfaces appear to be very close together or even superimposed at M_1 and M_2'.

*The distinction 'Division of Wavefront' for the methods of section 14.4 and 'Division of Amplitude' for those of section 14.9 and following is commonly used.

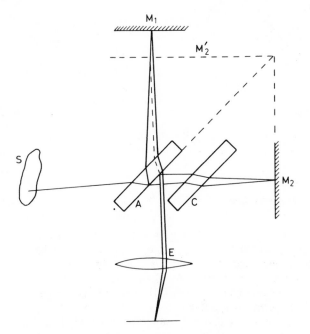

Figure 14.10 Michelson's interferometer

The light to M_1 passes twice through the glass support for the partial reflecting surface. Chromatic errors due to this may be cancelled out by inserting an identical block of glass to compensate at C. When the positions of M_1 and M_2 are such that the light paths are identical the two effective sources superimpose. With a small angle between the mirrors interference fringes can be seen even with white light. The different wave-lengths produce fringe patterns with spacing proportional to λ (14.02) but with a central fringe coincident for all colours where the phase difference is zero for all colours. On either side a few highly coloured fringes occur which rapidly merge to an even illumination as the fringe patterns of the different wave-lengths overlap. This action with white light can be very useful in positively identifying the point of effective coincidence of the mirrors and was used by Michelson to relate the wave-length of various monochromatic light sources to the unit of length, the standard metre.

A considerable number of interferometers are based on partial reflection techniques and some are described in sections 14.13 and 14.14.

14.9 THIN FILMS

In *Figure 14.11* light from an extended monochromatic source is incident upon a thin transparent film. A proportion of this light is reflected at each surface and the eye superimposes those coming from the same point on the film.

The actual ray bundles to the eye may be separated slightly before the action of the film and eye superimposes them on the retina but when the film is very thin this separation is very small and the bundles are largely mutually coherent. As they interfere the eye sees coloured fringes apparently localised in the plane

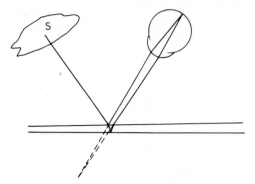

Figure 14.11 Interference with thin film

of the film. The phase difference causing the fringes comes from the path difference between the light reflected from the top surface of the film and that which enters the film and is reflected at the lower surface. The colours arise because different wave-lengths generate different phase differences from the same path difference. Thus one area of the film may generate a bright fringe for red-orange light but a dark fringe for blue violet.

If a soap film is caught across a wire it may be supported vertically in a glass container to protect it from draughts. The film gradually becomes wedge-shaped, due to gravity, and if illuminated with white light the reflection from it is seen to be crossed by broad horizontal bands of colour. After a time the upper region will get very thin and a perfectly black band will be formed at the upper edge. The same effect can be seen when two carefully cleaned glass plates are pressed into contact; where the air film between them is very thin compared with the wave-length of light a black patch will appear and the plates are said to be in **optical contact** at that place.

This dark region shows there is a phase difference between the two reflected waves for all colours. This cannot be due to the path difference which is very small and would produce various phase difference depending on wave-length. From electromagnetic theory it can be shown that a phase change of π occurs when a wave is reflected at a surface between a rarer and a denser medium. If the reflection is from a boundary between a denser and a rarer medium there is no change of phase. The two beams being reflected from the glass plates in optical contact come into each of these categories.

If crown glass is used for one plate and flint glass for the other the intervening space may be filled with a liquid of refractive index between the two glass types such as oil of sassafras. Under similar conditions of optical contact a bright patch is now seen as the light is being reflected from the same sort of boundary at each surface.

The path difference due to one beam traversing the film before reflection while the other is reflected from the first surface is required to be known so that the phase difference for a particular wave-length may be calculated. In *Figure 14.12* the difference in path length for the two beams which are subsequently superimposed by the eye is found by constructing DN perpendicular to the incident beams. There is then no path difference before that line and none after the point E where they combine. The difference then appears to be between

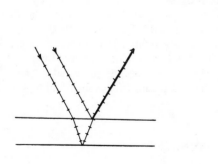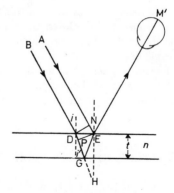

Figure 14.12 Reflection from a thin film. Conditions for a light or dark band

DGE and NE. However, the path NE does not lie in a medium of the same refractive index as DGE. The difference can most easily be accounted for by constructing PE perpendicular to DG. Then as was shown in Chapter 2 the time taken to travel along NE is the same as for DP.

Thus the waves are still in phase at P and E. The distance PGE can then be shown, for a parallel film of thickness t and index n to be given by $2t \cos i'$ using the construction to H. The phase difference due to this is given by equation (12.11) to be

$$\frac{4\pi nt \cos i'}{\lambda}$$

Remembering the phase difference due to reflection it is seen that for films bound by similar media a *dark fringe* will occur when:

$$\frac{4\pi nt \cos i'}{\lambda} = 2\pi k$$

or

$$2nt \cos i' = k\lambda \tag{14.07}$$

and a *bright fringe* when

$$\frac{4\pi nt \cos i'}{\lambda} = 2\pi (k + {}^1\!/_2)$$

or

$$2nt \cos i' = (k + {}^1\!/_2)\lambda \tag{14.07a}$$

The dark fringes so calculated will not be completely dark unless the reflection from the two surfaces produce interfering beams of equal amplitude. The fraction of incident light reflected at normal incidence at the boundary between media of index n_1 and n_2 is given by Fresnel's expression (see also section 16.4)

$$\left(\frac{n_1 - n_2}{n_1 + n_2}\right)^2$$

For a thin soap film in air or a thin film of air bounded by glass of identical type the intensity of the lower beam cannot equal that of the upper in *Figure 14.12* as this suffers two refractions as well as a reflection. However, there are other beams due to further reflections which provide interfering beams at larger phase differences. These give a multiple beam effect and provide for zero intensity in the dark bands. If the reflectivity of the two surfaces is increased a larger number of interfering beams occur giving the sharp dark fringes shown in *Figure 14.9*.

14.10 NEWTON'S RINGS AND FIZEAU FRINGES

Newton examined the effects produced on reflection from a thin film of air, by placing a convex lens of shallow spherical curve in contact with a glass plate (*Figure 14.13*). The enclosed air film will then be extremely thin at the point of contact and will gradually increase in thickness with increased distance from this point, lines of equal thickness being circles with the point of contact as centre. Illuminated with monochromatic light, the film, as seen by reflection, will therefore have a black spot at the point of contact, surrounded by concentric alternate light and dark rings. These are known as **Newton's rings** and a photograph of these is shown in Plate 2. If the film is illuminated with white light the rings will of course be coloured with colours which blur together 7 or 8 rings from the centre.

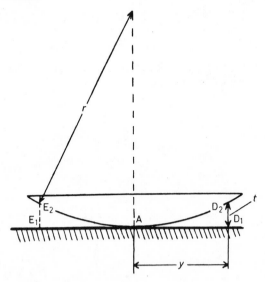

Figure 14.13 Newton's rings

Up to that ring the colours appear in the following order from the centre outwards.

1st order	Black, blue, white, yellow, red
2nd order	Violet, blue, green, yellow, red
3rd order	Purple, blue, green, yellow, red

4th order   Green, red
5th order   Greenish-blue, red
6th order   Greenish-blue, pale red
7th order   Greenish-blue, reddish-white

This is known as **Newton's scale of colours** and is shown in the colour Plate 6b.
From expression (14.07) the thickness t of the air film for any dark ring is

$$t = \frac{k\lambda}{2 \cos i'}$$

where k is the number of the ring (k = 0 for the central dark patch). If r is the
radius of curvature of the lens surface and y the radius of the kth dark ring, then
as t is small

$$r = \frac{y^2}{2t} \text{ (from equation 5.01)}$$

hence

$$r = \frac{y^2}{k\lambda} \tag{14.08}$$

for normally incident light when $\cos i' = 1$.

If the diameters of different rings are measured using a travelling microscope
then it is possible to assess the curvature of the lens. This may be in error if
the lens and plate are not in contact so a difference method is preferred using
the expression

$$r = \frac{y_1^2 - y_2^2}{(k_1 - k_2)\lambda} \tag{14.08a}$$

where y_1 and y_2 are the half-diameters of the k_1th and k_2th ring respectively.

If, instead of a lens, we use two flat glass plates with a thin piece of thin
paper between them at one edge, we then have a thin wedge of air which produces **Fizeau fringes**. Lines of equal thickness of the air film are generally
parallel to the edges of the glass plates which are in contact. Thus when the
arrangement is illuminated with monochromatic light a series of dark fringes will
be seen. If both plates have plano surfaces these fringes will be straight and
equally spaced. With normal illumination the spacing is given by x where

$$x = \frac{\lambda}{2\alpha} \tag{14.09}$$

where α, the angle between the plates is assumed to be very small. This expression may be calculated in a similar manner as (14.08).

In both cases the interference pattern is said to comprise **fringes of equal
thickness**. Any variation from flatness of the plates or sphericity of the lens will
change the thickness of the air gap and this will show in the fringe pattern. With
extended source illumination the fringes are localised in the plane of the film.
For thin films changes in the angle of viewing across the lens or plates only
introduces a fraction of a fringe error in the contouring action as $(1 - \cos \theta)$ is
less than 2% up to $10°$ from normal. The method constitutes a test of lens surfaces and is further discussed in section 14.12.

14.11 INTERFERENCE WITH THICK PLATES

Interference colours will be seen with white light only when the film is quite thin, that is, of only a few wave-lengths thickness. From expression (14.07a) it is seen that with a film of air illuminated with light at normal incidence, a maximum for a particular wave-length will occur when the thickness of the film is any *odd* number of quarter wave-lengths. Thus wave-lengths of 4 times, $^4/_3$ times, $^4/_5$ times, etc., the thickness of the film will be in the same phase on reaching the eye.

Suppose a film to have a thickness t = 0.0005 mm, the wave-length of green light, then the following wave-lengths will be in the same phase on reflection from the front and back surfaces, and will reinforce.

$\lambda = 4t\ \ \ = 0.0020$ mm $\qquad$ — infra red

$\lambda = {}^4/_3 t = 0.00067$ mm $\qquad$ — reddish orange

$\lambda = {}^4/_5 t = 0.00040$ mm $\qquad$ — violet

$\lambda = {}^4/_7 t = 0.00029$ mm etc. — ultra violet

Of these, only the second and third are in the visible spectrum and the reflected light is a mixture of these two wave-lengths and therefore distinctly coloured. As the thickness of the film is increased more of the wave-lengths of the visible spectrum are reinforced. Thus if the thickness of the film was 0.005 mm, proceeding as above, it will be found that twelve different wave-lengths in the visible spectrum, from $^4/_{27} t$ = 0.00074 mm to $^4/_{49} t$ = 0.00041 mm will be reinforced, and the mixture of all these wave-lengths cannot be distinguished by the eye from white light.

If the light reflected from a thin film is received on the slit of a spectroscope, the spectrum will consist of bright bands separated by dark spaces; thus with a film 0.0005 mm thick there will be two bright bands with their centres corresponding to wave-lengths of 0.00067 mm and 0.0004 mm respectively, while with the film of 0.005 mm thickness there will be twelve bands. This method may be used in determining the thickness of films.

As the separation of the surfaces providing the interfering beams is increased the visibility of the fringes is reduced with an extended source even if this is very monochromatic. This occurs because the pupil of the eye collects rays of different angles which give fringe patterns which spread across each fringe. An important exception occurs with a parallel-sided plate or air film for then the fringes are localised at infinity and the eye subtends a small angle. Such fringes are called **fringes of equal inclination** as the pattern then *depends* on changes of the angle of view as the eye looks to different parts of the plate.

However, when the plate or film is a wedge, fringes can only be seen using a point source of collimated illumination. In both cases the unaided eye can only see a small area of the fringe pattern at a time, but with collimated light fringes of equal thickness are obtained and the collimating lens can be used for the reflected light as well. In this case the eye can be placed to receive all the fringe pattern. *Figure 14.14* shows a **Fizeau Interferometer** using collimated light.

It will be seen that the source size and pupil size do not allow large wedge-angles to be used. In fact, a common method of adjusting this interferometer is to bring the two images of the source at S' into coincidence before placing the eye there to observe the fringes.

The fringes seen are 'cosine squared' fringes and the source must have a coherent length greater than the path difference used. Mercury isotope lamps provide sufficient coherence for several centimetres but the use of a laser source normally allows for many feet or, if the surfaces have high reflectance, for multiple beam fringes with a basic path difference of up to 2 or 3 feet. Best fringes are obtained when the path difference is equal to or a multiple of the resonant cavity length of the laser. With a laser source this interferometer is often called a **laser-Fizeau interferometer**.

14.12 CONTOUR AND INDEX MAPPING

In optical component fabrication it is important to know the precise curvature of any surface and also its figure, that is any variation in curvature over the surface. A traditional method, introduced by Fraunhofer uses **test plates** and is based on Newton's rings. Two glass plates are ground and polished to the exact curve required by the test surface one being convex and the other concave. The accuracy of the curves is tested by means of an accurate spherometer and by placing the plates in contact and, observing the interference pattern from the enclosed air film, any departure from the spherical can be seen. If both surfaces are

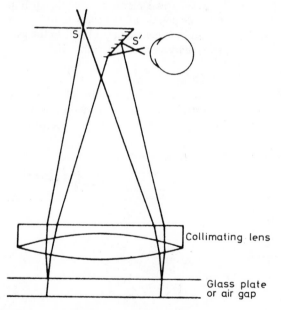

Figure 14.14 Fizeau interferometer

spherical and of equal curvature a uniform illumination is seen over the whole surface.

In testing a lens surface during manufacture the test plate of the opposite curvature is placed in contact with the lens surface, both surfaces being perfectly clean. If the lens is spherical but not of the correct curvature, circular fringes will be seen while any departure from spherical will show up in the shape of the rings. The rings do not immediately show whether the air film is thicker at the centre or at the edge, but in the latter case the lens can be rocked causing the fringe centre to move while in the former a slight pressure in the centre of the lens will cause the fringes to move outwards.

In spite of precautions it is all too easy to scratch the lens surface when a contact method such as this is used. Non-contact interferometers are discussed in sections 14.13 and 14.14.

For plano optical components Fizeau interference is commonly used. The fringes formed with an arrangement such as in *Figure 14.14* are fringes of equal thickness but more properly they constitute lines of equal *optical* thickness. Thus irregularities in the fringe pattern may be due to thickness variations or inhomogeneities of the material. In order to separate the two effects the interference fringes may be obtained from two known flats having between them the test component submerged in oil of equal refractive index. With such an arrangement irregular fringes must be due to inhomogeneities. Alternatively the fringes may be obtained with the component at a significant angle to the collimated beam. This will necessitate another collimating lens for the returning beam but if the effects at two angles are subtracted one from the other, that due to changes in refractive index may be found. This is normally done using automatic scanning techniques and a computer program to analyse the results.

When the plano optical component is intended to be an optical window its quality can be measured by inserting it between two plates forming the air film of a Fizeau interferometer. The use of a laser source allows the size of the air film to be increased and the component can be conveniently placed in the gap. The multiple-beam fringes show any homogeneity very clearly (Plate 10b).

14.13 EQUAL-PATH INTERFEROMETERS

The Michelson interferometer shown in *Figure 14.10* was modified by **Twyman and Green** to form a very useful testing instrument for optical shop use. When using collimated light from a monochromatic source the system can be considered

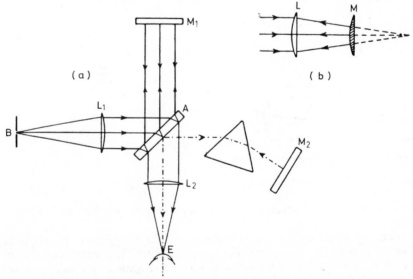

Figure 14.15 Interferometer method of testing prisms and lenses

to be a Fizeau interferometer when the reflected image of M_2 is close to M_1 but with the two beams separately available. The system as arranged in *Figure 14.15* can be used for testing prisms or optical windows. If mirror M_2 is replaced by a mirror under test that too may be interferometrically compared with M_1. For lenses the arrangement shown in (*b*) may be used where L is a lens producing a convergent beam. The test surface M is placed so that the beam is returned along its own path and therefore is re-collimated by L to interfere with the beam from M_1. A range of different radii of curvatures may be measured by this system and concave surfaces by replacing L with a divergent lens.

The results are seen as fringe patterns in which departure from circles or

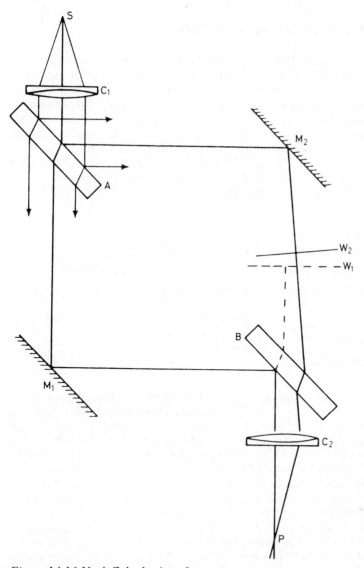

Figure 14.16 Mach-Zehnder interferometer

straight lines indicates a less-than-perfect component. The interpretation of the patterns can be difficult and interferometric testing is generally reserved for only the highest quality of optical component.

As has been indicated in section 14.12 variations in refractive index affect the fringe pattern and the **Mach-Zehnder** interferometer uses two partially reflecting elements (A and B) to provide the square arrangement in *Figure 14.16*. This has been used extensively in wind-tunnels where the different refractive index of air at different pressures shows up in the fringe pattern the airflow and shock-waves occurring in the test area of the tunnel which is crossed by one of the interfering beams.

14.14 UNEQUAL-PATH INTERFEROMETERS

Large path differences between interfering beams make heavier demands on the coherence of the light source particularly if the path difference is traversed many times as in multiple-beam interference. The availability of lasers as sources has stimulated considerable developments in this type although the **Fabry-Perot** was first conceived in the last century. This interferometer consists of nothing more than two glass or quartz plates with very plane surfaces held parallel a small distance apart by a hollow cylinder of invar or silica as in *Figure 14.17*. When the separation is fixed as is normally the case the instrument is called a **Fabry-Perot Etalon**. The inner surfaces of the plates are coated to give very high reflectivities and the plates themselves are made slightly prismatic to avoid any interference effects inside them.

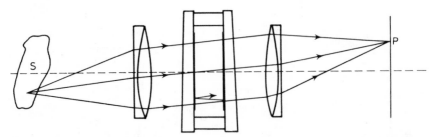

Figure 14.17 Fabry-Perot etalon

The prime purpose of the device is as a form of spectrometer. When the light from a monochromatic source is collimated and passes through the mirrors the multiple reflections result in very fine circular fringes, shaped as the sharpest in *Figure 14.9* so that the fine circles are bright against a dark background. Any other wave-lengths present in the source are easily seen even if their wave-lengths differ by considerably less than 0.1 nm from each other.

The **Fizeau interferometer** vies with the Twyman-Green for the testing of optical components. The collimating lens of *Figure 14.14* may be replaced by a multi-element lens based on aplanatic theory (section 18.3) as with a point source only spherical aberration has to be corrected. *Figure 14.18* shows the design of this system and different amounts of convergence of the beam may be obtained by taking out the component lenses starting with L_1 and working upwards. In the figure a convex surface is being compared with the exit surface of the lens

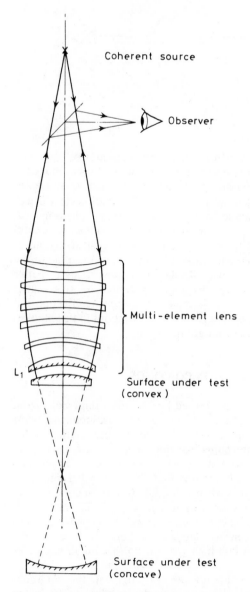

Coherent source

Observer

Multi-element lens

L_1

Surface under test
(convex)

Surface under test
(concave)

Figure 14.18 Fizeau interferometer for spherical surfaces

with a path difference of a few centimetres. A concave lens can be located at a position as shown for comparison with the same surface with a path difference of some feet. These path differences may be easily accommodated with a laser source which will even allow multiple-beam fringes.

A type of interferometer developed only since the advent of the laser is now in common use for measuring distances in workshops and laboratories (*Figure 14.19*). It may be considered as a very unequal path version of the Michelson interferometer. The system uses a stabilised laser (L) as its source. Light from

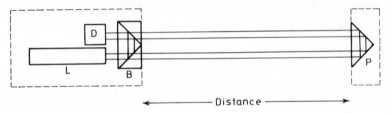

Figure 14.19 Distance measuring interferometer

this encounters a beam-splitter (B) which reflects a small amount of light. The main beam then traverses some distance before being returned by a corner cube prism (P) to another section of the beam-splitter which reflects most of it on to a photo-detector (D). The internal beam also reaches the photo-detector and broad interference fringes can be caused to occur. As the corner cube is moved, the path difference is changed and these fringes move rapidly across the photo-detector, which therefore generates electrical pulses as each bright fringe moves across it. If these fringes are counted and divided by the appropriate factor the distance moved by the corner cube can be displayed to an accuracy of better than 0.000004 inch. It is necessary that the counting system be such that it counts negatively for movement of the corner cube in the opposite direction and that the refractive index of the atmosphere is known as variations in this requires adjustment of the factor used in the computation.

14.15 SINGLE-LAYER ANTI-REFLECTION COATINGS

The interference effects of thin films as typified by Newton's rings and Fizeau interference have a usefulness beyond that of optical shop measurement. While the particular shape, spacing and order of the fringes convey information on the film thickness so it is possible by controlling that thickness to obtain a particular level of light reflectance which is uniform over a lens or other optical component if the film thickness is uniform. This effect was first observed by Fraunhofer as reducing the reflection at glass surfaces when a thin film or tarnish developed on exposure to the atmosphere of the glass used in lenses. Artificial methods of obtaining the tarnish were based on chemical etching and developed by Dennis Taylor in the early years of the century. Modern methods are based on the evaporation of particular substances in a high vacuum, so that no air molecules impede transfer to the optical surface.

For a single film to give zero reflection it is necessary to create the conditions for a dark fringe all over the surface. However, a further condition exists as the principle of superposition requires that the interfering beams must be of equal amplitude if they are to cancel out completely.

The fraction of the amplitude of the incident light reflected at normal incidence from a surface between two media of refractive indices, n and n' is given by Fresnel's expression: (see also section 16.4)

$$\left(\frac{n - n'}{n + n'}\right). \tag{14.10}$$

When a film is deposited on to glass, at the first surface of the film $n = 1$(air) and if n_f is the refractive index of the film, the fraction of the amplitude of the incident light reflected will be

$$\left(\frac{1 - n_f}{1 + n_f}\right) \qquad (14.10a)$$

At the second surface of the film in contact with the glass of refractive index n_g the fraction of the amplitude of the light reflected will be

$$\left(\frac{n_f - n_g}{n_f + n_g}\right) \qquad (14.10b)$$

Hence if the *intensities* of the light reflected from each surface of the film are to be equal

$$\left(\frac{1 - n_f}{1 + n_f}\right)^2 = \left(\frac{n_f - n_g}{n_f + n_g}\right)^2 \qquad (14.11)$$

from which we have $n_f = \sqrt{n_g}$ or the refractive index of the film should be equal to the square root of the refractive index of the glass. As reflection takes place each time at a surface between a rarer and a denser medium there will be no difference in phase between the two beams due to reflection alone. Therefore, in order that there shall be extinction of a particular wave-length at normal incidence, the thickness of the film must be such that

$$n_f t = \left(k + \frac{1}{2}\right)\frac{\lambda}{2} \quad \text{where k is an integer.}$$

That is, $n_f t$ the optical thickness, must be an odd number of quarter wave-lengths. Usually only one is applied and this is known as a **quarter-wave coating**.

Not all substances evaporate easily or deposit on a clean glass surface to form a hard durable film. The following is a short list of commonly used materials with their refractive indices.

Magnesium fluoride	MgF_2	1.38	Silicon monoxide	SiO 2.0
Silicon dioxide	SiO_2	1.45	Zinc sulphide	ZnS 2.3
Aluminium oxide	Al_2O_3	1.65	Titanium dioxide	TiO_2 2.35

For optical crown glass extinction requires a film of index $\sqrt{1.52} = 1.233$ but the lowest useful film index is magnesium fluoride which has a refractive index of 1.38. Using expression 14.10a, we find the amplitude reflected at the air/film interface is 0.159 (ignoring signs). Using the expression 14.10b, we find the amplitude reflected at the film/glass interface is 0.048. These amplitudes are in anti-phase for a quarter wave coating and so the resultant amplitude is given by 0.159–0.048 = 0.111. The square of this value gives the fraction of the intensity reflected which is 0.012 or 1.2%. Although this is not zero it is a considerable improvement over the 4% of the untreated glass. This latter value is found from equation 14.10 using $n = 1$ and $n' = 1.52$ and squaring.

Although the index of magnesium fluoride is too high for spectacle crown glass, for higher index glass types the performance of the single-layer coating improves. With a glass index of 1.9 it is theoretically possible to obtain zero reflectance as $\sqrt{1.9} = 1.38$.

These figures have been calculated for normal incidence and are effective for the wave-length at which the refractive indices are measured and for which the coating has a quarter-wave optical thickness. For other wave-lengths the result is less good so it is general practice to choose a design wave-length near to the centre of the visual range such as 510 nm. This means that most light is reflected in the red and blue extremes of the spectrum which gives rise to the purplish tinge on coated lenses and the old term, **bloomed lenses**. At angles other than normal incidence the optical thickness is affected by the cos*i* term and the Fresnel expressions change. Yet again any departure from the design wave-length gives a poorer result overall. *Figure 14.20* shows the reflectance values obtained at different angles of incidence for different wave-lengths.

The value of anti-reflection coatings is two-fold. In the first instance the reduction in reflection gives an equivalent increase in transmission as the absorption of the film is negligible. This may not seem large for a single surface but for a lens having six separate components there are 12 surfaces at which light is lost. If each surface transmits only 96% the total transmission of the lens is $(0.96)^{12}$ = 61%. For coated surfaces transmitting 98.8%, the total transmission is 87%.

The second advantage of coated surfaces is that the 40% portion of the light reflected in the example above is reflected about inside the lens and can end up in the image plane as stray light or flare spots. With coated lenses these flare spots are much reduced and are typically purple in colour.

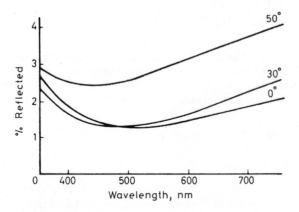

Figure 14.20 Single-layer anti-reflection coating

14.16 VACUUM COATING METHODS

The formation of a very uniform layer of material with a precise thickness in the range 0.1 μm to 1.0 μm on an optical component is not a simple process. The now traditional method comprises electrically heating the material in a boat of tantalum or molybdenum so that hot molecules of the material are ejected, and holding the object to be coated in the path of these ejected molecules which lose their heat as they strike the cold object and adhere to its surface. If this were done in an ordinary room most substances would react with the atmosphere and even those molecules ejected would be impeded by the molecules of air in their path. This means that the process must be done in a vacuum and values about 10^{-5} torr are commonly used.

As some materials liquify in the boat before evaporation the arrangement inside the vacuum chamber usually suspends the articles to be coated, commonly called **substrates**, over the point of evaporation so that they lie on the surface of a sphere. The density of the vapour stream follows the inverse square law and so it is necessary to have all the substrates to be coated at the same distance. If a large flat substrate is being coated or special accuracy is needed a rotating plate of metal may be positioned in the stream with a shape such that the stream is interrupted longer in the centre than at the edge and the propensity to deposit a greater thickness in the centre of the substrate is eliminated.

The material is evaporated at a rate depending on the temperature of the boat but this is too crude a relationship to give good control. If it is required to lay down a film with a quarter-wave thickness then the increasing thickness must be monitored during evaporation and the vapour stream interrupted as soon as the value is correct. This is done by continuously measuring the value of reflectivity from a particular substrate near the centre of the stream and having an iris diaphragm which can be closed when the reflectivity is at a minimum.

14.17 FRONT-REFLECTION MIRRORS

One of the easiest materials to evaporate is aluminium and when deposited on to a polished glass surface forms a very good mirror both for light incident first on the coating and for light which passes first through the glass. Reflecting components for use in the former fashion are called **front-reflection mirrors**. Their advantage lies in having no chromatic effects as the glass is not used optically and also that wave-lengths in the ultra-violet can be used which would otherwise be absorbed in the glass. Reflectance values of 90% are obtainable with freshly deposited aluminium.

Disadvantages stem from the tarnishing of the aluminium and that it is easily scratched. To avoid this an *overcoat* of silicon monoxide is evaporated on top of the aluminium although this causes a loss in reflectivity, particularly for blue light.

In spectrophotometers the light beams are generally focused by mirrors as they require no adjustment for variation of wave-length. If a number of mirrors are used in sequence, however, a reflectance of only 90% on each can rapidly reduce the total amount of light finally reaching the detectors in the same way that transmission values less than 100% for air-glass surfaces reduce the overall transmission of a lens. Increased reflectivity up to 99% can be obtained with pairs of half-wave layers on top of the aluminium. Here we are ·trying to get increased or *enhanced* reflection and the refractive indices of the layers are chosen to be as far apart as practicable. If MgF_2 (1.38) is deposited next to the aluminium and then ZnS (2.35) the total reflectivity increases from 92% to 97%. A further pair of layers gives 99%. This increase, as is typical with multilayer systems, is only achieved over limited wave-length ranges.

14.18 MULTI-LAYER ANTI-REFLECTION (AR) COATINGS

The disadvantage of single-layer A.R. coatings lies in the non-availability of materials of suitable refractive index for glass types of index lower than 1.9. A greater degree of freedom is found in design if two layers are used. Although a second coating material index may be even more incorrect, it now becomes possible to arrange for the larger amplitudes reflected to cancel to a more nearly zero value thereby obtaining a lower reflected intensity.

If, for example, a layer of refractive index 1.7 such as lead fluoride is interposed between the glass (1.52) and the magnesium fluoride (1.38) the three amplitudes of reflection are:

Air/MgF$_2$	0.17
MgF$_2$/PbF$_2$	0.105
PbF$_2$/Glass	0.06

If we arrange for the last two to be in anti-phase with the first the result is 0.005 giving a negligible reflectance value at the design wave-length. It should be remembered that the reflectance between air/MgF$_2$ and MgF$_2$/PbF$_2$ will undergo a change of phase of π but not that between PbF$_2$/glass so that two quarter-wave layers as in *Figure 14.21* will give the required phases.

The result for various wave-lengths (*Figure 14.21*) shows a complete extinction of reflection at the design wave-length but the reflectivity rises more sharply than for a single layer coating. It is possible with two layers to design not for zero reflection at the design wave-length but to obtain extinction at two other wave-lengths where the vector diagram forms a triangle but still gives a zero result.

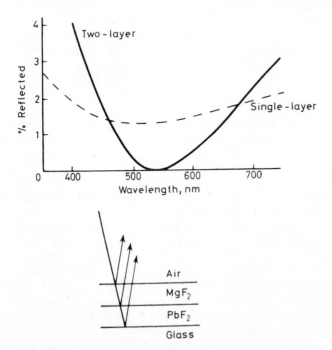

Figure 14.21 Two-layer anti-reflection coating

If three-layer systems are considered the number of degrees of freedom becomes still greater. This allows other constraints to be invoked. The durability of the coating may be considered important and the material chosen with this in mind. There is also the possibility that the coating can be 'tuned' for different substrate refractive indices by changing only the thicknesses of the layers and not the materials used. This means that the optical coating department of an optical company may meet most requirements without having to handle numerous coating materials. The actual materials and thicknesses used in these top-quality coatings tend to be proprietary information to the companies employing them.

14.19 MULTI-LAYER INTERFERENCE FILTERS

The narrower spectral response of the two-layer anti-reflection coating and the steeper sides of the three-layer types are related to the fine fringes found with multiple-beam interferometry. In general, the larger number of interfering beams, the sharper can be the effects on reflected or transmitted intensities for varying thickness or wave-length. These means that narrow spectral filters can be constructed with 10, 20 or even 40 layers of alternating materials so that the reflected light beams are alternately out of phase at a particular wave-length. This means that the transmission of the system is quite high (in the region of 60–70%) for that wave-length. However, for wave-lengths on either side of this the reflecting beams no longer cancel and high (99%) reflectance values can be obtained. The width of the transmitted light at the half-intensity points may be 5 nm or less with 20 layers. These systems are known as **band-pass filters**. Other transmission bands may exist at other wave-lengths but these can be cut out using a blocking filter (either multi-layer or absorbing). *Figure 14.22* shows a band-pass filter designed for use in colour television cameras for selecting the blue channel. The red transmission is cut out using an absorbing filter.

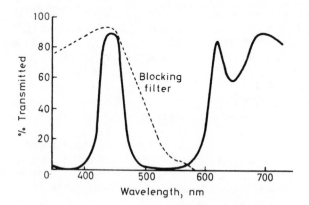

Figure 14.22 Band pass filter (blue) for colour television camera tube

Other multi-layer designs give high transmission for wave-lengths longer than a certain value and high reflectance for those shorter or vice versa. Such systems are known as **edge filters**. Both types are widely used in spectrophotometric instruments to assist in providing monochromatic beams, and in pollution-monitoring equipment where fixed wave-length beams are required.

The edge filter has found another application in the design of illumination systems in controlling the heat output of incandescent lamps. The heat is largely given out in the near infra-red part of the spectrum and would severely damage film material in high-powered projectors. Between the lamp and the film is placed an edge filter known as a **hot-mirror** which reflects the infra-red wave-lengths while allowing the visible light to pass through. Often lamps in these systems have reflectors behind them to concentrate their output on to the film. These reflectors often have edge filters deposited on them which reflect the visible light and allow the infra-red to pass through. They are naturally known as **cold mirrors** but the deeply curved shape of the reflector often makes a uniform coating difficult to achieve.

All multi-layer designs are very sensitive to angle of inclination. Temperature changes cause the layers to change in thickness and some critical designs must be held at constant temperature. Filters can be designed to work at non-normal angles of incidence but a problem arises in that the reflectivities of the interfaces vary with the polarization of the light. This can be useful when polarizing beam-splitters are required (section 16.4).

14.20 OPTICAL COATINGS AND INTERFERENCE—PERTINENT QUESTIONS

In the case of a single-layer anti-reflection coating the explanation of its action may well be as follows. The light wave approaches the first surface and due to the difference in refractive index across the interface a small fraction is reflected. On reaching the second interface a further fraction is reflected which interferes with the first reflected wave giving zero (or reduced) reflectivity.

Many students have asked the question, how does the light on reaching the first surface know of the presence of the second and not be reflected? The answer to this may be given in various degrees of erudition and mathematical rigour. However, the student is reminded of the points made in section 12.9 (developed further in section 15.12). The wave train builds up at the start in a gradual manner so that many wave crests have passed before full amplitude is reached. Thus between each half-wave (the phase difference of the 'reflected' portions) there is very little difference even during the early waves. Thus it is not a case that the reflection at the first surface has to be annulled in some way but rather that it is not allowed to form as the wave train reaches and crosses the interface system.

This whole question becomes less than hypothetical when the modern theory of light is adopted. This states that the wave motion associated with light is no more than a useful description of the fact that any given photon need not react with matter in a standard manner. It has a range of reactions available to it and the wave motion calculations are in fact determining the probability of particular events. Over fifty years ago at the University of Cambridge the Young's slits experiment was carried out with such a weak light source that most of the time only one photon was on its way from the source to the photographic plate which was the screen. Yet after an exposure of some months the interference pattern was found to be on the plate. Recent experiments with a laser source and sensitive photon detectors have shown that the expected pattern slowly built up from the apparently random arrival of each photon.

Thus the wave-motion is associated with each photon and not with the beam as a whole. In other words we can say that photons are components of light whereas waves are a description of it.

CHAPTER 14   *EXERCISES*

1. Explain the conditions necessary for interference to occur between two beams of light, and describe two methods of producing these conditions.

2. The effect at a given point of two beams of mutually coherent light depends upon the difference in phase between the two disturbances when they reach the point. Show that the difference in phase can be expressed in terms of the path difference by the relation

$$\text{phase difference} = \frac{2\pi}{\lambda} \times \text{path difference}$$

3. Two point sources emitting disturbances of 6 mm wave-length in a horizontal plane are 2 cm apart. Show in a diagram the lines along which there is no difference in phase and along which there are differences in phase of 1, 2 and 3 wave-lengths.

4. Describe a laboratory method of obtaining interference fringes in a manner based on Young's experiment.
Explain what measurement should be made and the calculation necessary to enable the wave-length of the light to be determined from the experiment.

5. Explain clearly why, to exhibit interference effects with incoherent sources, some arrangement such as Fresnel's bi-prism or mirrors is necessary.

6. A narrow slit illuminated with sodium light ($\lambda = 589$ mμ) is placed 15 cm from a bi-prism of angle 179° ($n = 1.5$); find the separation of the dark interference bands on a screen 1 metre from the prism.

7. A Fresnel bi-prism ($n = 1.5$) with angles of $^1/_2$° at its two edges is used to produce interference. It is placed 10 cm in front of a narrow illuminated slit and the interference fringes on a screen 1 metre from the prism are found to have a separation of 0.8 mm. What is the wavelength of light used? Explain what would happen if a very thin sheet of glass were placed in the path of the light from one half of the prism.

8. Two narrow slits 0.5 mm apart are illuminated with light of 600 nm wave-length, forming interference fringes on a screen 1 m from the slits.
(*a*) How far apart (centre to centre) are the dark bands in the pattern?
(*b*) A thin film 0.1 mm in thickness and of index 1.6 is placed over one slit. How far, and in which direction, are the fringes displaced on the screen?

9. Light from a narrow slit passes through two parallel slits 0.2 mm apart. The interference bands on a screen 100 cm away are found to be 3.29 mm apart. What is the wave-length of the light used? Describe a method other than the above for determining the wave-length of light.

10. Bi-prism fringes are produced with sodium light ($\lambda = 589$ mμ). A soap film ($n = 1.33$) is placed in the path of one of the interfering beams and the central bright band moves to the position previously occupied by the third. What is the thickness of the film?

11. In a Newton's rings experiment a curved glass surface rests in optical contact on a plano glass surface. The diameter of the 5th dark ring viewed in reflection is 9.96 mm for light of wavelength 628 mμ. The surfaces are then separated, coated to give a reflectivity of 90% and replaced in contact as before. Describe the new fringe pattern and derive the diameter of the 5th fringe. (Assume that the change of phase on reflection remains the same.)

12. Explain carefully the reason for the colours seen in thin films.
A convex lens is placed on a plane glass surface and illuminated with sodium light incident normally; if the diameter of the tenth black ring is 1.5 cm, find the curvature of the surface (wave-length = 589×10^{-7} cm).

13. Explain the appearance of coloured bands when light is reflected from a thin film of variable thickness. Describe a practical application of this phenomenon.

14. Newton's rings are formed between a plane surface of glass and a lens. The diameter of the third black ring is 1 cm when sodium light ($\lambda = 589 \times 10^{-7}$ cm) is used at such an angle that the light passes through the air film at an angle of $30°$ to the normal. Find the radius of curvature of the lens.

15. Circular interference fringes are formed by reflection at the air film enclosed between a plane and a spherical surface (Newton's rings). Show that the radii of the dark rings are proportional to the square roots of the natural numbers and the radii of the bright rings to the square roots of the odd numbers.

✓16. Explain with the aid of a diagram the formation of Newton's rings between two surfaces of approximately equal curvature.

A plano-convex lens ($n = 1.523$) of exactly one-eighth dioptre surface power is placed convex surface downwards on an optical flat. The rings are observed by means of a travelling microscope provided with a vertical illuminator, sodium light being used. Calculate the radii of the first and the tenth dark rings observed.

17. Newton's rings are formed in the case of an air film between an upper surface of radius r and a plane surface, normal incidence. Calculate the positions of the first six bright and dark rings for light of the three wave-lengths, viz.:

$\lambda = 680$ mμ, 589 mμ, 450 mμ

Plot the results by smooth (sine) curves showing the variation of light intensity from the centre outwards and hence deduce approximately the first series of Newton's colour scale.

18. Explain the production of Newton's rings by the light *transmitted* by a thin transparent film. Why are these rings of complementary colour to those formed by reflection from the same thickness of film?

19. Explain the difficulty in obtaining interference fringes by reflection at a film of appreciable thickness when the source emits white light. How is the difficulty affected by using (a) Light from a sodium lamp and (b) Light from a He-Ne Laser?

✓20. Two plane plates of glass are in contact at one edge and are separated at a point 20 cm from that edge by a wire 0.05 mm diameter. What will be the width between the dark interference fringes formed when light of wave-length 589 mμ falls normally on the air film enclosed between the plates?

21. Give a brief account, with an explanatory diagram, of the optical arrangement of a Michelson interferometer. Good fringes were observed with such an instrument with monochromatic light; when the movable mirror is shifted 0.015 mm, a shift of 50 fringes is observed. What is the wave-length of the light used?

22. A Michelson interferometer is adjusted to give the brightest possible fringes with a source of sodium light, which is emitting light of two wave-lengths $\lambda = 589.0$ mμ and $\lambda = 589.6$ mμ. On moving one of the mirrors a position is found where the fringes disappear, the maximum of one set falling on the minimum of the other. How far has the mirror been moved from its original position?

23. What are mutually coherent sources of light? Why are they needed in order to exhibit visible interference effects? How may they be produced from incoherent sources?

Interference fringes having been produced on a screen, a thin plate of transparent material is introduced into the path of the light from one of the sources. Explain briefly, with a diagram, what happens to the fringes, and why.

24. Interference fringes are formed by an arrangement such as a Fresnel bi-prism. When a thin flake of glass is introduced into one of the two interfering beams, the fringes are laterally displaced. Explain, with a diagram, why this is so and in which direction the displacement occurs.

If the refractive index of the glass flake is 1.5, the wave-length of the light used is 6×10^{-4} mm and the central bright band moves to the position previously occupied by the fifth, calculate the thickness of the glass.

25. Anti-reflection coatings are applied to the surfaces of lenses and optical parts in order to reduce the amount of reflected light and increase the transmitted light. Explain in

detail how this effect is brought about, making clear the role of the thickness and refractive index of the film. Give a diagram.

26. Explain how interferometry may be used to test the figure of a convex lens. Give diagrams of two methods.

27. Explain what is meant by a front-reflection mirror. How can a reflectivity of 98% be obtained?

28. Give an account of the method by which a very thin layer of material can be put on to a piece of glass.

29. A soap film illuminated by white light gradually becomes thinner as the liquid drains away. It is placed in front of the slit of a spectroscope so that the spectroscope receives the transmitted light. Describe what will be seen in the spectroscope eye-piece.

30. Show that the optical path difference between light reflected from the top surface and that undergoing a single reflection at the bottom surface of a parallel-sided film is $2 \mu t \cos r$, where μ and t are the refractive index and thickness of the film respectively and r is the angle of refraction.

When fringes are formed by interference in a thin film, explain why the condition is that $2 \mu t \cos r = n\lambda$ for a dark fringe when fringes are viewed in reflection, whilst for fringes viewed in transmission this condition is that for a bright fringe.

In a Newton's rings experiment the lower surface of the lens is not spherical but conical; contact is made with the glass plate over a very small flat area made by grinding flat the point or apex of the very shallow cone. What is the shape of the fringes formed and how are they spaced? When counting fringes radially outwards and viewing normally it is found that there are 10 fringes in a distance of 1 cm. What is the angle between the conical surface and the flat if the wavelength of the light used is 6000 Å?

31. Explain how the radius of curvature of the spherical surface of a plano-convex lens of low power, can be found using interference fringes.

Light, of wavelength λ, is incident normally on an air wedge of small angle, and the resulting interference fringes viewed by reflection. Derive an expression for the change in thickness of the wedge as the eye moves from one dark fringe to the next. If this wedge is replaced by a wedge of the same angle but made of a transparent plastic material of refractive index n, what change takes place in the separation of the fringes?

32. A simple interference filter is designed, using an air film bounded by parallel reflecting surfaces, to have a peak transmission at 546 nm. Calculate the thickness of film required for normal incident light and determine the shift in the wavelength of maximum transmission when light is incident at an angle of:

(a) $10°$ and   (b) $30°$ on the filter.

33. Magnesium fluoride of refractive index 1.38 is coated on to heavy flint glass of refractive index 1.7 to produce a non-reflecting surface for a wavelength of 500 nm. What thickness of coating is required? Why will magnesium fluoride not produce such an effective non-reflecting surface if it is coated on to glass of refractive index 1.5?

34. Anti-reflection coatings are applied to the surfaces of lenses and optical parts to reduce the amount of reflected light. Explain, in detail, how this is done, clearly showing the importance of thickness and refractive index of the film.

A non-reflecting layer is to be deposited on the surface of a lens having $n = 1.78$. Assuming that the coating layer has an index of 1.33 what would be the necessary thickness for zero reflection at $\lambda = 550$ nm?

35. A film index 1.4 is coated on to a glass of index 1.6 to give minimum reflection for a wavelength of 500 nm at normal incidence. Calculate the thickness of the film and the effective reflection coefficients for wavelengths of 500, 400 and 600 nm. Deduce the probable colour appearance of normally incident light reflected from the surface, assuming no change in index with wave-length.

✓ 36. A thin plano-convex lens of index 1.5 is placed curved side down on a plane glass plate and is normally illuminated from above by light of 600 nm wave-length. Newton's rings can be seen by reflection. The diameter of the 15th dark ring is 6 mm. What is the dioptric power of the lens?

Chapter 15

Diffraction and holography

15.1 INTRODUCTION

In Chapters 12 and 14 has been developed the concept that although the energy of light is propagated as photons the action of these can be most easily understood in terms of a wave motion. From an ordinary light source this wave motion is extremely complex as the motion comprises many waves in a random arrangement. The laser is a special source where the motion approaches that of a simple single wave. With wave motions such as water waves and sound it is clear that the waves bend round obstacles such as breakwaters and buildings and the discovery that the same is true of light held the wave theory of light in a position of pre-eminence for over a century.

This bending or **diffraction** of light has an important bearing on how precisely we can use light to measure size as the edges of shadows and images through lenses are found to have non-sharp forms with fringes. These effects will be described for incoherent sources and later for coherent sources where the fringing results in less precision. When these more pronounced effects can be utilised fully, the promise of **holography** can be realised as described in the final section of this chapter.

Previously we have made use of Huygen's principle in which every point on a wavefront is assumed to be the centre of a system of secondary waves or wavelets and the new wavefront is the common tangent of the secondary waves. Fresnel, by considering the mutual interference that takes place between these wavelets, succeeded in giving an explanation of diffraction effects. A more general theory was developed by Kirchhoff and applied to optical images by Fraunhofer and Lord Rayleigh. Abbe applied it to microscope images where the different parts of the object being illuminated by the same source can become partially coherent although a full understanding of imagery by coherent and partially coherent sources has only been developed within the last twenty years, notably by Born and Wolf. The simplified theory presented in the following sections assumes that the illuminating source is monochromatic *and uses very non-rigorous mathematics!*

15.2 HALF-PERIOD ZONES—SPHERICAL WAVES

Fresnel's treatment of diffraction was based on the idea of secondary wavelets emanating from all parts of the wavefront. *Figure 15.1* shows a wavefront arising from a point source at S. Fresnel maintained that the effect at P could be

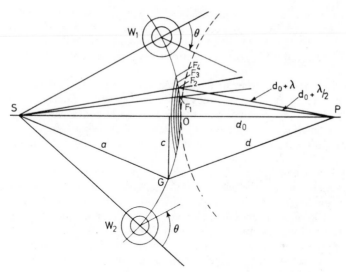

Figure 15.1 Wavelets and half-period zones

found either by allowing this wavefront to proceed until it reaches P or by dividing the wavefront into small areas so that each area can be assumed to generate a wavelet, as W_1 and W_2. Each of these wavelets then produces an effect at P which can be summed according to the principle of superposition to give the total effect of the wavefront at P. Obviously these two methods should yield identical results but the second will allow us to calculate the effect at P when parts of the wavefront are obscured by edges or apertures.

In his analysis Fresnel assumed that the *amplitude* of the effect at P of any wavelet is:

(1) Proportional to the amplitude of that part of the wavefront.
(2) Proportional to the area of the element of the wavefront generating the wavelet.
(3) Inversely proportional to the distance of the wavelet centre to P. (This is the inverse square law when intensities are considered.)
(4) Subject to an obliquity factor related to the angle θ which is zero when θ is 90° and rises to a maximum when θ is 0°. There is no wavelet in the backwards direction.

This simple theory encounters some difficulties of interpretation and justification particularly in the last assumption. However, a precise equation for this obliquity factor was not required in Fresnel's analysis as the major effect in the addition of the effects at P is the differences in relative phase of the disturbance due to each wavelet. This is correctly dealt with by this theory and remains an important contribution of Fresnel to the science of optics.

Referring to *Figure 15.1* we consider the shortest distance from the wavefront to P to be d_0 (= OP). Other parts of the wavefront will cause disturbances at P which are out of phase with that produced by the nearest part (at O) due to the longer routes giving a path difference. This may be most easily analysed by constructing a sphere centre P with a radius $d_0 + \lambda/2$. This sphere cuts the wavefront in a circle centred on SP passing through F_1 on the wavefront.

Within this circle the phase of the disturbances arriving at P varies between 0 and π. Another circle can be constructed using a sphere centre P of radius $d_o + \lambda$. This circle is also centred on SP and passes through F_2. The disturbances arriving at P from the annulus between these circles have phases varying between π and 2π.

Further annuli or zones can be constructed, each having an effect at P varying by $180°$ of phase. These hypothetical zones are known as **half-period zones** or **Fresnel zones**. It should be realised that the size of these zones will clearly depend on the wave-length of the light and the distance of the point P.

The reader may easily prove that, when the wavefront is spherical or plane and the wave-length is small compared with the distance d_o, the areas of the zones are all very nearly equal to $\pi d_o \lambda$, and the radii of their boundaries are proportional to the square roots of the natural numbers. With the short wave-length of light, these half-period zones will be very small. If d_o is 500 mm and the light of wave-length 6×10^{-4} mm the area of each zone is 0.942 sq mm and the radius of the first zone is 0.548 mm assuming a plane wavefront.

As the areas of these zones are equal they may be assumed to be sending out the same number of wavelets from equal elemental areas, but as the distance of P and also the obliquity are increasing as we go out from the central zone, the amplitude at P due to the zones is gradually diminishing from the central zone outwards. From the way in which the zones have been constructed it follows that the effect at P from any one zone is exactly opposite in phase from that of an adjacent zone. It is necessary therefore to find the resultant amplitude arising from a number of superimposed effects of very gradually decreasing amplitude and between every consecutive two of which there is a phase difference of π.

Let the amplitude at P due to successive zones be denoted by a_1, a_2, a_3, etc. As the average phase of the vibration due to adjacent zones differs by π, alternate amplitudes, a_2, a_4 etc. can be given negative values to show that the displacement is in an opposite direction to that of a_1, a_3, etc. Then the total amplitude at P will be

$$A = a_1 - a_2 + a_3 - a_4 + a_5 \text{ etc.}$$

As the amplitudes due to consecutive zones, although gradually decreasing, are nearly equal, we may say that

$$a_2 = \frac{a_1 + a_3}{2}, a_4 = \frac{a_3 + a_5}{2}, \text{etc.}$$

Then we may write A in the form

$$A = \frac{a_1}{2} + \left(\frac{a_1}{2} - a_2 + \frac{a_3}{2}\right) + \left(\frac{a_3}{2} - a_4 + \frac{a_5}{2}\right) + \text{etc.}$$

where each term in the brackets is equal to zero. The expression, if carried sufficiently far that the amplitude due to the outer zones is negligible, becomes

$$A = \frac{a_1}{2}$$

Hence, when a sufficiently large number of zones is considered, the total amplitude at a point, such as P, due to the wavelets from all points on a wave front is equal to half the amplitude due to the wavelets from the central half-period zone alone.

We have determined this result by dealing with the amplitude and phase of the effect due to each half-period zone taken as a whole. Actually, there is a continuous change in phase from 0 to π across the zone. This can be most easily analysed by using the graphical method of section 12.4. For this purpose each zone may be sub-divided into smaller rings having equal areas but differing slightly in phase. The amplitude of these zones will therefore decrease only gradually due to distance and obliquity factors.

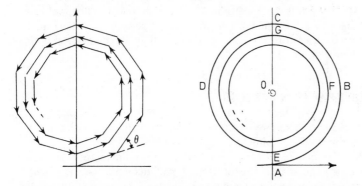

Figure 15.2 Vector addition of amplitudes–spherical wavefront

The graphical summation is shown in *Figure 15.2*, first for finite zones having equal phase differences and then in the limit as these zones become infinitely small. In this case the first half-period zone is represented by the arc ABC which is very nearly a semi-circle and the resultant amplitude is given by the length AC.

In the same way the second half-period zone is given by CDE and the resultant amplitude due to both zones is the very small value AE. If the construction is continued for a large number of zones the curve is seen to be a spiral and approaches O. Thus the resultant for all the zones from the wavefront is seen to equal AO which is closely equal to AC/2 as shown before.

Referring to any general point G on the wavefront the path difference of the effect at P is given approximately as

$$\text{P.D.} = \frac{c^2}{2a} + \frac{c^2}{2d} = c^2 \left(\frac{a+d}{2ad} \right)$$

using the sag formula in *Figure 15.1*. This gives a phase difference of

$$\delta = \pi \left(\frac{a+d}{ad\lambda} \right) c^2$$

which is sufficiently accurate as c is usually very small compared with a and d.

15.3 APPROXIMATE RECTILINEAR PROPAGATION OF LIGHT

It has been shown that in the case of light, the half-period zones will be very small, and a large number will be contained in quite a small area about the central point O. The effective portion of the wavefront may therefore be considered as confined to this small area, as the effect of the outer zones will be negligibly small as compared with that of a large number at the centre.

An object ordinarily considered small has nevertheless dimensions large as compared with the wave-length of light and will cover a considerable number of zones; the resultant amplitude of the vibrations at a point P (*Figure 15.1*) behind it from the remaining unscreened zones of the wavefront will consequently be inappreciable, the effect at P being *as if* the light travelled from the source *along the straight line OP*. Thus, except that there will be diffraction effects around the edge of the shadow, of a nature indicated in the first paragraph and to be investigated below, the statement that light travels in straight lines is seen to be approximately true; and it is quite legitimate in practice to accept the rectilinear propagation of light as a law upon which to base the actions of optical instruments and to employ therefore the idea of 'rays' of light in geometrical optics. (See sections 1.2 and 18.1.)

15.4 SIMPLE DIFFRACTION EFFECTS

The theoretical results obtained by considering the wavefront as in section 15.2 are confirmed by the results obtained by experiment.

(1) *Small circular aperture*: Light from a distant point source is passed through a perfectly circular aperture of 1 or 2 mm diameter, and the light is received on a screen, or better by means of an eyepiece. As the area of the half-period zones will depend on the distance d of the screen or eyepiece, the number of zones contained in the aperture can be varied by varying the distance d. At a certain considerable distance the aperture contains only one zone, and the illumination at the centre of the light patch is a maximum, since $A = a_1$. As the eyepiece is moved towards the aperture until two zones are included, the centre becomes dark, as now $A = a_1 + a_2 = 0$ (approx.). Moving still closer until three zones are included, the centre again becomes bright, since $A = a_1 + a_2 + a_3 = a_1$ (approx.). In this way a series of alternate light and dark centres is found as the eyepiece is moved towards the aperture.

(2) *Small circular obstacle*: One of the chief objections to the theory of Fresnel at the time of its publication was that put forward by Poisson, who showed that, according to the theory, there should be little loss of illumination at the centre of the shadow of a *small* circular object. As the areas of the zones are approximately equal, a small circular obstacle intercepting a few central zones should have little effect on the total disturbance reaching the point P from the whole wavefront, and there should be a bright spot at the centre of the shadow. Arago (1786–1853) showed by experiment that this was actually the case.

The experiment may be carried out as follows:

A circular opaque object, such as a smooth edged coin or a polished steel ball about 10 mm diameter, is suspended in the path of the light from a pinhole aperture 2 to 3 metres away. An eyepiece is mounted in the centre of the shadow at about an equal distance from the object, and a small bright spot of light will be seen in the field.* (Plate 8a.) Removing the object will make little difference in the brightness of the spot. It is important that the object shall be perfectly circular and have smooth edges, as otherwise irregular portions of a number of zones will be exposed and the appearance confused.

*About twenty half-period zones are covered in this case.

If the eye, placed in the centre of the shadow, views the object without the eyepiece, the edge of the object is seen as a brilliant luminous ring, showing that the light entering the shadow is travelling as if originating at the edge of the obstacle. An interesting case of the same kind can often be seen in mountainous districts. If one is just within the shadow of a fairly near mountain or hill before the sun has risen over the edge or just after it is set, trees on the sky line appear to be lit with an intense brilliance, while birds and even flies, far too small to be seen at such a distance in the ordinary way, appear as brilliant points of light.

(3) *The zone plate*: An interesting confirmation of Fresnel's theory is provided by the device known as a **zone plate**. It follows from our consideration of the effect of the various half-period zones (section 15.2) that, if the disturbance from alternate zones can be prevented from reaching the point P, that from the remaining zones, since it all arrives in the same phase, will add up and produce a greatly increased illumination at this point. To construct a zone plate a series of concentric circles with radii proportional to the square roots of the natural numbers are drawn on white paper, and the alternate rings blackened. The diagram is then photographed, considerably reduced, on to a glass plate so that the rings are alternately transparent and opaque. The rings on such a plate will correspond to half-period zones for a certain distance of the point P, depending of course, on their size; and if the zone plate is set up at the correct distance in front of a screen, there will be a concentration of light on the screen corresponding to each point on the object. The zone plate thus behaves in much the same way as a positive lens. Zone plates of any 'focal length' can, of course, be made by reproducing the diagram to the required size. Each zone plate has more than one 'focal length' as concentrations of light can occur at other distances where the path difference between neighbouring transparent rings is 2λ, 3λ, 4λ etc. giving progressively shorter f'.

A considerable increase in the concentration of the light could be effected if, instead of stopping out alternate zones, the phases of the wavelets from these zones were changed by π. The disturbances from *all* zones would then arrive at P with the same phase.

This can be done using photographic plates of dichromated gelatin. This material can be dissolved away when unexposed but the exposed areas remain to provide a transparent area giving a path difference from the clear areas. With a path difference of $\pi/2$ a **phase zone plate** is produced. These devices may be considered to be simple holograms (section 15.16).

15.5 HALF-PERIOD ZONES–CYLINDRICAL WAVES

Many of the diffraction phenomena are best seen when the source is in the form of a very narrow slit. In such a case the wavefronts may be considered as cylindrical, and the resultant amplitude along a line through a point, such as P, and parallel to the slit can best be found by dividing the wavefront into strips rather than zones. The construction of these half-period elements is similar to that of the half-period zones. If d is the distance of P from the nearest point of the wavefront, the distances of P from the outer edges of successive half-period elements will be $d + \lambda/2, d + 2\lambda/2, d + 3\lambda/2$ and so on.

The length of these strips being equal, their areas, unlike those of the half-period zones, decrease rapidly at first but more slowly as the distance from the centre increases, the more outer strips being practically equal in area. The amplitudes due to these outer strips, being approximately equal but of alternate phase, annul one another, and the effect of the whole wavefront is due to a few central elements. The effects of these central strips are not however equal, as was the case with zones, which had equal areas.

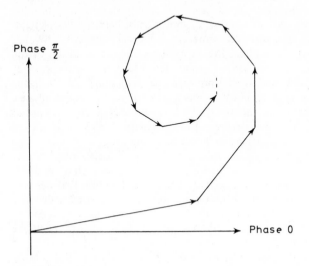

Figure 15.3 Vector addition of amplitudes—cylindrical wavefront

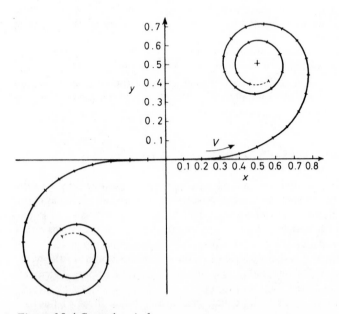

Figure 15.4 Cornu's spiral

15.6 CORNU'S SPIRAL, FRESNEL'S EQUATIONS

Because, for a cylindrical wavefront, the areas of the half-period strips are very unequal near to the axis we find that when the graphical method of section 12.4 is applied the spiral effect of section 15.2 still occurs but with a shift of centre. If narrow strips of equal phase difference are plotted as in *Figure 15.3* the amplitudes of the first few strips are large enough to give an offset before the spiral effect occurs. This can be reduced to a smooth curve when infinitely small strips are considered and the effects of those strips on the other side of the central strip show as a spiral to the opposite quadrant. The total curve known as **Cornu's Spiral** is given in *Figure 15.4*.

In this curve, the amplitudes emitted by each of the elemental strips are laid end to end but as the lengths of the strips are equal, distance along the spiral is related to distance across the wavefront. The resultant amplitude at the point P may be found from a line joining the ends of the available part of the spiral, the length of the line being proportional to the amplitude. Any part of this spiral has a slope given by the phase of the particular strip of the wavefront.

Thus from section 15.2 the phase,

$$\delta = \frac{\pi}{\lambda} \left(\frac{a+d}{ad\lambda} \right) c^2 = \frac{\pi}{2} \nu^2$$

where ν has been defined as

$$c \left(\frac{2(a+d)}{ad\lambda} \right)^{1/2}$$

It can be shown that this is the distance from the origin along the curve to the point having a phase difference δ. To obtain the equation of the curve it is seen that for a short element $d\nu$ of it the slope $dx/d\nu = \cos \delta$ and the slope $dy/d\nu = \sin \delta$.

These give

$$dx = \cos \delta \, d\nu = \cos \frac{\pi \nu^2}{2} \, d\nu$$

$$dy = \sin \delta \, d\nu = \sin \frac{\pi \nu^2}{2} \, d\nu$$

The co-ordinates of a particular point are then given by

$$x = \int_0^\nu \cos \frac{\pi \nu^2}{2} \, d\nu$$

$$y = \int_0^\nu \sin \frac{\pi \nu^2}{2} \, d\nu$$

These are known as **Fresnel's integrals** which cannot be solved as they stand, but yield infinite series which can be evaluated for specific values to give a table of the co-ordinates of the curve. In reality the equations of Fresnel were formulated first from a more rigorous analysis of his theory. The curve was developed later by Cornu in 1874 as an elegant description of the mathematics.

15.7 FRESNEL DIFFRACTION EFFECTS

The Cornu spiral (*Figure 15.4*) describes the action of a cylindrical wavefront at a given point P. Each part of the spiral relates to a part of the wavefront. The part at the origin of the graph is given by that part of the wavefront nearest to P which is the part intersected by the line SP at W (*Figure 15.5*). If all the wavefront has an effect at P the total amplitude of this effect is given by the line connecting the centres of the spirals. If some of the wavefront is obscured the total effect at P is given by a straight line joining the ends of that part of the curve related to the unobscured portion. Thus when a straight edge, X–X,

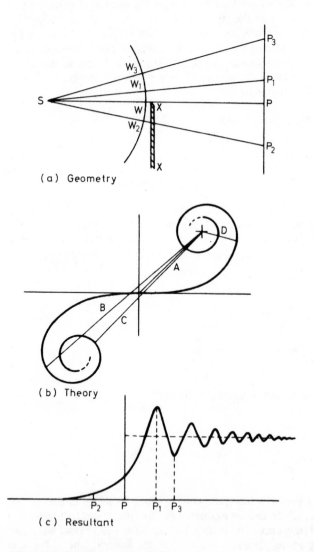

(a) Geometry

(b) Theory

(c) Resultant

Figure 15.5 Diffraction at straight edge

(*Figure 15.5a*) parallel to the source of the cylindrical wavefront obscures part of the wavefront, the effect at P is given by the line A on the Cornu spiral of *Figure 15.5b*. When squared, this gives the intensity for the position P as shown in *Figure 15.5c*.

For other points such as P_1 the nearest part of the wavefront becomes W_1 so that the unobscured part of the spiral is larger giving the line B (*Figure 15.5b*) as the resultant amplitude. This is longer than A and the greater intensity is plotted in *Figure 15.5c*.

For P_3 an even greater length of the curve is used but as this includes nearly a full turn of the lower spiral the resultant C is shorter than B. As more points

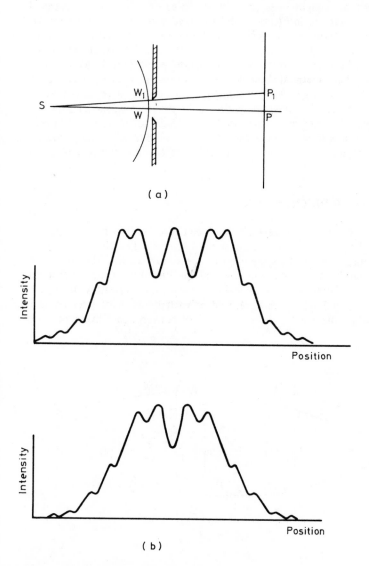

(a)

(b)

Figure 15.6 Fresnel diffraction effects at a slit

are considered the amount of the curve used travels round the lower spiral so that the straight-line resultants oscillate in length. This generates the wave-like structure to the graph of *Figure 15.5c* which shows as fringes when photographed with sufficient magnification (Plate 3).

If points inside the geometrical shadow are considered (i.e. below P in *Figure 15.5a*), the amount of the spiral used reduces. For example the point P_2 gives the resultant D which gives the intensity shown. Beyond P_2 the resultant straight line joins the centre of the upper spiral to points on the spiral giving a steadily reducing intensity, with no fringes, although the phase angle is varying rapidly.

Because the upper part of the wavefront in *Figure 15.5a* is completely unobscured the resultant straight lines all start at the centre of the upper spiral. If now a slit is considered as in *Figure 15.6a* this fixed point no longer occurs. The width of the slit determines the length of the curve to be used to obtain the resultant amplitude but the location of this on the spiral is dependent on the position of the point on the screen. For example the point P on the screen uses a length of the spiral symmetrical about the origin while the point P_1 uses a length having one end at the origin. The resultant patterns show fringes but these are critically dependent on the apparent width of the slit, that is, the length of the spiral used. Two examples are shown in *Figure 15.6b* for narrow and broader slits. If instead of a narrow slit a narrow wire is placed in front of the cylindrical wavefront, again parallel to its source, a similar effect is obtained and this is shown in Plate 4.

15.8 FRAUNHOFER DIFFRACTION

The effect of apertures on wavefronts has been discussed so far with the resultant pattern being received as a shadow on a screen. When the source of the wavefront is imaged by a lens system on to the screen we find that the image has a structure also, due to the lens imaging the diffracted light. In this case there is no need to compare the spherical or cylindrical wavefront with a spherical or cylindrical reference surface centred at each point on the screen. Any lens which images the source on to the screen converts the divergent wave-

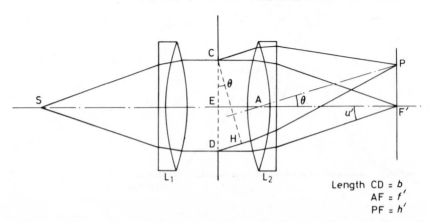

Figure 15.7 Fraunhofer diffraction. Single rectangular aperture

front into a converging one centred on the image point. For locations on the screen near to this point the comparison is between spherical or cylindrical surfaces which are tilted with respect to each other. The comparison becomes simpler if we assume that the diffracting aperture operates in parallel light between two lenses as in *Figure 15.7*. Any case where the screen is conjugate with the source is, however, known as **Fraunhofer diffraction**, and the images formed in optical instruments such as telescopes and the optical system of the eye are Fraunhofer diffraction patterns. The incident light does not need to be parallel but the instrument must be correctly focused.

15.9 FRAUNHOFER DIFFRACTION EFFECTS—RECTANGULAR APERTURES

In *Figure 15.7*, collimated light from a narrow slit S passes through the aperture, CD, and is imaged by the lens L_2 on to a screen. To develop the one-dimensional case, the aperture will be considered as the cross-section of a narrow slit parallel to the source slit S. According to geometrical optics a sharp image of the source should be formed at F' in the focal plane of L_2, but close inspection in an experimental arrangement will show that a diffraction pattern is formed on the screen consisting of a central bright band bordered by fringes of rapidly reducing intensity. As with Fresnel diffraction these fringes are coloured if a white-light source is used.

Considering first the point F' we note that this is the focus formed by L_2 of the plane wavefront emerging from L_1. If we consider the plane wavefront at CED, then the wavelets arising from all points on it will produce effects over most of the forward direction. Those that travel parallel to the lens axis and are focused at F' will all have the same path length as the lens creates the converging wavefront by making all the ray paths equal. These effects, therefore, arrive at F' with the same phase and this point, which is the centre of the pattern, is thus always a maximum and bright. In this important respect, the pattern will differ from that with Fresnel diffraction where the centre may be sometimes bright and sometimes dark.

For other points in the focal plane the lens will effectively scan the wavelets, taking into account their relative phases, over a plane tilted with respect to the axis. For the point P in *Figure 15.7* the ray paths CP and HP are equal as are the paths from all points on the line CH. With a single-slit source at S, the only wavelets available are in phase along the line CD. If they are compounded along the line CH there will be an increasing phase difference moving from C to H. If the aperture CD is divided into narrow strip elements of equal width parallel to the length of the slit, then the amplitudes due to each strip are equal and there will be a constant difference in phase between successive elements. This means that the graphical summation of these effects gives *Figure 15.8a* which in the limit reduces to the arc of a circle, *Figure 15.8b*. As different points in the focal plane are considered, the rate at which the phase difference increases will change and will give different curvatures to the arc. As the same total area of wavefront is being considered, the total amplitude and therefore the length of the arc, remains constant (neglecting the obliquity factor for small angles) but the resultant amplitude A, varies considerably, reducing to zero at times as shown in *Figure 15.8c*.

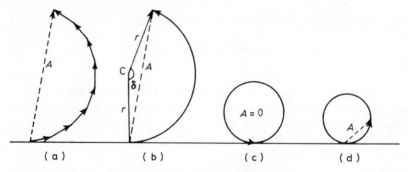

Figure 15.8 Vector addition for Fraunhofer diffraction

The resultant amplitude may be calculated more easily than under Fresnel conditions. In the simplest case, *Figure 15.8b*, the arc of the contributing vectors has an overall phase difference δ which is also the angle subtended at the centre, C, of the arc. If the length of the arc is l, the radius, r, is given by l/δ. The triangle gives the equation $A/2r = \sin \delta/2$ so that:

$$A = 2r \sin \delta/2 = \frac{2l \sin \delta/2}{\delta}$$

or, if the amplitude for zero is taken as unity (when $A = l = 1$) then

$$A = \frac{\sin \delta/2}{\delta/2} \tag{15.01}$$

The overall phase difference, δ, is generated by the path difference DH (*Figure 15.7*) so that

$$\delta = \frac{2\pi \mathrm{DH}}{\lambda} = \frac{2\pi b \sin \theta}{\lambda} = \frac{2\pi b \, h'}{\lambda f'} \tag{15.02}$$

where the angles $\angle$DCH and $\angle$PAF$'$ are equal for small values. Thus the equation for the amplitude becomes

$$A = \frac{\sin \left(\dfrac{\pi b \sin \theta}{\lambda} \right)}{\left(\dfrac{\pi b \sin \theta}{\lambda} \right)} \tag{15.03}$$

while the intensity is the square of this.

The full circles in the graphical addition diagrams occur when $\delta = 2\pi, 4\pi, 6\pi,$ etc. and give zero amplitude. This is also shown in equation 15.03, by putting the numerator equal to zero which occurs when $(\pi b \sin \theta)/\lambda = \delta/2 = k\pi$. Thus the locations of P for zero intensity are given by $\sin \theta = k\lambda/b$

or

$$h' = \frac{k\lambda f'}{b} = \frac{k\lambda}{2 \sin u'} \tag{15.04a}$$

The first dark band occurs at $\lambda f'/b$ so that the width of the central bright fringe is seen to be $2\lambda f'/b$ which varies inversely with b.

The locations of the maxima are not so straightforward. The opposite phase value, $\delta/2 = (k + \frac{1}{2})\pi$ gives

$$\sin \theta = \frac{(k + \frac{1}{2})\lambda}{b} \qquad (15.04b)$$

but this only gives the *approximate* positions as the denominator of equation (15.03) has an effect. By differentiating equation (15.01) and setting to zero we have $\tan \delta/2 = \delta/2$ for the maxima. This gives $\delta/2 = 0, 1.43\pi, 2.46\pi, 3.47\pi, \ldots$ which are different from $(k + \frac{1}{2})\pi$ although approaching it for the further locations. The amplitudes at these maxima may be calculated by putting these values of $\delta/2$ into equation 15.01 which was obtained by assuming the central amplitude to be unity*.

The values of maximum amplitude and intensity for the first three bright fringes are given in the table, along with the phase differences, δ, across the slit, which generate the maxima and minima.

Phase difference (δ)	Amplitude	Intensity	Fringe no.
0	1	1	zero
2π	0	0	minimum
2.86π	0.217	0.047	1
4π	0.	0	minimum
4.92π	0.215	0.017	2
6π	0	0	minimum
6.94π	0.091	0.008	3
8π	0	0	minimum

It is thus seen that the central maximum is of much higher intensity than the fringes and contains most of the light. The curves for amplitude and intensity against phase difference are given in *Figure 15.9*, and their appearance is shown in Plate 9a.

The above analysis has been restricted to one dimension, CD of the rectangular aperture between the lenses. If the other dimension, say CB, is considered it can be shown that this produces a fringe pattern orthogonal to the first where the locations of the maxima and minima are determined by the size of CB. The amplitude is given by

$$A(y) = \frac{\sin\left(\dfrac{\pi a \sin\varphi}{\lambda}\right)}{\left(\dfrac{\pi a \sin\varphi}{\lambda}\right)}$$

*In equation 15.01, $\sin \delta/2 = 0$, and $\delta/2 = 0$ for the central maximum. As the sine term approaches zero more slowly than the angle, the value of unity may be assumed.

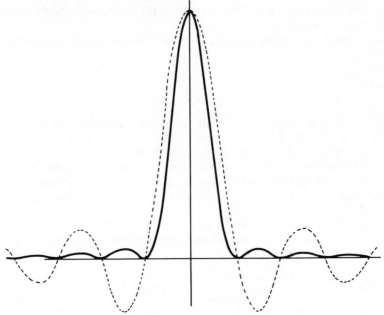

Figure 15.9 Amplitude and intensity curves for Fraunhofer diffraction from a slit. ---- Amplitude; ——Intensity

and the total pattern by

$$A(xy) = \frac{\sin\left(\dfrac{\pi b \sin\theta}{\lambda}\right)}{\left(\dfrac{\pi b \sin\theta}{\lambda}\right)} \times \frac{\sin\left(\dfrac{\pi a \sin\varphi}{\lambda}\right)}{\left(\dfrac{\pi a \sin\phi}{\lambda}\right)} \qquad (15.05)$$

Thus when one pattern gives zero amplitude, the total must also be zero so that the pattern has zero intensities along two sets of lines parallel to the sides of the rectangular aperture. Within the rectangles formed by these lines the intensity rises to a maximum.

15.10 FRAUNHOFER DIFFRACTION EFFECTS–CIRCULAR APERTURES

The most common aperture in optical instruments has a circular rather than a rectangular shape. When this is divided into strips to calculate the effect at off-axis points, their lengths are no longer constant so that the graphical addition is no longer a circle. This makes the summation extremely difficult as a double integral is involved necessitating the use of Bessel functions. However, it may be noted immediately that the system has symmetry about the optical axis so that the pattern must be circular and indeed takes the form of a bright central disc surrounded by alternate light and dark rings (Plate 8b) which will be coloured when the source is white. As with the slit fringes, the diameters of these rings are inversely proportional to the diameter of the aperture and are unrelated to the size of the source although this must be a small pinhole for best fringe visibility.

With apertures usually found in lenses, the rings will be so small as to be visible only when the pattern is viewed with magnification. Also, unless the lens is of high quality, its aberrations will mask the diffraction effects. When a lens has residual aberrations which are negligible compared with the diffraction effects it is said to be **diffraction limited**. With poor lenses, the pattern may always be artificially enlarged by placing a small aperture over the lens.

The Bessel function mentioned above is usually indicated by the symbol* $J_1(x)$. With a circular aperture, equations (15.01) and (15.03) become

$$A = \frac{2 J_1 (\delta/2)}{\delta/2} = \frac{2 J_1 \left(\dfrac{\pi b \sin \theta}{\lambda} \right)}{\left(\dfrac{\pi b \sin \theta}{\lambda} \right)} \qquad (15.06)$$

where b is the diameter of the circular aperture and θ the angle to a circle on the screen, centre F' and passing through P (as in *Figure 15.7*). The function $J_1(\delta/2)$ gives zero values for particular values of $\delta/2$ showing that the graphical summation of the amplitudes from the strips still gives a closed circuit at particular phase differences. However, the location of these zeros and the maxima between them is no longer open to the simple analysis of (15.04a) and (15.04b). The calculated values for the first three bright rings are given in the following table.

Phase difference (δ)	Amplitude	Intensity	Total fringe intensity	Fringe no.
0	1	1	1	Zero
2.44π	0	0		Minimum
3.27π	0.132	0.0175	0.084	1
4.47π	0	0		Minimum
5.36π	0.065	0.0042	0.033	2
6.48π	0	0		Minimum
7.40π	0.040	0.0016	0.018	3
8.48π	0	0		Minimum

The fourth column shows the total intensity in each ring, and it may be deduced that about 87% of the available light appears in the central disc. The radius of this disc can be considered to be the radius, h', of the first dark ring. From equation (15.02) and the table above we have that

$$\delta = 2.44 \, \pi = \frac{2 \pi b \sin \theta}{\lambda} = \frac{2 \pi b \, h'}{\lambda f'} \qquad (15.07)$$

so that

$$h' = \frac{1.22 \, \lambda f'}{b} = \frac{0.61 \, \lambda'}{\sin u'} \qquad (15.08)$$

where u' is the aperture angle (see section 17.2) on the image side and λ' the wave-length in the image space. This is usually, but not always, the same as that

*The value of $J_1(x)$ for various values of x is available in tables in much the same way that $\sin\theta$ is given for various θ although the Bessel function is rather more complex.

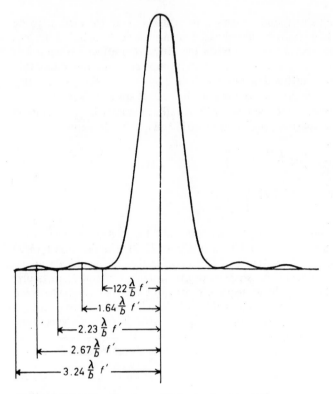

Figure 15.10 Distribution of light in the Airy Disc

in the object space; it is not so in the eye or with the immersion microscope objective.

This disc of light may be considered as the image of a point as formed by the lens. It is known as the **Airy Disc**, after the astronomer Airy who first investigated in 1834 the distribution of light at the focus of a lens. The variation in intensity is shown in *Figure 15.10* and a photograph of the pattern is given in Plate 8b.

It is thus seen that, whereas in geometrical optics we assume that a perfect lens will bring light from a luminous point to a *point* image, actually the image will be a disc of finite diameter depending largely on the wave-length of the light and the aperture of the lens. While the diameter of the Airy disc will be extremely small, it will affect the quality of images formed by lenses as described in section 15.11.

15.11 LIMIT OF RESOLUTION, RESOLVING POWER

Suppose B and Q (*Figure 15.11*) are two distant luminous points, such as two stars, subtending an angle w at a lens, then their images B$'$ and Q$'$ subtend the same angle at the second nodal point of the lens, and the separation of the images depends on this angle and on the focal length of the lens. Each image will consist of a diffraction pattern and when the separation of the images is large as

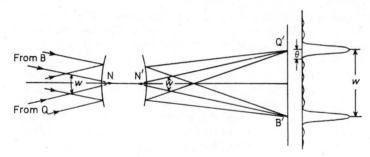

Figure 15.11 Images of two separated points

compared with the diameter of the Airy discs the distribution of light is as shown in *Figure 15.11*.

Under these conditions there will be no difficulty in seeing the two images as separate, that is, in 'resolving' the objects. As the angle *w* is reduced the images come closer together, while the diameter of the discs remains the same, and when the angle is reduced below a certain value the two discs will overlap to such an extent that the eye sees them as a single patch and is no longer able to

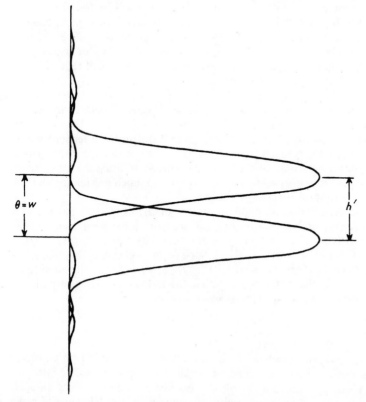

Figure 15.12 Intensity curves of two images just resolved

interpret the image as that of two object points. The smallest value of the angle *w* for which the images of two points can be detected as double is the **limit of resolution** or **resolving power** and is of very great importance in instruments, such as the eye, the telescope and the microscope, where objects with fine detail or point objects separated by small angles, such as double stars, are concerned.

In order that an expression may be obtained for the limit of resolution, it is necessary to adopt some standard for the separation of the Airy discs, such that the two object points shall be resolved. It is usually assumed that with a perfect optical system two points can be resolved if the centre of one diffraction pattern falls on the first dark ring of the other, that is, the angle *w* equals the angle θ, and the distribution curve is as shown in *Figure 15.12*. Hence it follows from expression (15.07) that the smallest angle subtended by two point objects that can just be resolved—the limit of resolution is:

$$w = 1.22\frac{\lambda}{b} \quad \text{or} \quad h' = \frac{0.61\lambda'}{\sin u'} \tag{15.09}$$

It is seen therefore that when an instrument, such as telescope or microscope, is required to have a high resolving power its aperture must be as large as possible, and this is the chief reason for the very large apertures employed in telescopes for astronomy. The large refracting telescope at the Yerkes Observatory in America, has an objective of 40 inches diameter, and it will be found from expression (15.09) that this will resolve two points subtending an angle as small as 6×10^{-7} radians or one-eight of a second in the case of green light of $\lambda = 500$ nm. As the eye is unable to separate with comfort points subtending an angle of less than about 1½ minutes, the image formed by the objective will require to be magnified about 720 times by the eyepiece.

The affect of aperture on resolution may be easily demonstrated by the following simple experiment. A piece of fine wire gauze is held in front of a brightly illuminated background and slowly brought up towards the eye, until the position is found at which it is just possible to distinguish the separate wires. If now a card with a very small pin-hole is placed before the eye, the separate wires will no longer be visible, and it will be necessary to bring the gauze much closer, so that the spaces between the wires subtend a much greater angle at the eye, before the separate wires are again seen.

In the foregoing consideration of the nature of the diffraction pattern at a focus and hence the resolving power of an instrument, it has been assumed that the optical lengths of all rays coming to the focus are equal, that is, that the emergent wavefront is perfectly spherical. Such perfection cannot be attained in practice, but Lord Rayleigh (1842-1919) in 1878 showed that the conditions would be satisfied if the difference between the longest and shortest paths to the focus did not exceed one-quarter of a wave-length.

Example:

A telescope with an objective of 40 mm diameter is focused on an object 1000 metres away; what is the smallest detail that could be visible with the instrument? If the eye can comfortably resolve two points subtending an angle of 100 seconds, what magnifying power is required for the telescope in order that the resolved detail shall be comfortably seen? (Take the wave-length of light as 500 nm).

From equation (15.09), the angle subtended by smallest detail resolved

$$w = \frac{1.22\lambda}{b} = \frac{1.22 \times .00055}{40} = 0.0000168 \text{ radians}$$

As the angle is small, $w = h/l$, where h is the size of the object and l its distance. Then

$$\frac{h}{1000} = 0.0000168$$

and so

$$h = 0.0168\text{m} = 16.8\text{mm}$$

The image of this object formed by the telescope is required to subtend an angle of 100 seconds = 100/206000 radians.

$$\text{Therefore required magnifying power} = \frac{100}{206000 \times 0.0000168} = 29$$

In considering the resolving power of a microscope it will be convenient to consider the cones of light entering the objective from object points. From expression (15.08) the smallest distance between two image points that can be resolved, $h' = 0.61\lambda'/\sin u'$. Assuming that the objective satisfies the sine condition (see equation 18.30) i.e. $n h \sin u = n' h' \sin u'$, then the minimum resolvable interval in the object space

$$h = \frac{n' h' \sin u'}{n \sin u} = \frac{0.61 n' \lambda'}{n \sin u} = \frac{0.61\lambda}{n \sin u}$$

where λ is the wave-length in air.

As proposed by Abbe, $n \sin u$ is known as the **numerical aperture** of the objective and is represented by N.A.

In the foregoing discussion the resultant of the curves in *Figure 15.12* has been obtained by adding the intensities of the two diffraction patterns and it has therefore been assumed that the two luminous points are incoherent. This is by far the most usual case. If, however, the two sources are coherent or have a high degree of mutual coherence we must add the amplitudes of their respective patterns before squaring to obtain the intensity. When this is done it is found that at the just resolved position for incoherent sources, the pattern for coherent sources is not resolved. A comparison between the two patterns is given in *Figure 15.13* where the curve for two coherent sources also assumes that they are in phase. For two coherent sources in antiphase the resultant intensity pattern is easily resolved.

In the case where a telescope is being used to view two stars there is no question that the analysis of the diffraction patterns will be on an incoherent basis. The case of a microscope is more difficult, however, as the objects on the stage are being illuminated by a single light source which although incoherent can give rise to partially coherent conditions for objects close together, when their wave-fronts will tend to be in phase.

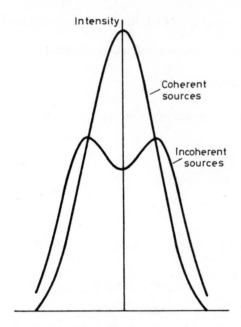

Figure 15.13 Resultant intensity curves for close objects

15.12 DIFFRACTION AND OPTICAL PROCESSING

We have seen that the location of the first dark ring (or fringe) in a Fraunhofer diffraction pattern is inversely proportional to the diameter (or width) of the aperture used. The pattern is centred on the image of the source so that if the aperture is very small compared with the lens in an arrangement similar to *Figure 15.7*, the pattern will be larger (and fainter) but still centred on the image of the source. If now a large number of equal apertures are irregularly distributed over the screen carrying the original aperture then each will form an identical pattern in the same place. Thus the intensity of the pattern will increase but, except for the central disc where the effect is modified by the spacings of the apertures, the rings will have the same diameters for a large number of circular holes as for one circular hole.

It has been shown, first by Babinet (1774–1862), that if the circular apertures in the opaque screen are replaced by circular opaque particles of the same size on a transparent screen, the diffraction effect is exactly the same, and examples of this phenomenon can be easily produced. On looking at a white luminous point on a dark background through a plate of glass covered with lycopodium dust, which consists of minute spheres of equal size, brilliant coloured rings will be seen around a bright centre. Similar rings will be seen through a dried blood film, a steamy glass and also close round the sun and moon in hazy weather. In these last two cases the rings will not be so brilliant as with lycopodium and blood, as the particles of moisture causing the diffraction are not all of the same size. These rings are known as **coronas** or **halos**, although the latter term is also applied to the larger rings often seen round the sun and moon, and which are due to refraction.

If we can measure the angle subtended by a given ring for a particular colour, we can find the radius of the particles from expression (15.07).

Young devised a simple piece of apparatus, which he called an **eriometer**, for this purpose. This consists of a metal plate having a small circular hole about ½ mm diameter drilled in the centre and around this at a distance of about 10 mm a circle of smaller holes. The plate is attached to a graduated bar and is illuminated from behind, preferably with monochromatic light. On looking at the central luminous point through a diffracting screen of small particles, the diffraction rings are seen, and by adjusting the distance between the screen and the plate a ring can be made to coincide with the circle of small holes. The distance is read on the scale; then if the angle is small, the distance will be approximately inversely proportional to the angle, and therefore proportional to the diameter of the particles. In this way Young compared the diameters of a great number of small particles, blood corpuscles, wool fibres, etc. The principle has been applied to the examination of the halos seen in glaucoma,* and to the measurement of fibre diameters in textiles.

Although very simple in use, Young's apparatus is, in fact, an example of what is now called **optical processing**, and for an understanding of this and the remaining sections in this chapter it is necessary to consider the concepts of **Fourier transforms**. When the summation of simple harmonic motion was developed in section 12.4 it was seen that two sine waves of differing amplitude and frequency could generate a waveform of complex shape. Fourier demonstrated that any form, complex or simple, could be mathematically analysed in terms of sine and cosine wave forms of varying amplitudes and frequencies. This applies both to wave motions and to wave forms. A single perfect sine-wave motion can be described as a single frequency, as we have seen. If this wave motion is modified in any way other frequencies are required. This applies even to such simple modifications as starting and stopping the wave motion. Thus

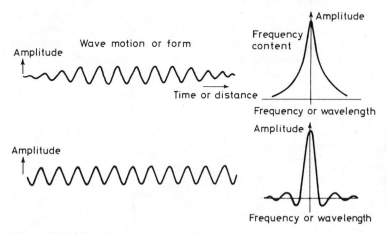

Figure 15.14 Fourier transforms—diagrammatic

*H. H. Emsley and E. F. Fincham. *Diffraction halos in normal and glaucomatous eyes*. Trans. Opt. Soc. 23, 1921–22.

when light comprises short wave trains we find that many frequencies are required to define them and the light is described as 'white'. The longer the wave trains, the fewer frequencies are required and so we find that best interference effects occur with light having a long coherence length, which means highly monochromatic.

Two particular examples of Fourier transforms are given in *Figure 15.14*. The upper diagram shows that for wave motions that start and stop gradually the mathematically derived frequencies are spread about the main frequency, and this is the shape of most spectral lines. If the wave form is started and stopped abruptly the frequencies required are shown in *Figure 15.14b*. Here it is seen that some frequencies are required to be negative or out of phase with the main frequency, and the shape bears a striking resemblance to the Fraunhofer diffraction patterns of section 15.9.

Considering *Figure 15.7* again it can be seen that for a wavefront arriving at the aperture at an angle the wave form within the aperture is a sine wave abruptly terminated at each end. It is found that Fraunhofer diffraction patterns are related to the shapes of apertures in the same way as Fourier transforms, each a unique description of the other. From this, optical systems can be devised which use diffraction patterns to perform mathematical processes or pick out particular types of object from general scenes.

15.13 FRAUNHOFER DIFFRACTION EFFECTS–MULTIPLE APERTURES

If in place of the slit CD (*Figure 15.7*) we have two equal narrow slits parallel to one another, each of these slits by itself would give rise to a diffraction pattern of the form described, the various maxima and minima due to one slit occupying the same position in the focal plane as those due to the other slit, if the incident light is parallel. There will, however, be interference between the light from the two slits. Since portions of the same wavefront pass through each slit, we may consider the slits as collections of coherent sources in the same way as the virtual sources of the double mirror or biprism experiment (Chapter 14).

If the width b of the slits is equal, the amplitude at the centre of the pattern will be twice, and the intensity therefore four times, that for a single slit. At any positions in the focal plane such that the difference in path $C_2 G_2$ (*Figure 15.15*)

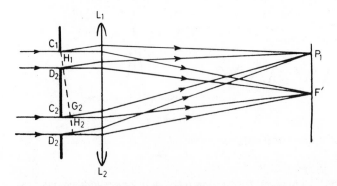

Figure 15.15 Fraunhofer diffraction. Two equal slits

of light from corresponding points on the two slits is an odd number of half wave-lengths, the intensity drops to zero, since the disturbance from any point in one slit annuls that from a corresponding point in the other. Thus a series of dark interference bands is superimposed on the diffraction pattern. The separation of these bands will depend on the separation $(b + c)$ of corresponding edges of the two slits, c being the width of the opaque space, and a number may lie in the central maximum of the diffraction pattern. For example, when $(b + c)$ is equal to $2b$, the new minima are half the distance apart of those for a single slit of width b.

Between these minima there will be maxima at positions where the difference in path $C_2 G_2$ is an even number of half wave-lengths. The resultant amplitude at these positions is the sum of the amplitudes due to each slit along the corresponding directions, and the intensity is therefore four times that at the same position with a single slit. Thus the maxima will be absent when they fall in positions corresponding to the minima of the diffraction pattern of the single slit, for in these positions the intensity due to each slit is zero. The intensity curve for two slits is shown in *Figure 15.16,* and the appearance in Plate 9b.

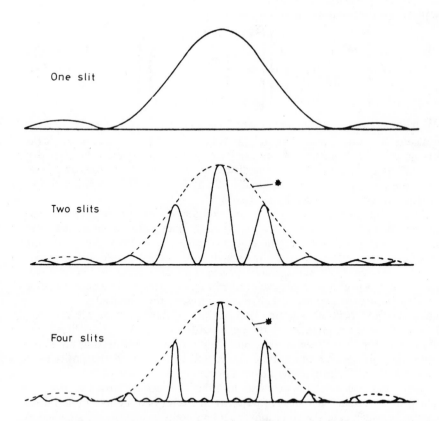

Figure 15.16 Intensity curves of Fraunhofer diffraction from one, two and four narrow slits. The heights of the curves for two and four slits are on a much smaller scale than those of the single slit.

The curve of the interference is that given for two-beam interference in section 14.6:

$$A^2 = 4a^2 \cos^2 \epsilon/2 \qquad \text{from (14.05a)}$$

where a is the amplitude of the interfering beams and ϵ is the phase difference between them. The curve of the diffraction pattern is

$$A^2 = \frac{\sin^2 \delta/2}{(\delta/2)} \qquad \text{from (15.01)}$$

where δ is the phase difference across each slit. The resultant curve is found from the product of these. However, both ϵ and δ are dependent on the angle θ, subtended by P_1 (*Figure 15.15*):

$$\epsilon = \frac{2\pi(b + c) \sin \theta}{\lambda}$$

$$\delta = \frac{2\pi b \sin \theta}{\lambda}$$

so that the total intensity curve is given by

$$I = 4a^2 \left[\frac{\sin^2 \left(\dfrac{\pi b \sin \theta}{\lambda} \right)}{\left(\dfrac{\pi b \sin \theta}{\lambda} \right)^2} \right] \left[\cos^2 \left(\frac{\pi(b + c) \sin \theta}{\lambda} \right) \right] \qquad (15.10)$$

which has a peak value four times that of a single slit.

As the first diffraction zero and the third interference zero coincide in the two-slit curve of *Figure 15.16* it can be shown from equation (15.10) that $c = 1\frac{1}{2}b$. If $c = 2b$ the zero of the diffraction pattern falls on the third interference maxima which is then virtually absent from the pattern although two very low intensity fringes mark the place.

When four equally spaced slits are considered the interference becomes multiple-beam but without the reducing amplitudes assumed in section 14.7. Thus although the fringes become sharper their profile is more akin to the diffraction patterns of single wide slits which gave circular graphical summation curves in section 15.9. More simply, principal maxima occur in positions such that the difference in path between the light from corresponding edges of adjacent slits is an even number of half wave-lengths, as the amplitudes along these directions from each slit add up. The intensity of these maxima will therefore be sixteen times the intensity at the same positions due to a single slit, that is, N^2 times, where N is the number of slits. Thus, as before, the maxima are absent where the positions in which they should fall coincide with the positions of minima in the diffraction pattern from a single slit.

Dark bands are formed in positions such that the path difference between the light from corresponding edges of adjacent slits is an odd number of half wavelengths, in which case the vibrations from any point in one slit are annulled by those from a corresponding point in an adjacent slit. There are, in addition, dark bands in directions such that the path difference from corresponding points on adjacent slits differs by an odd number of quarter wave-lengths, for in this case the path difference between alternate slits is an odd number of half wave-lengths, and the vibrations from the first slit cancel those from the third, and the vibrations from the second cancel those from the fourth. There are thus three minima between every two principal maxima. Subsidiary maxima occur between every two minima but their intensity is very much less than that of the principal maxima. The intensity curve for four slits is of the form shown in *Figure 15.16*, and a photograph of the pattern is shown in Plate 9c.

In the same way it may be shown that with any number of equally placed slits there will always be $(N - 1)$ dark bands and therefore $(N - 2)$ subsidiary maxima between every two principal maxima. When the number of slits is very large the intensity of the subsidiary maxima becomes so small as compared with that of the principal maxima that they will not be seen. The diffraction pattern then consists of the principal maxima in the form of very narrow bright lines which may be considered as images of the slit source, these being separated by comparatively broad dark spaces. These principal maxima will still, of course, be absent where their positions would correspond to the minima of the pattern for a single slit.

15.14 THE DIFFRACTION GRATING

As with multiple-beam interference, the sharper fringes associated with the four-slit pattern of the previous section means that the measurement of their position can be made more precisely. In general terms, the more slits the sharper the fringes, which leads to one of the most important applications of diffraction, the **diffraction grating**. This consists, in its simplest form, of a large number of extremely narrow equal parallel slits separated by equal opaque spaces that are usually of the same width as the slits.

The first diffraction gratings, made by Fraunhofer, about 1820, were formed of very fine wires closely and equally spaced. Later Rowland (1848–1901) produced gratings by ruling fine lines close together on the surface of glass with a fine diamond point. In these the diamond scratch may be considered as the opaque space between the transparent slits. Rowland succeeded in this way in making very accurate gratings containing as many as 20 000 lines to the inch. Owing to the very great difficulty of ruling such fine gratings, they would be too expensive for ordinary use, but replica gratings can be produced by a process invented by Thorp. A very thin cast is taken in polyester resin of the surface of a grating ruled on glass or metal, and the cast is mounted on a plane plate of glass. Such a replica gives effects very little inferior to the original ruling. The gratings in common use have rulings of approximately 10 000, 15 000 and 20 000 lines to the inch.

From what has been said in the preceding section it will be clear that when parallel monochromatic light from a distant slit or from the slit of a collimator passes through a diffraction grating and is focused by a lens on a screen or by a telescope, a central bright image of the slit is formed together with fainter bands on either side of it separated by dark spaces. As the light is passing through such a great number of slits and as these are so extremely narrow and close together, the various maxima will be sharply defined and widely separated.

The positions of the maxima, except the central one, are dependent upon the wave-length, and therefore if the light is white or any other compound light, the maxima for different wave-lengths occupy different positions, and a spectrum is formed. The different spectra corresponding to the various maxima from the centre outwards are termed the first, second, third, etc., order spectra respectively*.

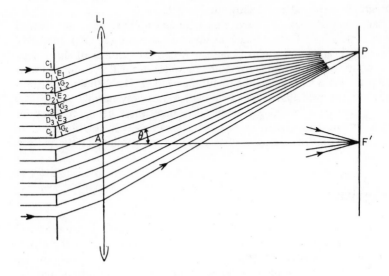

Figure 15.17 Diffraction grating

Figure 15.17 represents the section of a grating on which parallel light is incident normally. The diffracted light is received by lens L and imaged on to its focal plane. The intensity at point P is governed by two effects: the diffraction of the light from each individual slit which would, if the light from the slits were incoherent, give rise to a single diffraction pattern in the focal plane; and the multiple interference effects between light from all the slits if this is mutually coherent. When illuminated by a parallel beam from a collimated slit parallel to the slits of the grating this mutual coherence is normally very high, but the overall intensity of the interference pattern is given by the diffraction from each slit as in the two- and four-slit cases of section 15.13.

If the width of the slits is given by b then the diffracted amplitude from each slit at P is

$$a = \frac{\sin \delta/2}{\delta/2} \tag{15.11}$$

*From the previous section it will be clear that certain order spectra will be absent, this being the case when their positions would correspond to the minima of the diffraction pattern for the single slit.

where δ is the phase difference and the amplitude at F' is assumed to be unity (section 15.9). δ is given by the path difference between the sides of each slit, $D_1 E_1$, $D_2 E_2$, etc. (*Figure 15.17*) which are identical, and so

$$\delta = \frac{2\pi b \sin \theta}{\lambda}$$

The interference effect of all the slits can be calculated in a similar manner to that used in section 15.9 but with some important differences. The phase difference, ϵ, between each slit is given by the path differences $C_2 G_2$, $C_3 G_3$, etc. which are equal and so

$$\epsilon = \frac{2\pi(b + c) \sin \theta}{\lambda} \tag{15.11a}$$

where c is the opaque space between the slits. The distance $b + c$, that is, the width of one slit and one opaque space, is known as the **grating interval**.

Although the phase of the light from each slit varies across the slit, the phase difference between respective parts of slits is a constant and so we can show the phase of each slit as a single value in the summation. The length of each slit in a rectangular grating is also constant and so the amplitudes are equal. Thus the graphical summation is as shown in *Figure 15.18* where ϵ and the length, a, of each vector are constants. This is very similar to *Figure 15.8*, but the presence of

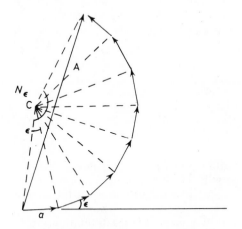

Figure 15.18 Grating—vector addition

the opaque strips means that polygon never becomes a circle. The calculation of the resultant amplitude follows from constructing bisectors of the angles between the vectors. As the polygon is regular, these all intersect at C. The angle subtended by each line at C is ϵ while the length of each bisector may be designated by r. The resultant A subtends the angle $N\epsilon$ at C where N is the number of slits. Then from the figure:

$$A = 2r \sin N\,\epsilon/2$$

$$a = 2r \sin \epsilon/2$$

Hence

$$A = \frac{a \sin N\,\epsilon/2}{\sin \epsilon/2} \tag{15.12}$$

Where (15.11) is inserted for a, the equation becomes

$$A = \frac{\sin \delta/2}{\delta/2} \times \frac{\sin N\,\epsilon/2}{\sin \epsilon/2} \tag{15.13}$$

where the amplitude at F' is taken as unity.

The expression for intensity is, of course, the square of this and writing in the full terms for δ and ϵ we have

$$I = \left[\frac{\sin \dfrac{\pi b \sin \theta}{\lambda}}{\dfrac{\pi b \sin \theta}{\lambda}} \right]^2 \left[\frac{\sin \dfrac{N \pi (b+c) \sin \theta}{\lambda}}{\sin \dfrac{\pi (b+c) \sin \theta}{\lambda}} \right]^2 \tag{15.14}$$

Putting N = 1, 2 and 4 gives the curves of *Figure 15.16*, where the overall intensity is controlled by the first term and the sharpening of the fringes by the second term. This latter expression is a maximum when $\epsilon/2 = k\pi$ at which points its value equals N* (k any integer).

This means that for the maxima

$$\frac{\pi (b+c) \sin \theta}{\lambda} = k\pi$$

or

$$(b+c) \sin \theta = k\lambda \tag{15.15}$$

In words, this means that the path differences $C_2 G_2$, $C_3 G_3$ in *Figure 15.17* are equal to a whole number of wave-lengths. The first, second, third, etc., order maxima occur in positions such that $C_2 G_2$ is equal to one, two, three, etc., whole wave-lengths respectively.

In contrast with the spectrum formed by a prism, the spectrum formed by a diffraction grating has the different colours directed into maxima whose angle θ is proportional to the wave-length. Thus the spectrum formed by one grating will be identical to that formed by another except in scale which will depend on the grating interval. The diffraction spectrum, for this reason, is often termed a **normal spectrum**. For values of k greater than 2 the maxima from different orders may overlap for different wave-lengths as the spread of the spectrum is

*Although both $\sin Nx$ and $\sin x$ becomes zero for $x = k\pi$ the sines become very nearly equal to the angles when x approaches $k\pi$ and it may be shown that the quotient equals N.

proportional to the order. When attempting to separate two wave-lengths close together it is preferable to use as high an order as possible as the width of the maxima does not change markedly with the order. However, the intensity of light in the higher orders is extremely small due to the overall diffraction pattern from a narrow slit. When the incident light is not at normal incidence it can be shown that 15.15 becomes

$$(b + c)(\sin\theta + \sin i) = k\lambda \tag{15.15a}$$

where i is the angle of incidence.

It is possible to rule transmission gratings so that each slit is in fact a prism. This has the effect of tilting the overall diffraction pattern away from the zero order and aligning its maximum intensity with one of the higher orders to one side. Such gratings are known as **blazed gratings**. It is also possible to produce reflection gratings where the slits become long narrow mirrors which can be tilted to give the blazed effect. These can be made by vacuum coating the replica gratings described earlier. A reflection grating can also be made on a polished concave surface so that no lenses are required to give sharp spectra, the light being focused by the concave mirror. Such gratings were first made by Rowland in 1880 and are commonly used in commercial spectrophotometers operating in the UV or IR where achromatic or even transparent lenses are impossible to construct. Gratings may also be made using the techniques of holography (section 15.16).

15.15 GRATING IMAGES AND THE OPTICAL TRANSFER FUNCTION

The diffraction gratings described in the last section were considered as elements in a beam of light but we may also investigate the problem of imaging such very fine objects. This was first considered by Abbe (1904) when deriving his theory of the microscope. In this, he took a transmission grating illuminated by a parallel beam as an object for lens L (*Figure 15.19*), to image on to a screen. The advantage of using a grating as an object lies in being able to apply the results of

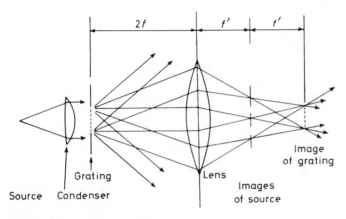

Figure 15.19 Grating images

interference and diffraction previously considered. As this parallel illuminating beam is coherent the grating will give first, and higher order beams as shown in the figure.

If the imaging lens L is of limited aperture not all these diffracted beams will pass through it. Provided the first orders are accepted by this lens, it will then form three point images in its focal plane. The images act as coherent sources to provide interference effects in the final image plane which constitute the image of the grating. If the first order diffracted beams fail to pass through the lens no image of the grating is seen as the zero order 'source' at F' has nothing with which to interfere. As the object grating is made finer the angle of the first order beam increases so that at some frequency of lines per inch the image of the grating is lost. With coherent illumination therefore it is found that the image of the grating quite quickly fades away as gratings of finer and finer structure are used as objects.

In the study of this fading it is normal to define the fineness of the grating in terms of its **spatial frequency**, that is the number of slits in a given length. This is commonly measured in cycles per millimetre or line-pairs per millimetre. Only if the grating has a sine-wave variation in transmission can it be said to have one frequency, according to Fourier's theory and even then the lack of *negative* transmission makes this less than exact. However, it is often a sufficient approximation to consider line/space gratings as having just one spatial frequency.

When such a grating is imaged by a lens it is found that diffraction and aberrations combine to reduce the clarity of the lines in the image. The ratio between the contrast of the image to the contrast of the object is generally called the **Optical Transfer Function (OTF)** or **Modulation Transfer Function (MTF)** of the lens (section 18.10). This parameter is usually assumed to be unity at coarse grating frequencies and a graph of the variation in modulation against grating frequency for coherent illumination will look like curve A of *Figure 15.20*.

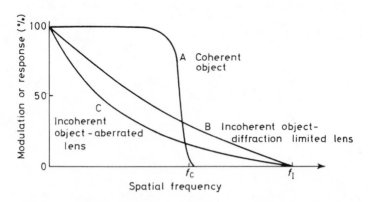

Figure 15.20 Image quality curves

From the Abbe concept that the 1st orders must be imaged by the lens we have that

$$(b + c) \sin \delta = \lambda \qquad \text{from (15.15)}$$

where $(b + c)$ is the period of the grating and θ may be set equal to u (section 17.2), the acceptance angle of the lens. So that with coherent light, we find that the maximum grating frequency ($f_C = 1/(b + c)$) which can be resolved is given by:

$$\frac{1}{(b + c)} = f_C = \frac{\sin u}{\lambda} = \frac{\text{N.A.}}{\lambda} \tag{15.16}$$

The above has assumed that the light from the individual slits is mutually coherent. This is much the case with a microscope but if the grating is self-luminous or illuminated by having an extended source imaged on to it then the calculation becomes somewhat different and rather more complex. An idea can be gained from the minimum separation distance h' for just resolved points given by (15.09) and (for rectangular apertures) by (15.04a). Although not exactly correct, the grating frequency may be regarded as equal to $1/h'$ and the maximum resolved frequency, f_I, is then $2 \sin u'/\lambda$ in the image plane or $2 \sin u/\lambda$ in the object plane. However, with incoherent light the change of modulation with frequency is much more gradual and the curves B and C of *Figure 15.20* show typical results for a perfect diffraction-limited circular lens and a lens with aberrations respectively.

Various systems are now available for measuring the optical transfer function of lenses by presenting various gratings to the lens under test and measuring electronically the modulation of the image (see section 18.11). When, as often happens in very sophisticated systems, the objects are not completely coherent or incoherent the calculation and measurement of the resolving action of lenses is very difficult.

15.16 HOLOGRAPHY, DIFFRACTIVE OPTICS

The basic concepts of holography were developed by Dr. Dennis Gabor in 1948, long before the laser was invented. Using mercury isotope lamps as his coherent sources he endeavoured to improve the resolution of the electron microscope and called his approach 'microscopy of reconstructed wavefronts.' This description is the essence of the holographic process. Normal photography recreates an object in the light and dark tones of an emulsion, but Gabor appreciated that when an observer views a real object he receives only the wavefront of the scattered light leaving that object and reaching his eyes. All the optical information must be in the phases, intensities and wave-lengths of that wavefront which are extremely complex even for simple objects. To record this wavefront for later reconstruction is normally impossible.

In holography the simplification of coherent monochromatic illumination is essential and even then it is not possible to record directly the phase values of the wavefront as these are not seen by optical detectors such as photographic emulsions and image tubes. In order to record phase effects it is necessary to add a coherent reference beam into the system which will interfere with the scattered wavefront so that differences in phase will show up as differences in intensity.

Leith and Upatnieks in 1961 introduced this reference beam at a fairly large angle to the scattered beam. This simple move creates very fine interference fringes in the plane of the photographic plate. When this is exposed and processed and returned to its original position the reference beam alone 'reconstructs' the scattered beam from the object by diffraction between the beam and the processed interference fringes. A number of 'orders' of image are created as with a diffraction grating. Some are real images while some are virtual. The first order virtual can be seen by an observer located as in *Figure 15.21b* as an image having considerable clarity of detail and shading and in three dimensions. The angle of the reference beam means that the straight-through light and the 'opposite' image are out of the field of view.

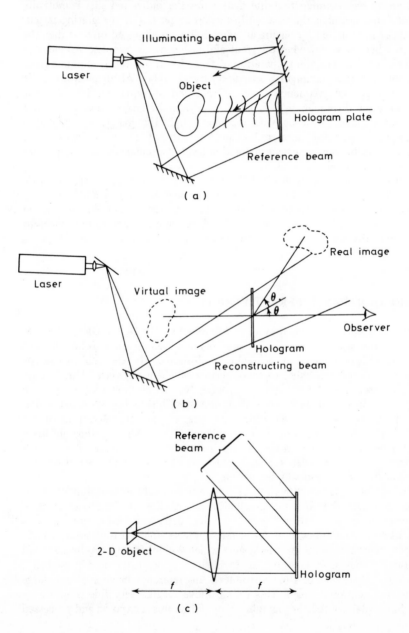

(a)

(b)

(c)

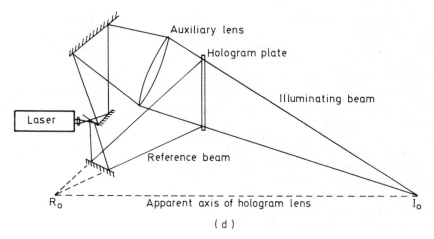

Auxiliary lens

Hologram plate

Illuminating beam

Laser

Reference beam

R_o

Apparent axis of hologram lens

I_o

(d)

Figure 15.21 (a) (b) Fresnel holography (c) Fourier-transform holography (d) Hologram lens

The three-dimensional aspect of the reconstruction is inherent in re-creating the wavefront rather than a flat representation of the object. The eyes only use two small areas of the wavefront at any one time. As the head is moved other areas are used and the object is seen 'in the round'. The wavefront is only re-constructed over the area of the interference fringes on the photographic plate so that the reconstructed object is only visible through the plate which acts as a window to the scene. The limited power of laser sources has restricted the size of hologram objects to a few feet although with special techniques holograms of rooms have been made. It is necessary to prevent incoherent illumination reaching the object or the plate.

Most holograms are made in darkrooms and it is generally necessary to mount the components on a vibration-free surface. If the exposed hologram is moved or tilted with respect to its original position in the reference beam the reconstruction distorts and detail is lost. If a reference beam of different wave-length is used the object may still be seen but at a magnification given by the ratio of the wave-lengths. Holograms recorded from radar signals or ultrasonic fields may be reconstructed using visible light and observed directly. These holograms are called **Fresnel holograms**. As might be expected it is possible to produce **Fraunhofer holograms**. These are made when the photographic plate is distant from the object compared to its size either in actuality or because a lens is used to collimate the scattered light from the object. *Figure 15.21c* shows a general arrangement using a lens. In the special case where the recording plane and the object are at the principal foci of the lens the hologram so formed is called a **Fourier-transform hologram**. This is because the intensity in the hologram at a given distance from the axis is found to be related to the amount of a given spatial frequency in the object (*Figure 15.17*). This can be used to process transparencies so that particular spatial frequencies are enhanced and others rejected—another form of optical processing.

Another form of hologram can be made when the lens of *Figure 15.21c* is positioned to form an image of the object close to the plane of the hologram.

Such **focused holograms** are more tolerant on coherence especially in reconstruction and are used in display applications. If a thick emulsion is used on the photographic plate it is possible by virtue of interference effects within the emulsion to obtain a hologram which responds differently to different colours in the reconstruction beam thus allowing colour holograms to be made.

It is seen that the image results from the reconstruction of the wavefront. If the object remains in place when the exposed hologram is returned to its original position, interference is possible between the actual object and the reconstructed image provided that both are illuminated from the same laser: thus if the object is moved slightly or deformed, two-beam interference fringes will be clearly seen. The techniques of **interference holography** have become standard in mechanical testing and vibration analysis.

When a hologram reconstructs a complex image from a reference beam it is found that a change in the angle or the collimation of the reference beam moves and aberrates the complex image. If, instead of a complex scene, a single coherent point of light is used it is found that, after processing, the hologram acts upon the reference beam to focus it as an image of the point. In other words the hologram is acting like a lens, but using diffraction rather than refraction. It is found that the illuminating beam of *Figure 15.21a* can be shone directly onto the hologram plate in the same manner as the reference beam (*Figure 15.21d*). After processing it is found that the **hologram lens** will image a point source placed at the origin, R_0, of the reference beam so that it appears as a point source placed at the origin, I_0, of the illuminating beam, neither beam being present. The point source does not have to be coherent and so we have a perfect lens *but* only for these particular conjugates, off-axis though the lens may appear to be.

The image of any other point will be aberrated in the ways described in Chapter 18. Furthermore, the use of diffractive effects (this technology is called **diffractive optics**) means that the process is very wave-length dependent. Unlike glass where the refractive index change from red to blue is a few percent at most, the power of a hologram lens for blue light will be only 50% of that for red light! This technology has been used very successfully for making diffraction gratings. It also has application in optical display systems where monochromatic light may be used. As with zone plates (section 15.4) other orders of image may occur. A great deal of current development work is directed to changing the form of the processed fringes so that all the light goes into one image. It will then be possible to use a system of holographic lenses which minimise the chromatic effects.

CHAPTER 15   *EXERCISES*

1. What must be the size of a circular opening in an opaque screen in order that it may transmit two Fresnel zones to a point 2 metres away? What will be the approximate intensity of light at this point? ($\lambda = 589$ mμ).

2. Describe, with a diagram, the Huygens-Fresnel principle of the approximate rectilinear propagation of light.

3. Write a short essay explaining how and why the geometrical laws of image formation are to be considered only as approximations from the standpoint of the wave theory of light.

4. Show that with a plane wavefront, when the wave-length is small the areas of the half-period zones for a point at a distance d from the wavefront are all very nearly equal to $\pi d\lambda$.

5. Find the diameters of the first four clear zones (the centre being opaque) of a zone plate that will 'focus' parallel incident light of $\lambda = 600$ mμ at 50 cm from the plate.

6. The diffraction bands at the end of the shadow of an opaque object decrease in width from the shadow outwards; explain why this is. When using white light these bands are coloured; state and explain the order of the colours.

7. Explain briefly the diffraction bands produced on a screen by light from a line source passing a straight edge parallel to it. Find the approximate intensity of illumination at the geometrical shadow edge and at the first bright band and first dark band; draw a graph to illustrate this variation of intensity.

8. When a thin wire is placed in the light from a slit source of monochromatic light its shadow on a screen is seen to be crossed with alternate light and dark bands parallel to the length of the wire; explain the formation of these bands. In an experiment the distance between the bright bands was 0.7 mm when the screen was 1 metre from the wire. What was the thickness of the wire? ($\lambda = 650$ nm).

9. Derive the expression $\sin \theta = \lambda/b$ for the position of the first dark band in the case of diffraction by a rectangular aperture in front of a telescope objective.

Calculate the resolving power of a telescope objective of 1½ inches (circular) aperture, assuming $\lambda = 560$ nm and that two stars can be resolved when the central maximum of one image falls upon the first dark ring of the other.

10. Explain how Cornu's spiral may be used to calculate diffraction patterns. To what type of pattern is it applicable?

11. Explain carefully the diffraction rings observed when a small light source is observed through a diffracting screen such as a glass plate covered with lycopodium powder.

In an experiment, the angular diameter of the rings was measured by using a plate containing a small central hole surrounded by a circular ring of smaller holes (Young's eriometer). Observed through lycopodium powder the distance from the diffracting screen to the hole was found to be 32.4 cm when a certain diffraction ring coincided with the ring of holes. Observed through a screen of blood corpuscles the distance for the same ring was 8.92 cm. If the mean diameter of the lycopodium grain is 0.029 mm find the diameter of the blood corpuscles.

12. Fraunhofer diffraction is observed using a square hole 2 mm side. Explain qualitatively what differences can be seen when this is replaced by a circular hole 3 mm in diameter.

13. Explain, giving a diagram, the formation of the diffraction pattern at the focus of, say, a telescope objective. What do the dimensions of this pattern depend upon and how are they associated with the resolving power of the instrument?

14. What is meant by the resolving power of a telescope? How may it be expressed and on what is it dependent?

A telescope with an objective of 2 inches diameter is focused on an object two thousand yards away: what would be the least size of the detail in the object which could be rendered visible with the instrument? ($\lambda = 570$ nm).

15. Assuming that two stars can be resolved when the central maximum of one image falls upon the first dark ring of the other, calculate the resolving power of a telescope objective, the aperture of which is 3 inches, and focal length 30 inches. Assume $\lambda = 560$ nm.

If the limit of resolution of the eye for comfortable working be taken as 100 seconds, find the focal length of the eyepiece necessary for clear resolution of the image when using the telescope.

16. Treating the eye as a perfect optical system of power 60D, aperture 3.5 mm and refractive index of vitreous as 4/3, calculate the radius of the first dark ring in the Airy diffraction pattern formed of a distant point source. ($\lambda = 560$ nm).

Taking the diameter of the foveal cones as 0.0025 mm, comment upon the eye's known capability of resolving two stars or point sources separated by less than one minute of arc.

17. Explain how diffraction limits the detail that can be seen with an optical instrument. Why does a microscope using ultra-violet light offer a better theoretical performance?

18. The angular separation of two stars is 1½ seconds. Find the minimum aperture that must be given to the objective of a telescope in order that the stars shall be just resolved by the instrument. Assume $\lambda = 560$ nm.

19. The focal length of a telescope objective is 6 inches and the magnifying power of the whole instrument is eight. Taking the resolving power of the eye as 100 seconds for comfortable working, find the angular magnitude of the finest structure the eye and eye-piece are capable of resolving and hence the minimum aperture of objective that will suffice (λ = 560 nm).

Why in practice would a larger aperture probably be used?

20. Draw carefully to scale the intensity curves of the diffraction patterns, as far as the third maximum, formed by two stars in the focal plane of a telescope objective of diameter 1½ inches and focal length 15 inches. Place the central (first) maximum of one curve over the first minimum of the other. Assume λ = 560 nm.

21. Calculate the radius of the first dark ring in the diffraction image produced by a telescope of 28 inch aperture, focal length 26 feet, λ = 560 nm.

On the assumption that two stars can be resolved when the centre of the image of one falls upon the first dark ring of the other, calculate the resolving power of this objective.

By how much would the image require to be magnified by the eyepiece in order that the eye may resolve it? The limit of resolution of the eye may be taken as 1½ minutes for comfortable vision.

22. Explain the action of the diffraction grating. State the advantages and disadvantages of using a grating to produce a spectrum as compared with the use of a prism.

23. Explain the production of spectra by a diffraction grating. Describe an experiment for determining the wave-length of light by means of a grating, stating clearly what quantities are measured and how the calculation is made.

24. What will be the angular separation of the two sodium lines (λ = 589.0 nm, λ = 589.6 nm) in the first order spectrum produced by a diffraction grating having 14,438 lines to the inch, the light being incident normally on the grating?

25. Parallel light from a mercury vapour lamp falls normally on a plane grating having 10 000 lines to the inch. The diffracted light is focused on a screen by a lens of 15 inches focal length. Find the distances in the first order spectrum between the lines corresponding to the wave-lengths.

579.1 nm	577.0 nm	546.0 nm	and	435.8 nm

26 A grating illuminated by imaging a source on to it is just resolved by a microscope. The condensing system is now changed to a lens which collimates a point source. Explain why the grating is no longer resolved.

27. Explain briefly what is meant by the diffraction of light. Give an example, to be met with in everyday life, of diffraction in the case of (a) water waves, (b) sound waves, (c) light waves. In what respects does the spectrum produced by a diffraction grating differ from a prismatic spectrum?

28. If you look through a piece of fine gauze (40 wires to 1 cm) at a narrow source, emitting light of wave-length 600 nm, placed 4 metres from the gauze, what will be the linear separation between the central and the first diffracted image?

Describe the advantages and disadvantages of using a diffraction grating to produce a spectrum as compared with the use of a prism.

29. Explain why it is possible to study much of the theory of lenses and of optical instruments generally on the basis of a ray theory rather than a wave theory of light propagation.

In what circumstances would the neglect of the wave theory introduce serious errors?

Explain why the image of a star viewed through a telescope appears smaller as the aperture of the telescope objective is increased.

30. Explain the term 'Spatial Frequency'. How can the performance of a lens be expressed using it?

31. Derive an expression which determines the positions of the spectral lines produced by a transmission diffraction grating, assuming normal incidence.

What influences the width of these lines? What effect has the width of each of the clear spaces on the lines?

A grating has 600 lines per mm. If the visible spectrum extends from 400 to 700 nm, find the linear separation of those wave-lengths in the focal plane of a telescope objective of focal length 25 cm in the second order. Assume normal incidence, first order.

32. Define Fraunhofer diffraction.

A parallel beam of monochromatic light is incident upon a plane diffraction grating at an angle of 30° to the normal. If the grating has 3000 lines per cm and the wave-length of the radiation is 632.8 nm determine the angles of all the transmitted orders.

33. Discuss briefly the physical properties of a laser light beam which make it suitable for demonstrating the interference and diffraction patterns formed by single and multiple apertures.

Show that the intensity I_P at a point P in a double-slit interference pattern is proportional to $\cos^2 v$ where $2v$ is the phase difference between light arriving at P from corresponding points in the two slits.

Sketch the Fraunhofer diffraction/interference pattern formed by a double slit, pointing out the 'missing order'. Assume $d = 4a$, where d is the separation between slit centres and a is the slit width. Sketch the resultant pattern on the same scale when (a) the slit width is doubled while the slit centre separation remains unchanged (b) the split separation (between centres) is doubled while the slit width remains at its original value.

34. Fraunhofer diffraction at a double slit may be explained as follows:
light from the two slits undergoes interference to produce fringes of the type obtained with two beams, but the intensities of these fringes are limited by the amount of light arriving at a given point on the screen by virtue of the diffraction occurring at each slit.

Explain this statement, sketch the resultant fringe pattern produced by such a double slit experiment and comment on the importance of dimensions b (slit width) and c (slit separation).

The Fraunhofer pattern from a double slit composed of slits each 0.5 mm wide and separated by $d = 20$ mm is observed in sodium light ($\lambda = 593$ nm) on a screen. How many fringes will occur under the central diffraction maxima?

Chapter 16

Polarization

16.1 INTRODUCTION

A vibratory disturbance may be propagated in one of two ways, (1) as a transverse vibration in which the vibration takes place perpendicularly to its direction of propagation; the simplest example of this type is seen in ripples on water, where the particles of the water vibrate vertically up and down, and successive vibrations are transmitted over a horizontal surface; and (2) as a longitudinal vibration in which the vibration takes place in the direction of its propagation, as in sound, where the vibrations consist of alternate compressions and rarefications of the air taking place along the direction in which the sound is transmitted. The various phenomena already described could be satisfactorily explained on the supposition that light was either one of these forms of wave motion, and the effects of interference and diffraction are found to occur in the case of both water waves and sound. There is, however, a series of phenomena, due to what is known as the polarization of light, that can only be satisfactorily accounted for on the assumption that light is a form of transverse wave motion. Some of these effects will be dealt with in this chapter.

The action of polarizing sunglasses is that of reducing reflected sunlight more than other light in the general scene. If two such sunglass lenses are put together, then it is possible, by rotating one about their common perpendicular axis, to vary the total amount of light passing through them (*Figure 16.1*). This phenomenon requires that the light between the two lenses has a preferred orientation

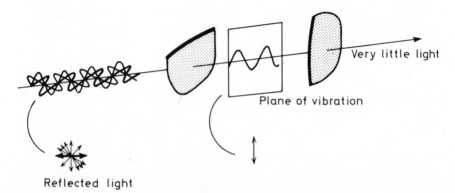

Very little light

Plane of vibration

Reflected light

Figure 16.1 Action of polarizing sunglasses

which the second lens can accept or reject. In the full theory of light, the wave motion used to describe it has a transverse varying electric field and transverse varying magnetic field which are orthogonal. Usually light is unpolarized, that is, it contains a great many wave motions with their vibrations at all angles (although each electric vector and its associated magnetic vector are always at right angles). The polarizing sunglass lens selects a particular orientation, and allows this to be transmitted while absorbing the rest (see *Figure 16.1*).

One form of polarized light is **linearly polarized light** which occurs when the light has its electric field varying in a single fixed orientation. The term 'plane of polarization' was originally used to describe the orientation of the magnetic vector but (of course) it has sometimes been used referring to the electric vector. To avoid this confusion the term **'plane of vibration'** is now used and this refers to the electric vector, being that plane containing the electric vector and direction of propagation of the light. The term 'plane of polarization' should not be used. In the paragraphs which follow we will think of light in terms of its electric field vector only.

The action of the sunglass lens described above is not that of selecting a particular plane of orientation from amongst *all* others.

The action of a light beam polarized in one plane of vibration is not restricted to that plane as it is possible to calculate its components in other planes. Thus, a disturbance a orientated at an angle θ to the x axis has a component $a \cos \theta$ in the x direction and $a \sin \theta$ in the y direction. For disturbances in all orientations between $0°$ and 2π, the intensity will be $2\pi a^2$. For the components of these disturbances in the x direction the intensity will be

$$\int_0^{2\pi} a^2 \cos^2 \theta \, d\theta = \pi a^2$$

while in the y direction,

$$\int_0^{2\pi} a^2 \sin^2 \theta \, d\theta = \pi a^2$$

Thus the intensities are divided equally between the orthogonal orientations. The perfect linear polarizer would transmit 50% of the incident **unpolarized light** in which the planes of vibration are randomly orientated. Quite often the incoming light is not completely random in this respect and exhibits **partial polarization**. The more monochromatic light becomes the more likely it is to be polarized in some way. Strictly monochromatic light always exhibits some form of polarization as it has insufficient wave trains to obtain randomness during normal observation times. When unpolarized light is divided into two orthogonal linearly polarized beams it is found that the two beams are incoherent and no interference effects can be obtained between them. This remains the case even when their planes of vibration are subsequently made parallel.

An optical system which generates polarized from unpolarized light is called a **polarizer**. When used to examine polarized or partially polarized light it is called an **analyser**. When two polarizers are used orthogonally as in *Figure 16.1* they are said to be **crossed**.

16.2 LINEAR POLARIZATION BY ANISOTROPY–CRYSTALS AND GRIDS

The earliest form of polarizers were **dichroic crystals** like tourmaline. This family of substances has crystals with long narrow lattice structures which orientate parallel to each other. This means that the material has an action on any light beam going through it which depends on the direction of the beam and on the orientation of the plane of vibration with the axis of the crystals. The material is **anisotropic**. It is useful to define a direction within the crystal where the effect *does not* occur–this is known as the **optic axis**. If a plate is cut from a tourmaline crystal with the optic axis parallel to the surfaces of the plate then an unpolarized light beam passing through the plate will have one component of its electric vectors in the optic axis and the other perpendicular to it. In tourmaline the electric vector perpendicular to the optic axis is very strongly absorbed while the other is less so. The emergent beam is thus strongly polarized provided the plate is a few millimetres thick. These effects are very colour dependent and so tourmaline changes colour depending on the angle from which it is viewed, hence the term *dichroic* which merely means two-coloured. If the plate was cut so that the light passing through it was going *along* the optic axis no effect would occur, because the absorption would be the same for the two components.

Other crystals exhibit similar properties. The earliest form of Polaroid used large numbers of dichroic herapathite crystals aligned on a plastic sheet. Once again the unpolarized light has its electric vector component strongly absorbed when it is perpendicular to the optic axis. This selective absorption occurs because some of the electrons in the crystal are partially free to move in a direction perpendicular to the optic axis but not at right angles. The electric vectors of the incident light beam try to stimulate electric currents in the material but only those perpendicular to the optic axis succeed and the resultant electron energy is absorbed into the lattice.

An alternative to long crystals is long parallel wires in the form of a grid. When the incident unpolarized beam reaches the grid the components of the electric vectors that are parallel to the wires induce currents in the wires by driving the conduction electrons along the wires. These conduction electrons lose energy either by collision with the lattice in which case the light is absorbed or by reradiating in the backwards direction in which case the light is reflected. The components of the electric vectors which are perpendicular to the grid are not able to generate electric currents across the narrow wires (if they are sufficiently narrow) and are therefore transmitted (*Figure 16.2*).

Notice that when electric vectors are considered the component *against* the gaps in the grating is transmitted. The efficiency of this device as a polarizer is critically dependent on the ratio of the wave-length of the incident light to wire spacing. *Figure 16.3* shows that d, the wire spacing needs to be at least 4× smaller than the wavelength of the light.

The first grid polarizers were made for radio waves using actual wires wound between spacers by Hertz in 1888. The spacing was 1 mm. In 1963 Bird and Parrish evaporated gold obliquely onto a plastic replica of an optical diffraction grating so that the metal only reached the tips between the grooves. This gave a grid with a period of 0.463 μm which polarized infra-red radiation of wave-length 2 μm and longer. Experimental gratings have been made (at the National Physical

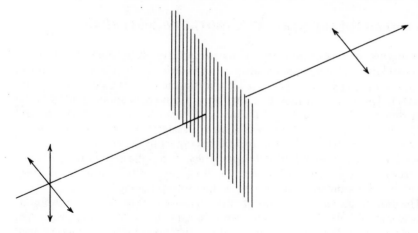

Figure 16.2 Action of grid polarizer

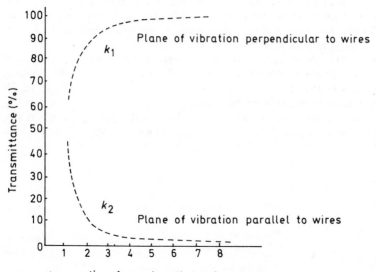

Figure 16.3 Polarizing effect of wire grid

Laboratory) with periods of 0.22 μm which clearly polarize in the visible but nothing is commercially available for these wavelengths.

The most common linear polarizer is not the dichroic Polaroid described above but a later development which is a molecular analogue to the wire grid. This E-Sheet Polaroid is formed from a sheet of polyvinyl alcohol which is heated and stretched in *one* direction. This process aligns its long hydrocarbon molecules. When the sheet is dipped into an iodine solution the iodine attaches to the long molecules forming long aligned conducting chains which act like wires. These are separated by distances of molecular dimensions so the grid polarizes in the visible although it is less good at the blue end of the spectrum.

16.3 LINEAR POLARIZATION BY ANISOTROPY–BIREFRINGENCE

Birefringence is another property of anisotropic crystals in which the electrons can move with different amounts of freedom in different directions. An example of this is Calcite or Iceland spar which is a crystalline form of Calcium Carbonate ($CaCO_3$). This cleaves readily into the form of rhombohedra or rhombs in the form shown in *Figure 16.4a*. All the sides are parallelograms of angles 78° and 102°. All the corners except two contain both angles. The two 'blunt' corners contain angles of 102° only. The two blunt corners in the equally-sided crystal shown are joined by an axis of *symmetry* which is an optic axis. In this direction the crystal appears symmetrical and therefore apparently isotropic. All other directions are asymmetric and the crystal is anisotropic along these.

The anisotropy takes the form of *two* refractive indices. When a *near* object, such as a small hole in a card, is viewed through a rhomb of calcite, two images are seen and on rotating the rhomb about the direction of view one image is seen to remain stationary while the other moves round it. The line joining the two images will always be parallel to the optic axis direction. If the blunt corners are cut off by plane faces so that it is possible to view the object along this direction it will be found that there is then only a single image and therefore no double refraction of the light travelling along the optic axis.

Because calcite has only one direction of zero effect it belongs to a class called **uniaxial** crystals. The two types of ray which give rise to double imaging are called **ordinary** rays (o-rays) and **extraordinary** rays (e-rays). *Figure 16.4b* shows how a beam of unpolarized light falling normally onto one surface of the rhomb is doubly refracted into two beams. The ordinary ray behaves as if the material were isotropic, as in the case of glass. The extraordinary ray behaves

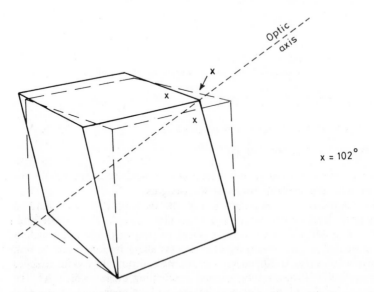

Figure 16.4a Rhombic calcite crystal showing blunt corner, optic axis and reference cube (dashed). The top and lower faces of the reference cube are co-planar with the rhomb faces

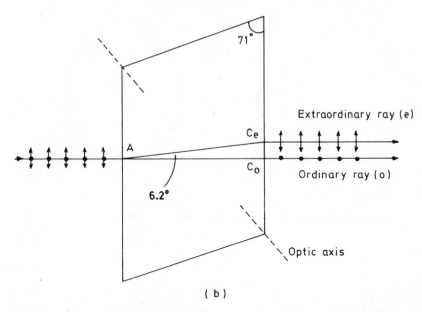

71°

Extraordinary ray (e)

C_e

C_o

Ordinary ray (o)

A

6.2°

Optic axis

(b)

*Figure 16.4b Double refraction by calcite of unpolarized beam incident normally
in section containing the optic axis*

quite differently changing its direction on entering the crystal even though the
incident light is normal to the surface. It is found that these rays are linearly
polarized in planes orthogonal to each other.

By measuring the angles of refraction of the ordinary and extraordinary rays
for different angles of incidence we find that $\sin i/\sin i'$ is a constant for the
ordinary ray but varies with angle of incidence for the extraordinary ray. Hence
we conclude that the velocity of light and the refractive index of the crystal are
constant in all directions for the ordinary ray but differ in different directions
for the extraordinary ray. Only along the optic axis of the crystal will the two
velocities be equal.

The principle of wavelets developed by Huygens (see section 1.2) and used
later by Fresnel (section 15.2) may also be applied to the propagation of wavefronts in anistropic media. In uniaxial crystals such as calcite, quartz, tourmaline
and ice, the wavefront associated with the ordinary ray is constructed from wavelets having the same velocity in all directions, that is, they are spherical wavelets.
For extraordinary rays the wavelets are ellipsoids of revolution which coincide
with the sphere of the ordinary wavelets along the optic axis, where their velocities are equal. *Figure 16.5a* shows that the velocity of the extraordinary wavelets
may be smaller or larger than the ordinary ray giving rise to the terms positive
uniaxial crystal and negative uniaxial crystal as shown. The general sphere-
ellipsoid wavelet shape for a positive uniaxial crystal is shown in *Figure 16.5b*,
which also shows the wavelet shape for a **biaxial** crystal, that is a crystal having
two axes. These wavelets are both ellipsoids with four distortions on them. The
inner ellipsoid is pinched outwards while the outer ellipsoid is dimpled in to
coincide in the same plane as the optic axes although not co-incident with them
as the axes define identical velocities and *directions*. Strongly biaxial crystals
are not often used and will not be dealt with here.

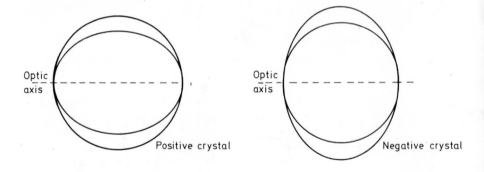

Figure 16.5a Wavelet sections for uniaxial crystals

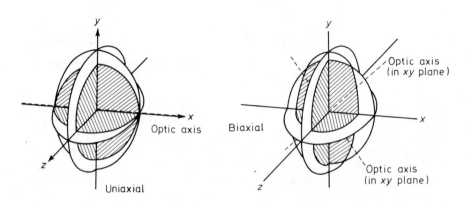

Figure 16.5b Crystal wavelets

Positive crystals				Negative crystals		
	n_o	n_e			n_o	n_e
Quartz	1.5442	1.5533		Calcite	1.6585	1.4864
Ice	1.309	1.313		Tourmaline	1.669	1.638
Mica	1.561	1.594		Sodium nitrate	1.5874	1.5361

Figure 16.5c Table of indices for birefringent crystals; Mica is slightly biaxial

In *Figure 16.6* parallel light is shown incident normally on one face of a crystal of calcite, the optic axis of which is in the plane of the paper. When the plane wavefront meets the surface AD, wavelets of the form described above travel into the crystal from each point on the surface; those originating at A and D are shown in the figure. The refracted wavefronts of the ordinary light $M_o N_o$

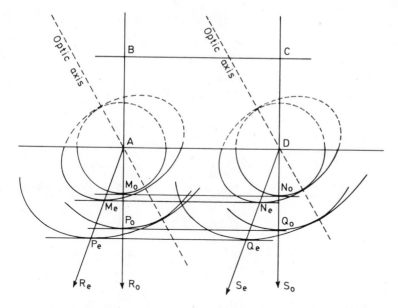

Figure 16.6 Huygens' construction for ordinary (o) and extraordinary (e) wavefronts and ray directions

and P_oQ_o are tangential to the spheres, while those of the extraordinary light M_eN_e and P_eO_e are tangential to the ellipsoids. It is seen that, in this case, the two sets of wavefronts are parallel to one another, but the rays of the *extraordinary beam are not perpendicular to the wavefronts*, and may not always lie in the plane of incidence; the extraordinary beam is therefore deviated from the normal. The rays of the ordinary beam are perpendicular to the wavefronts as in an isotropic medium. Other examples with the incident light reaching the surface at various angles should be solved by similar graphical constructions. It will be obvious that when the light is incident in the direction of the optic axis the ordinary and extraordinary beams coincide, since the spheres and ellipsoids are in contact along this direction.

An important case occurs when the optic axis is perpendicular to the plane of incidence and parallel to the face of the crystal. In the graphical construction of this (*Figure 16.7*), as the optic axis is perpendicular to the plane of the paper, the sections of both ordinary and extraordinary wavelets will be circles. Therefore, in this plane the velocity of the extraordinary light is the same in all directions, and the value $\sin i/\sin i'$ is a constant. For this reason the refractive index n_e of a crystal for the extraordinary beam is defined as the ratio of the velocity of light in air to the velocity of the extraordinary beam in a plane perpendicular to the optic axis. The values of the ordinary and extraordinary refractive indices for the D line of a few of the more important uniaxial crystals are given in *Figure 16.5c*. The **linear birefringence** of uniaxial crystals is defined as the difference between the extraordinary and ordinary refractive indices, $n_e - n_o$. Few naturally occurring crystals have a linear birefringence greater than calcite (negative).

As shown in *Figure 16.4b*, the two emergent beams of light are found to be linearly polarized in orthogonal directions. The ordinary ray has a vibration

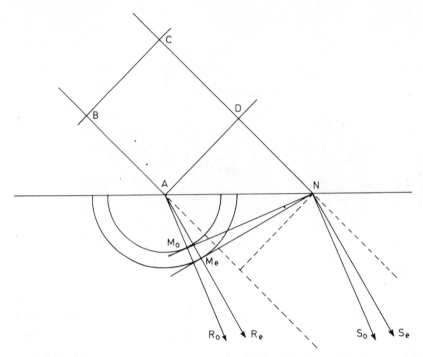

Figure 16.7 Huygens' construction for double refraction. Optic axis perpendicular to plane of incidence

orientation that is *always* perpendicular to the optics axis. On the other hand, the extraordinary ray plane of vibration contains the transmitted ray and the optic axis. This plane is sometimes called a **principal section**—*Figures 16.4b* and *16.6* are principal sections. The polarization is complete, that is each of the two beams is completely linearly polarized. Even with the high linear birefringence of calcite the angle between the beams is not very large so that some extra technique is needed to separate them. The **Nicol prism** was invented by the Scottish physicist William Nicol in 1828 and was the most widely used linear polarizer for over a century. Its form is shown in *Figure 16.8* where the normal angles of a principal section of calcite have been modified by grinding and polishing the input and output faces. The 'extra technique' is introduced by cutting the crystal diagonally and then cementing together again with Canada Balsam ($n = 1.526$). Light entering one of the end faces of the prism is double refracted into ordinary and extraordinary beams. On reaching the cement the o-ray, for which the calcite has a refractive index of 1.6585, is totally internally reflected. The e-ray, for which the calcite has a *minimum* refractive index of 1.4864, cannot be totally internally reflected until the effective refractive index exceeds that of the cement *and* the angle of incidence is greater than the critical angle.

The allowed range of angles of the incoming light is about 28°, limited by the total internal reflection conditions for each type of ray and for the need to get the refracted rays to traverse this long crystal. The change in angle of the end faces is designed to assist this by setting the incoming ray parallel to LS in the centre of the 28° acceptance angle (*Figure 16.8*). Another form of birefringent

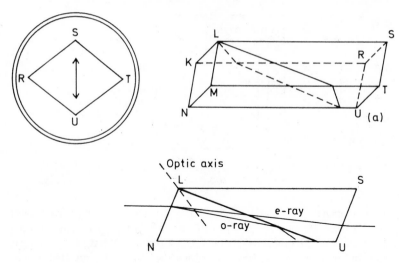

Figure 16.8 Nicol prism construction

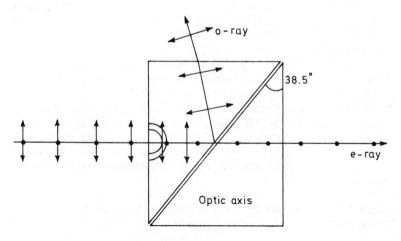

Figure 16.9 Glan–Foucault prism polarizer in calcite

polarizer is known as the **Glan-Foucault Prism**. This uses prisms cut from the natural calcite crystal so that the optic axis is perpendicular to the prism section (*Figure 16.9*). This means the maximum difference between the index for the e-ray and o-ray is utilised and the total internal reflection occurs at an air interface. Although this prism is much shorter than the Nicol, its acceptance angle is only about $10°$. If the prisms are cemented together the **Glan-Thompson prism** is produced with an acceptance angle of about $30°$ but with a performance limited by the quality of the cement.

As an alternative to using total internal reflection it is possible to divide the two beams by emergent angle. In the **Wollaston double-image prism** (*Figure 16.10*), the first prism is cut from a birefringent crystal so that its optic axis is

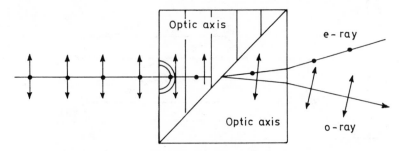

Figure 16.10 Wollaston double-image prism polarizer in calcite. The rays are designated with regard to the second *prism*

parallel to the entrance surface AC and lies in the prism section, while the second prism has its optic axis parallel to the exit surface, but perpendicular to the prism section as in the Glan-Foucault prism. On entering the first prism the light divides into an e-ray and o-ray, travelling along the same path but at different velocities experiencing different refractive indices. On reaching the angled interface, which must be well cemented or in very good optical contact to avoid total internal reflection, the designations of the two rays exchange. Because the optic axes of the two prisms are perpendicular to each other, the o-ray of the first prism becomes the e-ray of the second and vice-versa. Therefore, one ray sees an interface between n_o and n_e while the other ray sees n_e to n_o. Thus, one ray is refracted towards the normal to the interface while the other is refracted away from it. Each ray undergoes a further refraction at the final surface.

In this way considerable angular separation of the two beams can be obtained, while the dispersion of one prism is almost neutralised by the other. Prisms of this kind have been used for doubling the image in measuring instruments, where the fact that the light is polarized (as in *Figure 16.10*) is of no importance, for example the Javal-Schoitz Ophthalmometer or Keratometer. The Wollaston prism is commonly made from quartz or calcite and separation angles from $15°$ to $45°$ are commercially available.

16.4 LINEAR POLARIZATION BY ISOTROPY–SCATTERING AND REFLECTION

This section deals with polarization effects that do not depend on asymmetric crystal structures. Here the asymmetry needed is provided by the geometry of the situation. Light is scattered by atoms, molecules and particles that can vibrate in resonance or near-resonance with the frequency of the light. The interaction is generally between the light energy and the electron cloud in the atom or molecule. The characteristics of the electron clouds at the frequency of the incident light are very variable in the case of solids and these determine whether the light is reflected, refracted or absorbed.

In the case of gases the most likely method by which the atomic electrons lose the absorbed energy is by re-radiating or **scattering**. The nearer the light frequency to resonance the greater the scattering. Most gases resonate in the ultra-violet and therefore the atmosphere scatters more at the blue end of the spectrum, giving rise to the blue sky and, because the blue is scattered *out* of the beam, to red sunsets.

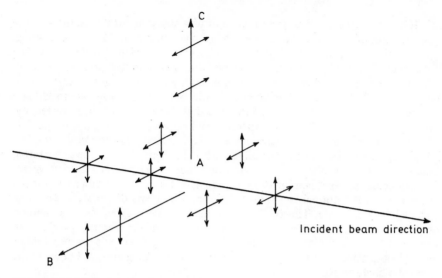

Figure 16.11 Polarization by scattering

If we consider a beam of unpolarized light reaching a scattering medium at A (*Figure 16.11*) the near-resonant vibrations in the atoms of the media are transverse to the incident direction. Consider an observer at B whose awareness of the vibrations at A is by means of the scattered light. It is apparent from the geometry that there are no vibrations in the direction of the incident beam—the scattered light is therefore, linearly polarized (similarly at C).

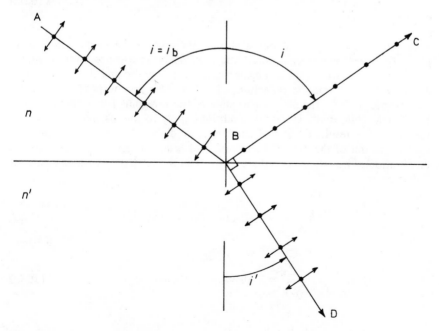

Figure 16.12 Polarization by reflection

If an analyser is used to view a patch of blue sky at about $90°$ to the sun, the polarization of the scattered light can be seen by rotating it. Multiple scattering, that is further scattering of the once scattered light tends to depolarize the beam but 70% to 80% polarization can be found in favourable circumstances.

The geometry of *Figure 16.11* is also applicable when light is reflected. In *Figure 16.12* the angle of incidence of the unpolarized beam approaching a dense medium has been chosen so that the angle between the light refracted and the light reflected is $90°$. The reflection and refraction takes place in the first few molecular layers of the dense material and viewed from the position C the light has no vibration in the BD direction so that only light of the polarization shown is reflected; akin to the scattering situation. This has its electric vector perpendicular to the plane of incidence, that is, the plane containing the incidence ray and the normal to the surface. The other plane of vibration would be parallel to this plane. The two orientations have been given many symbols*. Probably the most common are p and s derived from the German 'parallel' and 'senkrecht'. The light reaching C is the s type having its electric vector vibrating in a plane perpendicular to the plane of incidence. The angle i_b is known as the **Brewster angle** or polarizing angle.

From *Figure 16.12* we have

$$n \sin i = n' \sin i' = n' \cos i \qquad (16.01)$$

if $\qquad i + i' = 90°$

Therefore

$$\tan i_b = \frac{n'}{n} \qquad (16.02)$$

This is the angle at which the polarization of the reflected beam is the most complete. At other angles some polarization occurs in accordance with equations which can be derived from a consideration of the electric and magnetic vectors at the point of refraction/reflection. There must be no abrupt changes or *discontinuities* in these fields. The reasoning of the foregoing is a simplification of the E field conditions. Different conditions apply to the magnetic field which yield the same result. **Fresnel's Equations** were derived by him to give the relative *amplitudes* of the reflected and refracted beams in the two principal planes of vibration. When these are squared we obtain the relative intensities which for reflection are

$$\left(\frac{I_R}{I_I}\right)_p = \frac{\tan^2 (i - i')}{\tan^2 (i + i')} = \text{Reflection Coefficient, } R_p \qquad (16.03)$$

$$\left(\frac{I_R}{I_I}\right)_s = \frac{\sin^2 (i - i')}{\sin^2 (i + i')} = \text{Reflection Coefficient, } R_s \qquad (16.04)$$

*Others are p, s; TM, TE; ||, ⊥; l, r.

It is apparent that (16.03) gives $R_p = 0$ at the Brewster angle given by (16.02). For normal incidence $i = i' = 0$ and the equations become indeterminate. However, if we expand the sine terms and remembering that sin = tan at small angles we have

$$R_p = R_s = \left[\frac{\sin i \cos i' - \cos i \sin i'}{\sin i \cos i' + \cos i \sin i'} \right]^2 \tag{16.05}$$

Using Snell's law and the fact that cosines go to unity for small angles we get

$$R_p = R_s = \left[\frac{n' - n}{n' + n} \right]^2 \tag{16.06}$$

When i exceeds the critical angle, these simplified equations cannot be directly applied and a more comprehensive treatment including the relative phases of the vibrations is needed*.

The graphs of *Figure 16.13* show the reflection coefficients for three refractive indices and for external or internal reflection. The intensities of the transmitted beams are found by the coefficients

$$T_p = 1 - R_p \tag{16.07}$$

and

$$T_s = 1 - R_s \tag{16.08}$$

assuming no absorption and limiting the measurements to air (i.e. after two surfaces).

Where there is absorption in the material the reflected light as well as the transmitted light is affected. Most importantly the reflectivity for the p beam does not fall to zero, although the general shape of the curves remains the same. With metallic coatings for mirrors, unwanted polarization effects are normally not too serious but with multi-layer dielectric coatings as described in section 14.19 serious problems can occur at non-normal angles.

On the other hand, if polarization effects are required then this method may be used. With a single surface of water the reflected light is more than 80% polarized over a range of incidence angles from about $40°$ to $70°$ which explains the value of polarizing spectacles in marine or rain conditions (see *Figure 16.1*). The axes of any polarizer can be found by looking through it at a scene reflected from a liquid or glass surface near the Brewster angle and rotating the polarizer for minimum transmission.

With a parallel plate of glass two surfaces are in action. The internal and external Brewster angles are linked by Snell's Law and so maximum polarization occurs at both surfaces for the same beam angle. The reflection of the s-polarization at this angle rises from about 7.5% for the single surface to 13% for the plate ($n = 1.52$).

For two plates it is 20% and 28% for four. When a large number of plates are stacked together with small air gaps between a maximum reflection of about 35% may be obtained. The transmitted beam also becomes more polarized.

*See Longhurst, *Geometrical and Physical Optics*, Longman (1973) or Hecht and Zajac, *Optics*, Addison-Wesley (1974).

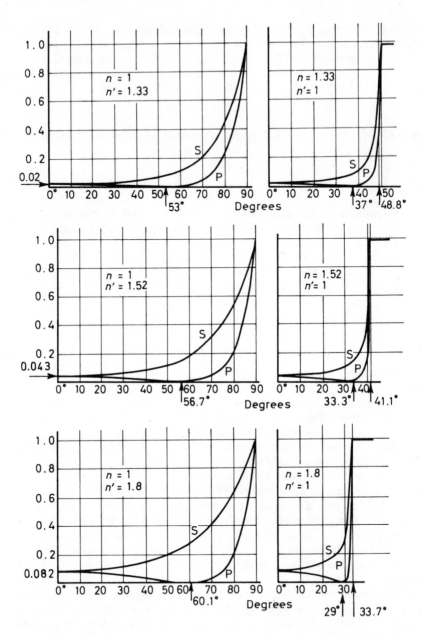

Figure 16.13 Reflection coefficients at boundaries of non-absorbing media

This concept may be realised using optical thin films whose refractive index with respect to a glass cube may be chosen to give a Brewster angle of 45°, for a chosen wavelength. At other angles the effect is less complete as it is at other wave-lengths. *Figure 16.14* shows the design and performance of such a polarizing beamsplitter.

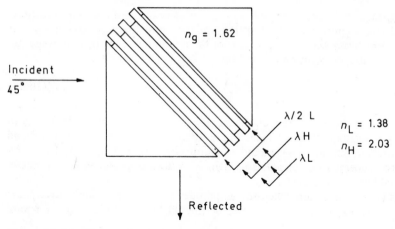

Figure 16.14a Design of polarizing beamsplitter—7 layer

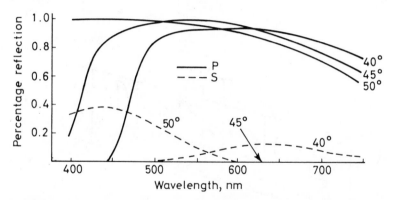

Figure 16.14b Performance against (external) angle of incidence. Design wavelength 540 nm (Design by Dr P.B. Clapham, National Physical Laboratory, acknowledged with thanks)

16.5 RETARDERS AND CIRCULARLY POLARIZED LIGHT

The Wollaston prisms described in section 16.3 produced linearly polarized light from an incident unpolarized beam. The first component of the prism was birefringent crystal cut with its incident face parallel to the optic axis. In this direction the crystal displayed the maximum difference between the o-ray and the e-ray refractive index and, therefore, light velocity and phase. If a parallel-side plate of the crystal is cut in this way the two rays emerge without change in direction but with part of the energy *retarded* behind the other part. No effect of this can be seen when the two beams are incoherent as the summation of the two random vibrations remains random.

If the incident beam is linearly polarized, then the two emergent beams will be coherent but the electric vector of one will be delayed with respect to the other. These are the conditions described in section 12.4. The resulting emergent light depends on the relative amplitudes and relative phase of the o-ray and e-ray. The relative amplitudes depend on the orientation of the plane of vibration of the incident beam with the optic axis of the retarder, and the relative phase $\Delta\varphi$, depends on the index difference and thickness (t) of the retarder

$$\Delta\varphi = \frac{2\pi}{\lambda_0}(n_0 - n_e)t \qquad (16.09)$$

λ_0 is the wavelength in vacuum as n_0 and n_e are the refractive indices with respect to vacuum. In most cases the resultant will be an ellipse as was found with (12.07) (section 12.4) and indeed **elliptically polarized** light may be regarded as the general form of polarized light with **linearly polarized** and **circularly polarized** being special cases.

This latter case occurs when the amplitudes are equal and the relative phase difference is 90° i.e. $\pi/2$. A retarder of this thickness is called a **quarter-wave plate**.

$$\frac{\lambda_0}{4} = (n_0 - n_e)t \qquad (16.10)$$

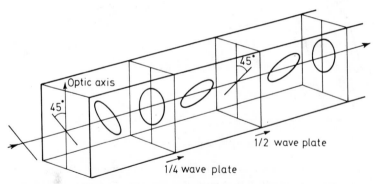

(a) Incident plane of vibration at 45° to optic axis

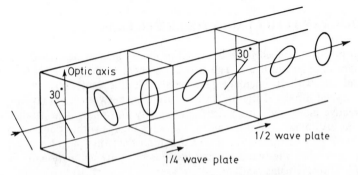

(b) Incident plane of vibration at 30° to optic axis

Figure 16.15 Passage of polarized light through retarder

Clearly it is not possible to make a simple plate retarder which is a quarter-wave plate for all wavelengths. *Figure 16.15* shows how the form of the polarization changes as the light passes through a retarder for linearly polarized incident light with a direction of vibration at 45° to the optic axis (giving equal components) and at 30°. It is seen that a plate of double thickness produces linearly polarized light of a different orientation. Such a **half-wave plate** introduces a phase difference of 180° between the components and so for any orientation with respect to the optic axis the emergent beam has the reverse orientation as shown.

Circularly polarized light may be produced from unpolarized light by mounting a linear-polarizer in front of a quarter-wave plate with its plane of transmitted vibration at 45° to the optic axis of the plate. Such a combination is called a **circular polarizer** and it is evident that it only works for light incident from one side.

16.6 DEFINITIONS AND TYPICAL VALUES*

Degree of polarization

A beam of partially polarized light may be considered as comprising a completely polarized component of intensity I_a and a completely unpolarized component of intensity I_b. The **degree of polarization**, P is defined by

$$P = \frac{I_a}{I_a + I_b}$$

assuming no coherence between the beams.

Principal transmittances and ratios

When the direction of vibration of an incident linearly polarized beam is orientated so that the transmittance of a polarizer is a maximum, then the ratio of transmitted to incident intensities is defined as the **major principal transmittance**, k_1. This direction of vibration also defines the **transmission axis** of the polarizer.

When the orientation is such as to give a minimum transmittance, the ratio of the intensities is defined as the **minor principal transmittance**, k_2. The **principal transmittance** ratio is defined as k_1/k_2 while the **extinction ratio** is the inverse of this.

Polarizance

In the particular case where the incident light is unpolarized the degree of polarization produced by a polarizer is sometimes called **polarizance**. In terms of principal transmittances the polarizance is equal to $(k_1 - k_2)/(k_1 + k_2)$.

If the incident light is partially polarized, the degree of polarization produced by the polarizer will not be equal to its polarizance.

*See NPL Report MOM25 *Polarized Light: Definitions and Nomenclature* R. J. King, National Physical Laboratory, Teddington, England. (July 77) on which this section is based, with permission.

Extinction ratio

The inverse of the principal transmittance ratio is defined as the **extinction ratio,** k_2/k_1. If two identical polarizers are used together the transmittance when their transmission axes are parallel is

$$\tfrac{1}{2}\, k_1 k_1 + \tfrac{1}{2} k_2 k_2 \simeq \frac{k_1{}^2}{2}$$

when k_2 is small. When the axes are crossed the transmission is

$$\tfrac{1}{2} k_1 k_2 + \tfrac{1}{2} k_1 k_2 = k_1 k_2$$

This crossed transmittance is often referred to as the extinction and the ratio

$$\frac{k_1 k_2}{k_1^2/2} = \frac{2k_2}{k_1} \tag{16.11}$$

is sometimes (erroneously) called the extinction ratio.

When polarizers are used with visible light the values of k_1, k_2 etc. must be weighted to match the visual response. The performance of polarizers in the IR and UV will be different. When a polarizer is used with unpolarized light the total visual transmittance is

$$k_\mathrm{v} = \frac{k_1 + k_2}{2}$$

which is usually very nearly equal to $k_1/2$. For sheet polaroid the type numbers used refer to the value of k_v. Types HN-38, HN-32 and HN-22 have k_v values of approximately 0.38, 0.32 and 0.22 respectively. The transmittances of crossed pairs $(2k_2/k_1)$ of the above types are about 5×10^{-4}, 5×10^{-5} and 5×10^{-6} respectively.

The best Glan-Thompson prisms have an extinction ratio, (k_2/k_1), less than 1×10^{-6}. The extinction ratio of a single reflection may not be as small as the zero value of the curves of *Figure 16.15* might suggest. In practice the incident beam would show some angular spread and the surface will not be perfect as regards cleanliness and structure. For a thin-film polarizer designed along the lines of *Figure 16.14* an extinction of about 10^{-3} in transmission is found at the design angle and wavelength.

The degree of polarization of a single reflection varies with incident angle and refractive index as shown in *Figure 16.16*. Naturally, these curves apply to specular reflections only. Light diffusely reflected from textured surfaces has a much lower degree of polarization but this light contains information which is normally required. The value of polarizing sunglasses in reducing glare from specular reflections can be judged from *Figure 16.16*.

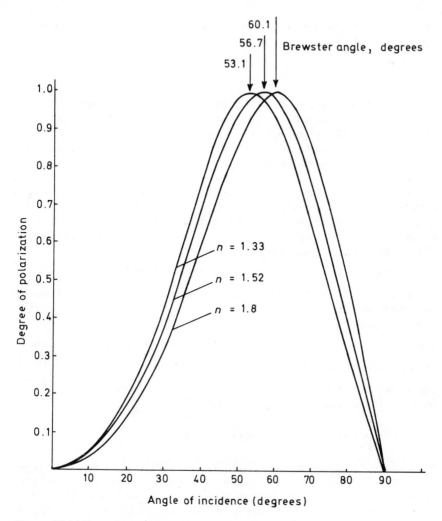

Figure 16.16 Degree of polarization by reflection

16.7 METHODS OF CALCULATION

When light is transmitted by an efficient linear polarizer the extinction ratio is at least 10^{-4} and so the degree of polarization may be taken as unity. Subsequent components of this linearly polarized beam will be coherent and so we must work with amplitudes. If a second linear polarizer is inserted in this beam with its transmission axis orientated at angle θ to the first the light transmitted by the system will be a maximum when θ is zero and a minimum when θ is $90°$. At intermediate angles the second polarizer selects the component amplitude $a_1 \cos \theta$ where a_1 is proportional to $\sqrt{k_1}$ for the first polarizer. The transmitted intensity is then proportional to $k_1 k_1' \cos^2 \theta$ where k_1' applies to the second polarizer. If the incident light is unpolarized a factor $^1/_2$ must be applied as k_1 is the principal transmittance for polarized incident light. Thus, the system transmittance is given by

$$\frac{k_1 k_1'}{2} \cos^2 \theta$$

If now a retarder is introduced between the linear polarizers the situation is complicated to the extent that the optic axis of the retarder must be defined with respect to the transmission axes of the polarizers and the phase difference ϕ, introduced by it must be known. If the angle between the first polarizer and retarder is α and between the retarder and second polarizer is β we may proceed as follows.

The light reaching the retarder has amplitude components $a \cos \alpha$ and $a \sin \alpha$ with respect to its optic axis. These components will have further components $ab \cos \alpha \cos \beta$ and $ab \sin \alpha \sin \beta$ with respect to the transmission axis of the second polarizer, but they will differ in phase by ϕ by the action of the retarder. The transmitted disturbance is therefore

$$A' = ab \cos \alpha \cos \beta \, [\sin \omega t] + ab \sin a \sin \beta \, [\sin (\omega t + \phi)] \tag{16.12}$$

where these are not orthogonal components but both in the transmission axis of the second polarizer.

By (12.04) we have

$$I' = a^2 b^2 \cos^2 \alpha \cos^2 \beta + a^2 b^2 \sin^2 \alpha \sin^2 \beta + 2a^2 b^2 \sin \alpha \cos \alpha \sin \beta \cos \beta \cos \phi$$

$$= a^2 b^2 \cos^2 (\alpha - \beta) - a^2 b^2 \sin 2\alpha \sin 2\beta \sin^2 \phi/2 \tag{16.13}$$

because we have calculated with respect to the transmission axes of both polarizers. The transmittance of the system is given by

$$k_1 k_1' \left[\cos (\alpha - \beta) - \sin 2\alpha \sin 2\beta \sin^2 \frac{\phi}{2} \right] \tag{16.14}$$

Obviously the problem becomes much more complex when k_2 is not negligible, and when more complicated systems are to be analysed. A number of different methods have been invented to cope with this. Each one requires the application of a very specific set of rules under stringent conditions in an empirical procedure designed as an analogue computing scheme. The **Poincaré sphere** uses a sphere as a map with linearly polarized light (of varying orientation) around the equator and circularly polarized light at the poles. Everywhere else is elliptically polarized and calculations are made by changing position. **Jones' vectors** and **Stokes' vectors** are similar approaches with complementary advantages and disadvantages. In matrix form the former are 2×2 while the latter use 4×4 matrices (Mueller matrix). The form of these have been calculated and tabulated in W. A. Shurcliff *Polarized Light*, (1962) Harvard/Oxford.

16.8 APPLICATIONS OF POLARIZED LIGHT

The most obvious use, in the form of polarizing sunspectacles, has been covered in section 16.1 and their value noted in section 16.6. The polarizing effect by reflection can also be used in the study of surface structure. This is particularly useful on materials which absorb light but which can be optically polished. The reflection of linearly or circularly polarized light is accompanied by a change in phase of the components giving an elliptically polarized reflected beam. Instruments called **ellipsometers** are used and can measure surface effects a small fraction of a wavelength thick.

Many organic tissues have polarizing properties and a polarizing microscope is used to study these. This is not very different from an ordinary microscope but has provision for polarizers and analysers together with a rotatable mount for the stage. The lenses used must be particularly free from strain.

If a normally isotropic glass plate is compressed it is found to act like a negative uniaxial crystal. The optic axis is in the direction of the stress. This means that for light incident perpendicular to this direction the glass acts like a retardation plate. This stress induced birefringence is called **photoelasticity**. For most purposes it is found that the birefringence is linearly proportional to the stress, the constant of proportionality being the **stress-optical coefficient**.

The effect of applied stresses may be seen by placing the material between linear polarizers, or between circular polarizers. Models of proposed structures may be built and loaded so that the distribution of the strain can be seen. (Stress is the applied load; strain is the response of the material). With optical glass a similar method may be used to check the absence of strain for good homogeneity and isotropism. On the other hand glass, which is very strong in compression, may be toughened by generating zones of tension and compression so that the compressive region is at all the surfaces while the central bulk is in tension. This is done either by quickly cooling the outer surfaces with a cold air blast or by chemical means.

A quantitative measure of the stresses can be obtained by adding a **compensator** between the polarizers. This has two thin prisms of different retarding materials (or different optic axis directions) arranged so that near their centre they compensate for each other and, with crossed polarizers, give a black fringe. Other fringes, where the retardation is a multiple of 2π, are coloured due to dispersion in the prisms. In Plate 7 these horizontal fringes can be seen either side of a toughened glass sample viewed edge-on. The compressive stress gives a new position where the compensator plus sample gives zero retardation. Likewise, the tensile stress moves the black fringe in the opposite direction. The fringes in Plate 7 show the typical parabolic form of thermally toughened glass. Chemical toughening gives a 'top-hat' shape to the fringes indicating a thinner compressive region. For toughened spectacle lenses the fringes seen when the lens is inserted face-on (as used) should be straight and undeviated as the compressive and tensile regions should everywhere compensate for each other.

16.9 POLARIZATION AND VISION

The human eye can just detect the polarization of light. This was first noticed in 1844 by Haidinger and the phenomenon is known as **Haidinger's Brush.** When a sheet of linear polarizer is held between the observer and a uniform white background the brush appears if the polarizer is quickly rotated through 90°. This faint pattern is yellow and double ended like two cones stuck point to point. It subtends an angle of about 2 degrees. Its orientation depends on the plane of vibration of the polarizer and as it fades away in a few seconds it is generally thought to be a fatigue effect. If the polarizer is rotated again it reappears and can be made to rotate if the polarizer is slowly rotated. A similar but fainter effect can be found with circularly polarized light. The phenomenon is confined to blue light—if this is excluded the brush fails to appear—so that the yellow occurs where the blue response has been fatigued.

Not all light reaching the retina is absorbed by it. The techniques of retinoscopy and ophthalmoscopy would be impossible if this were the case. Although less blue light than yellow and red (and infra-red) light is reflected it is found that the blue reflection is mainly specular while for the longer wavelengths the reflection is mainly diffuse. This can be shown by using linearly polarized light which retains its polarization with specular reflection. The location of the reflecting surfaces in the retina of the living eye is still open to argument. It seems clear that the longer wavelengths are reflected from more than one place while the blue has a single plane of reflection. Wavelengths shorter than 400 nm are hardly reflected at all.

16.10 OTHER ASPECTS OF POLARIZED LIGHT

Some materials, such as quartz, are not only birefringent, they also exhibit **optical activity**. When it is cut as a parallel plate with its optic axis perpendicular to the surfaces it has the ability to rotate the plane of vibration of incident linearly polarized light. Thus, if between a pair of polarizers crossed to give extinction, a plate of quartz cut as above is inserted, it will allow light to be transmitted. If the second polarizer is rotated, a new position of extinction can be found. It follows, therefore, that the light leaving the quartz plate is linearly polarized but the plane of the vibrations has been rotated.

With a given substance the angle through which the plane of vibration is rotated is proportional to the thickness of the plate and with a given thickness of plate, this angle varies inversely with the wavelength, being approximately proportional to λ^{-2}. Hence, with white light the extinction described above will be incomplete and different colours will be transmitted in turn as the analyser is rotated.

Some materials are birefringent without being optically active (calcite) while others are optically active without being birefringent. The last category includes liquids particularly sugar solutions. The rotation may be right-handed (positive) or left-handed (negative) defined as the required movement of the analyser to regain extinction as seen by an observer looking from it towards the polarizer. Fused quartz (fused silica) has no birefringence or optical activity.

Instruments which measure optical activity are called **polarimeters** and when they are specifically applied to measuring the strength of sugar solutions they are called **saccharimeters**. The theory of optical rotation given by Fresnel showed that linearly polarized light could be resolved into two circularly polarized beams. If these were transmitted at different velocities through a substance the resultant of them, the plane of vibration, was effectively rotated.

As light is an electromagnetic disturbance it may be expected that it should be influenced by electric and magnetic fields. This is particularly so when the field is acting on the source of the light. In this case the reaction is primarily between the field and the electron orbits of the source emitting the light. The

Zeeman effect relates to magnetic fields and the **Stark effect** to electric fields. In both cases each spectral line of the source is observed to divide into a number of closely spaced lines.

Related to the Zeeman effect is the **Faraday effect**. If a magnetic field is generated along a piece of glass, a linearly polarized beam directed parallel to the field has its plane of polarization rotated. This effect is greater with high index materials but occurs over a wide range of solids, liquids and gases. A related phenomenon, the **Voigt effect**, occurs when the light path is perpendicular to the magnetic field. Under these circumstances the material may become birefringent. Related effects with electric fields are also found.

The above effects are due to interactions within the atomic structure. Other, and usually larger, effects can occur when the magnetic or electric field affects the alignment of molecules within a material. The **Cotton–Mouton** effect occurs with liquids like nitrobenzene when anisotropic molecules are all lined up by a transverse magnetic field and bulk birefringence results. Nitrobenzene exhibits a similar action with a transverse electric field which is known as the **Kerr effect**. This is used in the **Kerr cell** which uses the fast switching of an electromagnetic field to alter the polarizing properties of the liquid. When placed between a polarizer and analyser the system comprises an **electro-optical shutter** which may be used for chopping light very quickly. This has been used for velocity measurements, optical radar, projection television, and Q-switch lasers. In this last application, the cell actually holds up the lasing effect with the energy stored in the inverted population. When the cell is opened the lasing action builds up very quickly and a very short powerful burst of laser light is created.

CHAPTER 16   *EXERCISES*

1. Describe two experiments which show that light is propagated as a transverse wave motion. How is the law of rectilinear propagation explained on the wave theory?

2. Give a short account of the nature of light, describing experiments to prove the truth of your statements.

3. Explain what is meant by plane polarised light and describe two methods by which it can be produced.

4. Explain why two crossed tourmalines or Nicols allow no light to pass through, and why interposing a plate of crystal between them may restore the light.

5. Describe the phenomenon of double refraction. The end face and the diagonal face of a Nicol prism are at right angles. Show on a careful diagram the approximate paths of the ordinary and extraordinary rays within the prism, the light being incident on the end face at an incident angle of $50°$. The refractive indices of Iceland Spar are approximately 1.66 (ordinary) and 1.49 (extraordinary) and of Canada balsam 1.53.

6. Show that when $i + i' = 90°$, $\tan i = n$. What is the difference between the incident and reflected light for this particular angle of incidence?

7. Describe Brewster's law concerning the polarisation of light by reflection. Describe a form of double image prism, making quite clear, with the help of a diagram, its construction; and explain carefully the reason for the course taken through the prism by an incident ray.

8. If a picture is hung so that it is badly seen owing to light reflected from its surface it can often be clearly seen through a Nicol prism or other analyser. Explain this.

9. Using Fresnel's law of reflection plot the curves showing the variation of the intensity of the reflected light with variation of the angle of incidence when light is reflected from the surface of a medium of refractive index 1.6, for vibrations taking place (*a*) parallel (*b*) perpendicular to the plane of incidence. From the curves determine the polarising angle.

10. The angle of incidence of a parallel pencil of white light on the plane polished surface of a block of glass of refractive index 1.760 is 60° 24′. What will be the angle between the reflected beam and the beam refracted into the glass?

Explain carefully any differences there will be in the nature of the incident, the reflected and the refracted light.

11. Explain what is meant by the term 'polarizing angle' when used in connection with light incident in air on the surface of a plane glass plate. Show that if the refractive index of the plate is n then the tangent of the polarizing angle is equal to n.

When polarizing lenses are used in sunglasses they are orientated to accept light polarized in a particular direction. What is the direction normally chosen and why is it chosen?

A man uses a piece of Polaroid to minimise the glare reflected from the still surface of a lake. At what angle of view relative to the surface will he obtain the best extinction of the glare?

12. A Wollaston double image prism is made of two 45° prisms of quartz, the refractive indices being 1.544 for the ordinary ray and 1.553 for the extraordinary. Find the angle between the two emergent beams if the incident beam of unpolarised light is normal to the first surface.

13. A polariscope consists of two sheets of Polaroid. If the transmitted intensity of the light is I_0 when the analyser axis is parallel to the polarizer axis, and is zero when the axes are crossed (i.e. perpendicular to each other) what is the transmitted intensity when:

(a) The analyser axis is 30° from the polarizer axis?

(b) A quarter wave plate is placed between the crossed polaroids with its optic axis inclined at 30° to the polarizer axis?

14. Plane polarised light ($\lambda = 600$ nm) is incident on a plate of quartz 0.5 mm thick cut with its faces parallel to the axis. Find the difference in phase in the two beams leaving the plate.

15. A thin plate of a doubly refracting crystal cut with the optic axis in the plane of the surfaces is placed between crossed Nicols. Explain carefully the appearances seen as the plate is rotated about a line joining the Nicols. When the plate of crystal is placed with its optic axis at 45° to the principal plane of the polariser what will be the effect of rotating the analyser?

16. A beam of plane polarised sodium light is incident on a plane plate of quartz cut with the optic axis in the plane of the surfaces, the plane in which the vibrations of the incident light are taking place being at 45° to the optic axis. Find the least thickness of the plate so that the resultant vibration produced by the ordinary and extraordinary beams immediately on their emergence from the plate is (a) plane, (b) circularly polarised.

17. Two Nicol prisms are placed in line to form a polariscope; compare the intensities of the transmitted beam when the principal planes of the Nicols are inclined at (a) 0°, (b) 30°, (c) 45°, (d) 60°, (e) 90°.

18. Explain what is meant by a quarter wave plate. The refractive indices of mica being 1.561 (ordinary) and 1.594 (extraordinary), what must be the thickness of a film of mica to produce a quarter wave plate for sodium light? ($\lambda = 589$ nm).

19. When an observer looks at an object through a Wollaston double image prism made of calcite ($n_0 = 1.66, n_e = 1.49$) he sees two images separated by 2°. What are the apical angles of the calcite prisms?

20. (a) Explain how circularly polarized light may be produced by a quarter wave plate. (b) Explain how a half wave plate may be used to give any desired rotation of the plane of vibration of linearly polarized light.

21. Along what direction in a uniaxial crystal must light travel in order that the velocities of the ordinary and extraordinary rays shall be equal? Define the refractive index of a crystal for the extraordinary ray.

A ray of plane polarised light, the vibration direction being vertical, is incident normally on a Wollaston prism made of quartz, the axis direction of the first component prism being vertical. Show on a diagram how a ray will pass through and emerge from the prism. Give a brief explanation. (N.B.—In quartz the extraordinary index exceeds the ordinary.)

22. A double image prism consists of a 30° prism of calcite cemented to a glass prism. The calcite prism is cut with the optic axis of the crystal in the plane of the first surface and parallel to the base-apex line. If light falls normally on the first face of the calcite prism, find the angle of the glass prism required in order that the extraordinary beam shall emerge undeviated. What will be the deviation of the ordinary beam?

Refractive indices: Calcite, ordinary   1.659
                     extraordinary   1.486
                 Glass   1.510

23. A beam of plane polarised light of wave-length 540 nm falls normally on a thin plate of quartz cut so that the optic axis lies in the surface. If the ordinary and extraordinary refractive indices are respectively 1.544 and 1.553, find for what thickness of the plate the difference in phase between the ordinary and extraordinary beams is π radians on emergence from the plate.

24. Plane waves of light are incident obliquely on the surface of a uniaxial crystal such as calcite. Give a careful diagram with explanation showing the passage of the light within the crystal when the optic axis lies perpendicular to the plane of incidence and parallel to the surface of the crystal.

Hence define the refraction index of the crystal for the extraordinary ray.

In what important respect do the ordinary and extraordinary beams emerging from the crystal differ from one another?

25. Explain the construction and action of a Nicol Prism. Describe two practical examples of the use of polarized light.

The effective intensity of a source is reduced by the use of a polarizer and analyser whose relative orientation is θ. Calculate the value of 'θ' which will reduce the intensity of transmitted beam to $1/7$ incident beam intensity.

26. Describe the construction of an apparatus for observations on photo elastic strain effects. Discuss the technique of using this apparatus and the theory involved in its operation.

27. Plane waves of light are incident obliquely on the surface of a calcite crystal. Give a clear diagram showing the passage of light within the crystal when the optic axis lies perpendicular to the plane of incidence and parallel to the surface of the crystal.

In what important respect do the ordinary and extraordinary beams emerging from the crystal differ from one another?

A ray of unpolarized light falls on a calcite crystal, the optic axis of which is parallel to the surface. The angle of incidence is 34° and the plane of incidence coincides with the principal section of the crystal. Find the angles of refraction of the O and E rays for sodium light (λ = 600 nm).

$$n_o = 1.658, n_e = 1.486$$

Hint: Regard the normal as the x-axis. The line $y = mx + c$ touches the ellipse

$$\frac{x^2}{a^2} + \frac{y^2}{b^2} = 1$$

when $m^2 a^2 + b^2 = c^2$.

28. A plane-parallel calcite plate is cut with its faces parallel to the optic axis. Describe what happens to a narrow pencil of light as it is transmitted through the crystal and into the air beyond when it is incident (*a*) normally, and (*b*) at any other angle on one of the parallel faces.

Two sheets of polaroid, each of which absorbs 24% of any polarized light incident on it, are used as a reduction filter in front of an unpolarized source of light. Calculate the angle at which the principal sections of the two sheets must be set in order that the overall transmission of the filter should be 25%.

29. What is the Brewster angle? Derive an expression for it in terms of the refractive index of the reflecting medium.

Distinguish between dichroic and birefringent materials, explaining the physical properties of each. Give examples of each type of material.

Describe one application of polarisers.

Explain the properties of a quarter-wave plate and show that its thickness is given by

$$d = \frac{\lambda}{4\,(n_o - n_e)}$$

When $n_o > n_e$ and n_o and n_e are respectively the ordinary and extraordinary refractive indices and λ is the wave-length of the light used. What are the uses of a quarter-wave plate?

Chapter 17

The effects of limitation of
beams in optical systems

17.1 INTRODUCTION

In the earlier chapters on geometrical optics little attention has been paid to the actual extent of pencils and beams passing through any optical system or instrument, except that most of the theoretical results obtained are limited, in the case of simple lenses, to the paraxial region, and therefore the pencils and beams are very narrow. Thus in the graphical constructions of Chapter 6 the actual paths of light by which the image is formed were not considered, and the rays used in the constructions are not usually directions of the light that actually passes through the lens or instrument. An example of this is shown in *Figure 17.1*, where, in addition to the rays used in graphically constructing the image, the path taken by the light from one end of the object through the lens is shown.

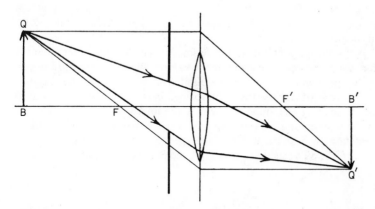

Figure 17.1 Graphical construction of image and actual path of rays from object point

Pencils and beams passing through any instrument are limited by apertures, which may be the actual apertures of the lenses, prisms, etc., or additional diaphragms or stops. These latter are used to cut off light not travelling in the required direction to form a perfect image, to cut off the outer parts of the image, where the definition is defective, that is, to limit the field of view, and to intercept stray light reflected from the interior of tubes, lens mounts, etc. The various apertures of a system will influence (1) the field of view, (2) the depth of focus, (3) the illumination of the image, and (4) the resolving power.

17.2 ENTRANCE AND EXIT PUPILS

One of the apertures of a system or instrument will be effective in limiting the width of the pencil of light from an object point on the axis, and this is known as the **aperture stop** of the system. This may be the mount of one of the lenses or a stop in front of, behind or within the system. When there are any lenses between the object and the aperture stop, the pencil in the object space will be travelling as though it passes through the image of the aperture stop as formed by any lenses in front of it, and in the image space the pencil is limited as if by the image of the aperture stop formed by any lenses behind it. In *Figure 17.2*, $S_1 T_1$ is the aperture stop of the system of two lenses at A_1 and A_2. It will be seen that the pencil from the object point B is incident as if passing through ST, the image of $S_1 T_1$ formed by the first lens at A_1, while the pencil going to the image point B' emerges from the system as though it had come through S'T', the image of $S_1 T_1$ formed by the second lens at A_2. Any rays which pass through an actual stop must pass through or towards corresponding points in the image of the stop, and any rays cut off by a stop will not pass through the opening in the image of the stop. In dealing with the limitation of pencils and beams in passing through a system it is very convenient to be able to consider all the apertures as being in the same medium, and this may be done by taking in the object space all the actual apertures in front of the system and the images of other apertures formed by any lenses in front of them. In the image space we have any actual apertures behind the system and the images of the other apertures formed by lenses behind them.

The stop or stop-image that subtends the smallest angle at the object point B is called, after Abbe, the **entrance pupil** of the system. When the aperture stop is in front of the system it will be the entrance pupil, but in other cases the entrance pupil will be the image of the aperture stop formed in the object space. The stop or stop-image subtending the smallest angle at the image point B' is called the **exit pupil** of the system, and will correspond to the aperture stop when this is behind the system or will be the image of the aperture stop formed in the image space. In *Figure 17.2*, ST is the entrance pupil and S'T' the exit pupil. Since any

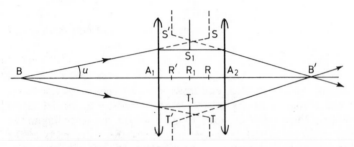

Figure 17.2 Entrance and exit pupils

ray in the object space passing through or towards a point in the entrance pupil, on leaving the system is travelling through or as though from a corresponding point in the exit pupil, the exit pupil is the image of the entrance pupil formed by the complete system, and the centres of the pupils R and R' are conjugate points with respect to the system. The angle subtended at B by the radius RS of the entrance pupil is known as the **aperture angle** u of the system for that particular object. This angle is used in the Sine Rule (section 18.5) and the definition of Numerical Aperture (section 15.11).

17.3 FIELD OF VIEW

In addition to limiting the width of the pencil from an axial object point and therefore the amount of light going to the image, the apertures of a system will restrict the extent of object of which an image can be formed by the system. This extent of object is called the **field of view** of the system, and is usually expressed as the angle that the field subtends at some given point in the system. In the case of an instrument, such as a telescope, used with the eye, the field of view is the extent of object that can be seen through the instrument by the eye rotating about its centre of rotation. As a simple example of field of view we may consider the case of a person in a room looking through a window. Here the aperture stop of the system is the pupil of the person's eye, and the entrance pupil is the image of this stop formed by the cornea. The field of view is obviously dependent upon the size of the window opening and the position of the eye, and the extent of the field of view may be determined by drawing straight lines from the centre of rotation of the eye past the edges of the window. The window here acts as the **field stop**.

Figure 17.3 shows the field of view given by a lens of circular aperture DE with a circular stop ST placed before it. ST in this case is both the aperture stop and the entrance pupil with respect to an object point at B. In *Figure 17.3a*, the full pencil of light passing through ST from the object point B is transmitted by the lens and goes to form the image at B'. It will be seen that pencils of the same width are transmitted from all object points between B and Q_1 (*Figure 17.3b*), and the image $B'Q'_1$ will have uniform illumination. If the two apertures are centrally situated with respect to the optical axis, there will be a circular field of radius $B'Q'_1$ in the image plane in which the illumination is uniform, corresponding to a circular field of radius BQ_1 in the object plane. With object points lying farther from the axis than Q_1, the whole of the pencil passing through the entrance pupil is not transmitted by the lens, and the illumination of the image is accordingly reduced; thus in *Figure 17.3c*, only one-half the incident pencil from Q_2 is able to pass through the lens. The width of the transmitted pencil is continually reduced as the object point is taken farther from the axis until as shown in *Figure 17.3d*, the incident pencil from Q_3 just misses the lens aperture entirely. Thus the complete field of view in this case will consist of a central portion over which the illumination is uniform, this is known as the **field of full · illumination**, and an outer 'vignetted' portion in which the illumination gradually decreases to zero. From the figure it is seen that the field of full illumination is determined by drawing straight lines through corresponding points in the edges of the two apertures on the same side of the axis, such as the points E and T, D and S (*Figure 17.3b*). The total field is determined by straight lines drawn

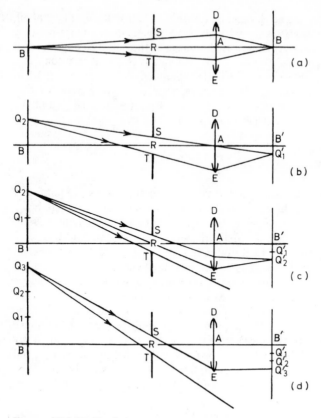

Figure 17.3 Field of view

through points on opposite sides of the axis, such as E and S, D and T (*Figure 17.3d*). This case represents the field of view obtained with a photographic lens fitted with a diaphragm in front, the so-called landscape lens.

From section 17.2 it is clear that a ray directed towards the centre of the entrance pupil will pass through the centre of the stop and emerge through the centre of the exit pupil. This ray is called the **principal ray**. It is used in Chapter 18 to establish the parameters of an off-axis object and image. Clearly the ray Q_2 R E Q_2' in *Figure 17.3* is a principal ray. The ray which defines the stop effect is called a **marginal ray**. This uses an axial object and is the ray at the maximum angle allowed by the stop. In *Figure 17.3* the ray B S G B' is a marginal ray.

The stop or stop-image in the object space that with the entrance pupil limits the field of view is called the **entrance port**, and the actual stop or lens mount corresponding to this is the **field stop**. In *Figure 17.3* the lens aperture DE is both the field stop and the entrance port. The entrance port will be the stop or stop-image that subtends the smallest angle at the centre R of the entrance pupil. The image of the entrance port formed by the entire system is the **exit port** and limits the field of view in the image space.

In order to determine the field of view of any instrument, we must first find the images in either the object or image space of all stops that are likely to behave as entrance and exit pupils and as entrance and exit ports. Then if we consider the object side, the entrance pupil is found as the actual stop or stop-image subtending the smallest angle at the axial point of the object and the entrance port as the actual stop or stop-image subtending the smallest angle at the centre of the entrance pupil.

The fields of full and partial illumination on the object side can then be determined by drawing straight lines through corresponding and opposite edges of the entrance pupil and port, as in *Figure 17.3*. When the object lies between the entrance pupil and entrance port, as in section 17.8, the field of full illumination is determined by lines joining opposite edges of the entrance pupil and port, and the total field by lines joining corresponding edges. Frequently the field of view of an instrument is defined as the field over which the illumination does not fall below half the illumination at the centre. This, as will be seen from *Figure 17.3c*, is an angular field equal to the angle subtended by the entrance port at the centre of the entrance pupil, and is therefore independent of the size of the entrance pupil.

Similar constructions to the above, using the exit pupil and port, will determine the field of view on the image side, or this may be determined directly as the image of the object field. A few examples will illustrate the method.

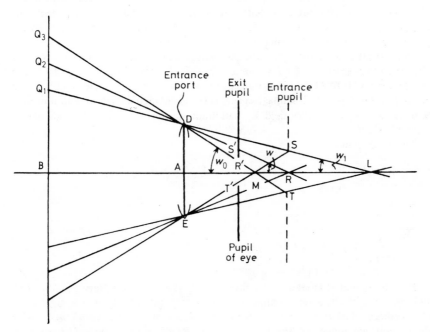

Figure 17.4 Field of view of magnifier

Convex lens used as magnifying glass: The aperture stop in this case is the opening in the iris of the eye, and the exit pupil $S'T'$ (*Figure 17.4*) is the pupil of the eye seen through the cornea. The entrance pupil ST is the image of $S'T'$ formed by the magnifying lens, and if, as is usually the case, the eye is fairly close to the lens, that is, inside its principal focus, the entrance pupil is a virtual image behind the lens. The lens rim is both the field stop and entrance port in this case. Then the angle $BLQ_1 = w_1$ is the semi-angular field of full illumination, the angle $BRQ_2 = w$ is the semi-angular field of half illumination, and the angle $BMQ_3 = w_0$ is the semi-angular total field.

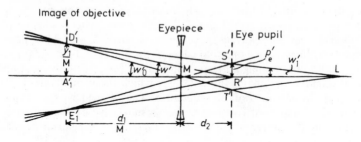

Figure 17.5 Field of view of Galilean telescope

Galilean Telescope (Figure 17.5): In this case it will be more convenient to consider the field of view referred to the image space. The aperture of the objective will be the entrance port and its image formed by the eyepiece, the exit port, this will be a virtual image situated between the objective and eyepiece (see Chapter 10). As the aperture of the eyepiece is always of much larger diameter than that of the eye pupil, the latter will act as exit pupil of the complete system consisting of telescope and eye. In *Figure 17.5* $D'_1 E'_1$ is the image of the objective formed by the eyepiece—the virtual Ramsden circle—and $S'T'$ the eye pupil. Then w'_1, in the image space, is the semi-angular field of full illumination, w' the semi-angular field of half illumination and w'_0 the semi-angular complete field.

From the figure

$$\tan w' = \frac{D'_1 A'_1}{A'_1 R'}$$

$$\tan w'_1 = \frac{D'_1 A'_1 - S'R'}{A'_1 R'}$$

$$\tan w'_0 = \frac{D'_1 A'_1 + S'R'}{A'_1 R'}$$

If y_1 is the semi-diameter of the objective, d_1 the separation between objective and eyepiece, p'_e the semi-diameter of the eye pupil, and d_2 its distance from the eyepiece, then the semi-diameter of the Ramsden circle is y_1/M, where M is the magnifying power of the telescope, and its distance from the eyepiece is d_1/M.

Hence

$$\tan w' = \frac{\dfrac{y_1}{M}}{\dfrac{d_1}{M} + d_2} \tag{17.01}$$

$$\tan w'_1 = \frac{\dfrac{y_1}{M} - p'_e}{\dfrac{d_1}{M} + d_2} \tag{17.02}$$

$$\tan w'_0 = \frac{\dfrac{y_1}{M} + p'_e}{\dfrac{d_1}{M} + d_2} \tag{17.03}$$

The field of view on the image side is called the **apparent field of view** of the telescope, and the field of view on the object side—the actual field of view—can obviously be determined by dividing the apparent field by the magnifying power of the instrument.

Astronomical Telescope: In the astronomical telescope the Ramsden circle is a real image of the objective and in the ordinary way will coincide in position with the pupil of the eye. In this case the rim of one of the eyepiece lenses would be effective in limiting the field of view. In practice, however, a stop is usually placed in the common focal plane of objective and eyepiece, its diameter being such that the outer vignetted portion of the field is cut off. The field therefore has a well defined boundary, the image of the stop seen through the eyepiece. The apparent field will be the angle subtended at the eyepiece by the stop diameter, and the angle subtended at the objective by the stop diameter will be the actual field.

The aim in many instruments is to obtain as large a field of view as possible together with magnification. This entails a high value for the apparent field, and such a requirement demands the highest qualities in the design of the lenses. Thus in a binocular magnifying six times with a field view of $8°$, the eyepiece is required to give good definition over a field of $48°$. If the magnification is increased to eight times, for the same apparent field covered by the eyepiece, the actual field is reduced to $6°$, or if the same actual field is required the eyepiece must be designed to give good definition over a field of $64°$. Therefore an increase in the magnification tends to lead to a decrease in the field of view. The apparent field of view is a useful value in comparing the fields of telescopes of different magnifications and expresses what might be called the efficiency of an instrument.

17.4 DEPTH OF FOCUS

The distance from a lens of any image point is dependent upon the distance of the corresponding object point, and there will be a solid image formed in space of any solid object. While the dimensions of the image in all directions perpendicular to the axis are the same fraction of the corresponding object dimensions, if aberrations are neglected, this is not so for dimensions along the axis, as was shown in Chapter 6. The image of a solid object formed in space is not, therefore, an exact scale reproduction of the object.

In practice, the image formed by an instrument is always received on a surface or screen, either directly, as in the case of photographic and projection apparatus, or on the retina of the eye in the case of the images formed by microscopes and telescopes. Only object points that lie in the same plane can, of course, have sharply defined image points on a receiving screen in a particular position, and object points in front of or behind this plane will be represented by patches of light that will be circular on the lens axis and elliptical in other positions. As these patches do not in the ordinary way depart far from a circular form they are usually referred to as **blur circles**. From *Figure 17.6* it is clear that the dimensions of these so-called blur circles depend on the position and diameter of the exit pupil and the distance of the true image position from the screen.

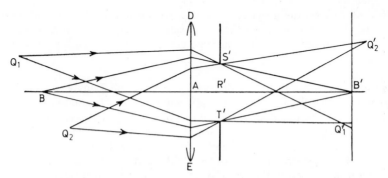

Figure 17.6 Depth of focus

It is well known that in practice it is possible to produce apparently sharp images of objects at different distances from the lens simultaneously on a plane surface. Thus in a photograph of, say, a landscape, object planes that may be separated by considerable distances are all reproduced with apparent sharpness on the film. The reason for this is that objects that subtend less than a certain angle, usually taken as one minute, at the eye appear to the eye as points, and if the blur circles are sufficiently small compared with the distance at which they are viewed, there will be no apparent difference between the portions of the image that are perfectly sharp and those that are blurred to this extent. A simple example of a similar kind occurs in the case of a picture printed in halftone. This is made up of light and dark dots, as can be seen with a magnifier, but when viewed at a suitable distance gives the effect of a sharp picture, the separate dots being indistinguishable.

The greatest distance the screen may be moved away from the theoretical image position without the image appearing unsharp is called the **depth of focus**. The corresponding distance on the object side, that is, between object planes that are apparently sharply imaged in one plane, is often called the **depth of field**. These distances are obviously largely dependent upon the standard of definition required, that is, the permissible diameter of the blur circles. In the case of an ordinary photograph to be viewed directly, 0.25 mm is frequently taken as the limiting value of the blur circle diameter, but a better standard is a blur circle diameter of one-thousandth the focal length of the lens. This will allow for the enlargement usually required by the small pictures obtained with short focus lenses.

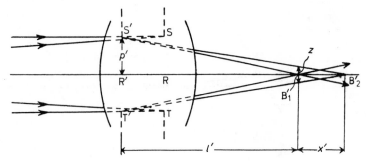

Figure 17.7 Depth of focus

Figure 17.7 represents a system having entrance and exit pupils ST and S'T' respectively. Light from an axial object point focuses at B'_1 and from some other closer point at B'_2; then if a receiving screen is placed at the image point B'_1, the second image is represented by a blur circle of diameter z. From the figure

$$\frac{z}{x'} = \frac{2p'}{l'+x'} = \frac{2p'}{l'}$$

since x' is usually very small as compared with l'. If B'_1 is at the principal focus of the system, l' will in most cases differ little from the equivalent focal length of the system, and we then have

$$\frac{z}{x'} = \frac{2p'}{f'} \tag{17.04}$$

$2p'/f'$ is the **aperture ratio** of the system, and is for most practical purposes the reciprocal of the f/No., as in most photographic objectives there is little difference in the diameters of entrance and exit pupils.

The distance $(f' + x')$ is the image distance of the nearest object sharply defined when the lens is correctly focused on a very distant object. This object distance is called the **hyperfocal distance**, and if the screen is placed at the correct focus for this distance, all objects from infinity to approximately half the hyperfocal distance have apparently sharp images on the screen.

17.5 THE TELECENTRIC PRINCIPLE

If the aperture stop of a system is so placed that the entrance pupil falls in the first focal plane (*Figure 17.8a*) the exit pupil is at infinity, and the rays through the centre of the entrance pupil from all points on the object are parallel to the axis in the image space. Similarly, if the exit pupil lies in the second focal plane (*Figure 17.8b*) the entrance pupil is at infinity, and the rays through the centre of the exit pupil are parallel to the axis in the object space. Such systems are said to be **telecentric** and have important applications in measuring instruments. Where the length of an object is to be measured by projecting its image on a scale, as, for example, in the various micrometer microscopes, errors frequently arise from the fact that, owing to depth of focus, the image may not be exactly

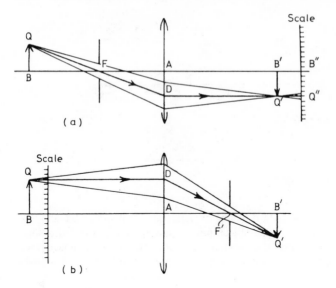

Figure 17.8 The telecentric principle

in the plane of the scale, although both appear in focus together. An error may then be introduced as can be seen from *Figure 17.6*, where if the scale is supposed to be in the position of the screen, the distance of Q_1 from the axis as measured by its image on the scale will be too large, while that of Q_2 will be too small. A similar error may occur when the scale is on the object side but cannot be placed in the object plane. This error can be avoided by the use of the telecentric system, for, as is shown in *Figure 17.8a*, the length $B''Q''$ projected on the scale is the same as that of the focused image $B'Q'$. The telecentric principle is applied in the Wessely Keratometer, an instrument used in ophthalmic work for measuring the distance of the back vertex of a spectacle lens from the eye's cornea.

17.6 THE ILLUMINATION OF AN IMAGE

If the loss of light due to reflection and absorption of the light passing through a system is neglected, it can be shown that the illumination of an image is proportional to the luminance of the object and the area of the entrance pupil. Let h be the area of an object of luminance B at a distance l from the entrance pupil of area a, then if a and h are small compared with l, the luminous flux reaching the entrance pupil will be Bha/l^2. Neglecting reflection and absorption losses this flux is transmitted by the system and goes to the image. The area of the image h' will be $l'^2 h/l^2$, and since the flux is distributed over this area, the illumination of the image will be Ba/l'^2. In the case of the eye l' is constant, and the illumination of the retinal image is dependent only upon the luminance of the object and the pupil diameter. With a given pupil diameter, an object will therefore appear equally bright no matter what its distance, neglecting, of course, absorption due to the atmosphere.

When an instrument such as a telescope is used with the eye, with the exit pupil or Ramsden circle in the plane of the eye pupil, the luminous flux entering the eye is governed by either the Ramsden circle or the eye pupil, whichever is the smaller. The luminous flux entering the instrument is proportional to the area of the entrance pupil, that is, to p^2, where p is the radius of the entrance pupil, and after passing through the instrument this flux is concentrated in the area of the exit pupil of radius p', if reflection and absorption losses are neglected. Providing the eye pupil of radius p'_e is not smaller than the exit pupil of the instrument, so that the whole of the light can enter the eye, the ratio of the flux received by the eye with the instrument to that received from the same object by the unaided eye will be

$$\left(\frac{p}{p'_e} \right)^2$$

The retinal image, however, is magnified M times, and the area over which the light is distributed is therefore increased M^2 times, hence the illumination on the retina of the image seen through the telescope will be

$$\left(\frac{p}{p'_e} \right)^2 \frac{1}{M^2} = \left(\frac{p'}{p'_e} \right)^2$$

times that of the image received by the unaided eye. This factor cannot be greater than unity, for if the exit pupil is greater than the eye pupil the light from the outer parts is cut off by the iris, and the effective diameter of the exit pupil is then the diameter of the eye pupil.

Thus an extended object seen through a telescope can never appear brighter than when seen without the telescope, and will actually always be less bright through losses of light by reflection and absorption. To obtain the maximum illumination with a telescope in viewing an extended object, it is necessary that the diameter of the exit pupil should be equal to the diameter of the eye pupil and therefore the objective diameter should be M times the diameter of the eye pupil. Manufacturers of binoculars frequently specify the **light transmitting power** of an instrument as the square of the diameter of the exit pupil in millimetres. It will be realised that this value has no significance when the diameter of the exit pupil exceeds the diameter of the eye pupil possible under any conditions.

When the object is a point, such as a star, the image is a disc, the size of which depends on the aperture of the objective, as explained in Chapter 15, and is independent of the magnification. Then, if the exit pupil of the instrument is smaller than the pupil of the eye, all light which passes through the entrance pupil—the objective aperture in this case—passes into the eye, and the luminosity of the image, as compared with the image seen by the unaided eye, is increased in the ratio of the area of the objective to the area of the pupil of the eye. When the diameter of the exit pupil exceeds the diameter of the eye pupil, the latter can be considered as the effective exit pupil and its image as formed by the instrument as the entrance pupil. The area of the effective entrance pupil is then M^2 times that of the eye pupil, and the luminosity of the image is increased in this ratio. Thus when the object is a point it is possible to increase the luminosity by means of a telescope, and in the case of large astronomical telescopes, where the diameter of the objective may be several feet, the increase in luminosity will be very considerable. Since the apparent brightness of a star can be increased in this way while the apparent brightness of the sky, this being an extended object, is unchanged or more often reduced, it is possible by means of a large telescope to see stars in full daylight.

The visibility of the stars will be less than at night due to the reduction in their apparent contrast by the presence of the light scattered from the atmosphere within the field of view and also from light from outside the field of view. This stray light can reach the image surface by multiple reflections from the lens surfaces or by reflections from the housing of the telescope. When faint objects close to bright objects are being studied this stray light generates **veiling glare**. To reduce it, **light baffles** and **glare stops** are introduced into the system. These may be introduced almost anywhere in the system; field stops and aperture stops also assist in reducing glare. The term glare stop is often used for a second stop, in a complex system, placed at the image of the aperture stop. In *Figure 10.16* a glare stop is placed at the image of the objective lens, the aperture stop.

17.7 THE NATURE OF A BEAM

Figure 17.9 shows the beam from a source QS and the lens DE. The individual pencils converge to form an image of the source $S'Q'$, and the edges of the beam are shown by the thick lines $DS'T'$ and $EQ'W'$. This beam has its smallest cross-section at $S'Q'$, and this position is termed the *waist* of the beam. Here the illumination will be greatest, since the flux is concentrated into the smallest space. Beyond the waist the beam is diverging from the point P. The position of the waist is often important, for example, in the projection lantern the objective should occupy approximately this position in the beam from the condenser, in order that the whole beam shall be transmitted by an objective of reasonable aperture. When the image of the source is larger than the lens aperture or when it is virtual, the lens aperture will be the waist of the beam.

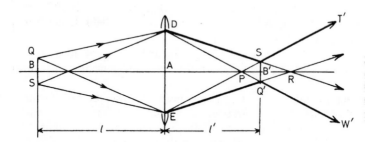

Figure 17.9 The nature of a beam

When an eye is situated in a beam between the positions P and R the whole lens is seen filled with light of uniform intensity as the out-of-focus image of the source subtends a greater angle than is subtended by the lens aperture. This appearance is known as **complete flashing** or the **Maxwellian view**. As the eye moves from this position, either towards or away from the lens, the area of the portion of the lens from which the light appears to be coming is reduced, a blurred image of the source being seen partially filling the lens aperture. The beam from a curved reflector may be considered in the same way, and the principle has important applications in dealing with motor headlights, searchlights and signalling lamps.

The principle of the Maxwellian view is sometimes useful in photometry, and is particularly applicable to the measurement of very small sources or extended sources of low intensity. A convergent lens forms an image of the source in the pupil of the eye, the eye being accommodated so that an image of the lens is formed on the retina. The eye then sees the lens as a field of uniform luminance. When the image of the source falls completely within the pupil of the eye, the apparent luminance B' of the field, neglecting any loss of light by reflection and absorption at the lens, is

$$B' = \frac{h\,B l'^2}{a_e\,l^2} = \frac{I\,l'^2}{a_e\,l^2} \tag{17.04}$$

where h is the area, B the luminance and I the candle-power of the source, and a_e the area of the eye pupil. When the source is of such dimensions that its image more than covers the pupil, the apparent luminance of the field is equal to that of the source.

$$B' = B$$

again neglecting losses of light at the lens.

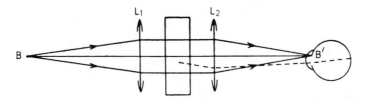

Figure 17.10 Test for striae. Dotted line shows direction of light, refracted by vein, entering the eye

A simple application of the principle of flashing is used in the examination of glass for veins or striae. Light from a bright point source B (*Figure 17.10*) is rendered parallel by the lens L_1, the plate of glass to be examined is placed in the parallel beam, and the light is brought to a focus at B' by the lens L_2. The eye is placed at this focus, but in such a position that the focus falls just outside the pupil and, if the glass is homogeneous, no light enters the eye. Any veins of different refractive index will refract the light, and some of the light from them will enter the eye, which then sees the veins as bright lines on a black background. A similar principle is used in the 'knife-edge' test of lenses and curved mirrors and in retinoscopy.

17.8 OBJECT ILLUMINATED BY TRANSMITTED LIGHT. CONDENSERS

In the foregoing considerations the object has been assumed as self-luminous or as one that reflects diffuse light in all directions. There are, however, a number of cases, particularly in optical projection and microscopy, where the object, usually in such cases called a *slide*, is more or less transparent and is illuminated by transmitted light from a source some distance behind.

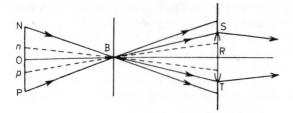

Figure 17.11a Point object illuminated by transmitted light

Let us consider first the case of an axial object point B (*Figure 17.11a*), illuminated by a source NP which is supposed to be of uniform luminance and circular in form. ST is the entrance pupil of the instrument, in this case a single lens, that forms an image of B. It will be seen that when, as in the figure, the source subtends an angle at the object equal to or greater than that subtended by the entrance pupil, the cone of light transmitted by the system is the same as that from a self-luminous point at B. When, however, the source, such as np, subtends a smaller angle, it governs the angular cone entering the system, and behaves as the entrance pupil.

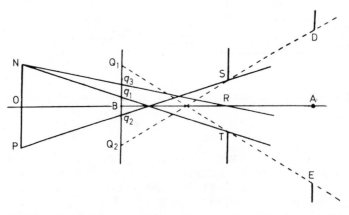

Figure 17.11b Object illuminated by transmitted light

With an extended object the extent of object from which full pencils can pass through the instrument is limited by the angle subtended by the source at the entrance pupil. This is shown in *Figure 17.11b* where ST is the entrance pupil and DE the entrance port of the instrument forming an image of an extended object placed at B. If the object was self-luminous or diffusely reflecting, $Q_1 Q_2$ would be the extent of object, of which the image would be evenly illuminated—the field of full illumination. If however, the object is transparent and illuminated by light from the source NP, the extent of object from which complete pencils can pass is limited to $q_1 q_2$; this is determined by connecting the opposite edges of the source and entrance pupil. Thus for a point on the object, such as q_3, no light is passing from the source through the lower half of the entrance pupil, and the illumination of the image of this point is consequently reduced to half that of the axial point. Hence we see that the source may behave as the entrance port of the instrument, as it does in the case just considered.

It is necessary therefore that if the full aperture and field of view available in the instrument forming the image are to be used, the source must subtend a sufficiently large angle both at the object and at the entrance pupil. In a few cases where the required luminance of the source is comparatively low, an evenly illuminated surface such as the white sky can be used as the source and will subtend a sufficiently large angle. Usually, however, because of the impossibility of producing an extended source of high luminance and also the necessary limitation of the distance at which the source can be placed owing to size of lamp and amount of heat radiated, the source cannot be made to subtend a sufficiently large angle directly. Under these circumstances we may make use of an image of the source formed by a positive lens or system, and this image may be made to subtend a sufficiently large angle at the object or at the entrance pupil. A lens used for this purpose is called a **condenser**. By making the condenser of large aperture as compared with its focal length, large angular cones of light may be collected from the source and passed through the system.

From the conditions to be satisfied in the illumination of a transparent object, which have been discussed above, it will be seen that the condenser may be used in two different ways according to the purpose of the instrument. When the instrument is required to form images of object points with pencils of large aperture, the angular extent of the object being small, the object should be situated at or close to the image of the source formed by the condenser, that is, in the waist of the beam. The condenser aperture should then be large enough to make use of the full angular aperture of the system. This is the arrangement used in the microscope with substage condenser and in the film projector. As an image of the source is formed at or close to the object and will therefore be more or less in focus in the final image it will be necessary that the source should be one that is uniform over its area as most quartz halogen lamps. When the instrument is required to form an image of a large angular extent of object, as in the slide projector, the image of the source should be formed at, or near, the entrance pupil. In this case the aperture of the condenser will act as the entrance port of the system, provided no other stop or stop image subtends a smaller angle at the entrance pupil, while the actual entrance pupil or the image of the source, whichever is the smaller, acts as entrance pupil. In the projection lantern the slide is placed close against the condenser and the slide mask is itself the field stop and entrance port (see *Figure 10.17*).

CHAPTER 17   EXERCISES

1. Explain the meaning of the terms: aperture stop, entrance and exit pupils, and field of view in relation to an optical instrument.

Find the position and magnification of the entrance pupil of the eye, taking the actual pupil to be 3.6 mm behind the cornea, which is assumed to be a single surface of radius 7.8 mm. Assume the refractive index of the aqueous to be $^4/_3$.

2. The eye (assumed to be the 60 D simple eye) with pupil diameter 6 mm is placed behind a thin positive lens of 5 cm focal length and 2 cm aperture, the distance between lens and eye pupil being 8 cm. Find by construction (paraxial rays) the position and size of the entrance pupil of the system consisting of the lens and eye.

3. A system consists of two thin lenses having focal length + 10 cm and + 15 cm respectively separated 5 cm; an aperture 2 cm diameter is placed coaxially between the lenses and 3 cm from the shorter focus lens which is turned towards the object. Find the positions and sizes of the entrance and exit pupils and ports.

4. If the lenses of the system in question 3 have diameters of 3 cm and the object is situated 50 cm from the first lens, find the field of view of full illumination and the total field.

5. An object 2 cm long is placed 12 cm from a positive lens of 6 cm focal length and 6 cm aperture. Make a careful diagram showing the pencils from each end of the object by which an eye with a pupil diameter of 8 mm sees the aerial image, the eye being 25 cm from the image.

6. An unaccommodated emmetropic eye having a pupil diameter of 4 mm is placed 20 mm behind a positive lens of 30 mm focal length and 15 mm aperture. An object is brought up towards the lens until its image is clearly seen. What will be the total extent of object that can be seen and over what extent of object seen will there be no diminution of illumination?

7. A telescope is made up of two thin positive lenses, the objective have a focal length of 6 inches and aperture 4 inches and the eyepiece a focal length of 1½ inches and aperture 2 inches. There are no diaphragms in the instrument. Find by drawing a diagram carefully to scale the field of view and position of the exit pupil.
In what respects would the performance of such an instrument be defective?

8. A Galilean binocular magnifying four times has an objective of 160 mm focal length and 40 mm aperture. The pupil of the eye is 5 mm diameter and situated 10 mm from the eyepiece. Find (a) the apparent and actual field over which the illumination does not fall below half that at the centre, (b) the apparent and actual field of full illumination, (c) the apparent and actual total field.

9. Discuss the limitation of the field of view in a binocular of the Galilean type.

10. It is required to make up a telescope of the astronomical type to have of field of 10°, magnifying power of eight, aperture ratio of object glass $^1/_6$, using a one-inch Ramsden eyepiece.
Determine the dimensions of the parts and draw a careful full-size diagram (to scale) showing the passage through the instrument of a pencil from the limit of the field.

11. What is meant by 'depth of focus'? On what will the depth of focus for any lens depend?

12. A + 6 D lens, having an aperture equal to one-twelfth the focal length, is correctly focused on a very distant object. Find the position of the nearest object which would have its image apparently well defined in the same plane, assuming the image to be well defined when images of points do not exceed 0.1 mm diameter.

13. A camera fitted with a lens of 6 inches focal length and aperture $f/8$ is correctly focused on an object 10 ft away. What will be the range of apparent good focus on either side of the object position if the image is considered sharp when blur circles do not exceed 0.005 inch diameter?

14. With reference to the photographic camera explain (a) the purpose of the adjustable aperture or stop; (b) the meanings and significance of depth of focus.
If the correct exposure with an aperture of $f/8$ is $^1/_{30}$ second, what would be the exposure with the $f/16$ aperture?
If the depth of focus with the $f/8$ aperture is 0.016 inch, what would be the depth of focus with the $f/16$ aperture?

15. Explain carefully why the image of an extended object seen through an optical instrument can never appear brighter than when seen by the unaided eye. How may the apparent brightness differ in the case of a point object, such as a star?

16. A circular source of 5 mm diameter is placed on the axis of, and 6.25 cm from, a + 21 D thin lens of 3 cm aperture. Find the positions on the axis between which an eye, considered as a point, will see the whole lens filled with light—complete flashing.

Explain how this phenomenon can be utilised in examining glass for defects; give a diagram.

17. An optical system consists of a + 10 D lens 40 mm in diameter and a − 40 D lens 20 mm in diameter, separated by 12.5 cm. Between the lenses is a 6 mm stop located 8 cm from the plus lens. For an object plane located 30 cm in front of the plus lens determine:
(a) The location and size of the entrance pupil of the system.
(b) The location and size of the entrance port of the system.
(c) The diameter of the field of full illumination.
(d) The diameter of the full field of view.

18. Describe and discuss the telecentric principle, and show that when this is used, the size of an out of focus image is the same as that of an image sharply in focus.

19. What functions can stops fulfil in an optical system?
An astronomical telescope is made from an objective of focal length + 50 cm and an eyepiece of focal length + 10 cm. If the objective has a diameter of 5 cm, how big is the exit pupil of the instrument and where is it situated?

20. Explain clearly the action of the following stops in an optical system:
(a) aperture stop
(b) field stop
(c) glare stop
A telephoto lens consists of two thin lenses the first of power + 5 D, the second of power − 2.5 D separated by 10 cm, their diameters being 20 mm and 8 mm respectively. Find the position and size of the aperture stop and entrance pupil, and determine the focal ratio (f/no.) of the complete system.

21. Explain the meaning of the terms aperture stop, entrance and exit pupils, and field of view in relation to an optical instrument.
A thin lens with a focal length of + 3.0 cm and aperture 4 cm has a 2.5 cm stop located 1.5 cm in front of it. An object 1 cm high is located with its lower end on the axis 6.0 cm in front of the stop. Find the position and size of the exit pupil and then locate graphically the image position by drawing the two marginal rays and the chief ray from the top of the object.

22. Discuss the roles played by stops and diaphragms in optical systems.
An optical system consists of two converging lenses of focal lengths 10 cm and 20 cm separated by 5 cm. An aperture stop, 1 cm in diameter, is placed midway between them. Calculate the positions and sizes of the entrance and exit pupils.

23. Explain the terms aperture stop and entrance pupil. Draw a diagram of a photographic lens system which consists of two separate lens elements with a variable stop between them. Indicate clearly the entrance and exit pupils of the system.
A thin lens with a focal length of +5 cm and an aperture of 6 cm has a 3.0 cm stop located 3.0 cm behind it. An object 3.0 cm high is located with its centre on the axis 12 cm in front of the lens. Find:
(a) by formula the position and size of the entrance pupil
(b) the image location graphically by drawing the two marginal rays and the principal ray from the top end of the object.

24. Explain the meaning of the following terms:
(a) aperture stop
(b) entrance pupil
(c) field stop
(d) entrance window.
A coaxial optical system consists of two thin lenses, with a stop midway between them, separated by a distance of 15 cm. The first lens is a converging lens of focal length 8 cm and diameter 2 cm. The diameter of the stop is 1.5 cm. For a distant object field determine the positions of the entrance pupil and entrance window and the magnitude of the semi-angular mean field of view.

Chapter 18
Aberrations and image quality

18.1 INTRODUCTION

An optical system or instrument is required to produce, in a predetermined position and of a certain size, an image of a given object. In order that this image shall be a faithful and clearly defined copy of the object, each point of the latter must be imaged as a point, or at least as a small diffraction disc, in the image, and the totality of these image points must build up a geometrical figure similar to the object. We learn from Chapter 15 that for the realisation of this it is necessary that the aperture of the system be appreciable; also that the emerging wavefronts corresponding in the several object points be spherical so that the 'rays' of each emerging pencil, axial as well as oblique, shall reunite in common image points. We have previously seen (Chapters 5, 6 and 7) that when the aperture is appreciable, the whole of the light from a given object point does not reunite in a point after reflection or refraction at a spherical surface or refraction by a single lens. Hence in practice images suffer from defects or **aberrations** unless steps are taken—as by combining a number of lenses, probably made of different glasses and of various thicknesses—to eliminate or reduce such aberrations, in which case we have a 'corrected' system.

Any complete study of aberrations and lens design is outside the scope of this book, but it is desirable to obtain an idea of the general nature of the more important aberrations and some of the principles involved in their correction. For a fuller account of aberrations and lens design the student should consult H. H. Emsley *Aberrations of Thin Lenses*, Constable (1956) which relates more to single lenses while Warren J. Smith *Modern Optical Engineering*, McGraw-Hill (1966) gives a wider coverage of optical systems and computational methods.

It has been shown that refraction at a lens surface follows Snell's law:

$$n \sin i = n' \sin i'$$

This formula does not give a simple relationship between i and i'. However, sine can be represented as a series by:

$$\sin i = i - \frac{i^3}{6} + \frac{i^5}{120} - \frac{i^7}{5040} + \dots$$

where i is measured in radians. These terms get smaller and smaller when i is less than 1 radian (about 57°) so for a lot of cases we can use a part of the series as a reasonable approximation. The assumption, $\sin i = i$, gives the paraxial approximation which has been used so far in this book. This is also called 1st order theory.

A closer approximation is

$$\sin i = i - \frac{i^3}{6}.$$

When this is used it is possible to derive five independent ways in which an image may be an imperfect rendering of an object. These were studied by Ludwig von Seidel who published the mathematical expressions in 1855-56. Consequently they are often called the Seidel Aberrations. Because the extra term in the approximation contains i^3, this analysis is called third-order theory.

An even closer approximation can be obtained by taking

$$\sin i = i - \frac{i^3}{6} + \frac{i^5}{120}$$

which gives fith-order theory. This yields a great many aberration types, most of which remain small. They are called secondary aberrations to distinguish them from the Seidel or primary aberrations. Fifth-order theory and secondary aberrations are not widely used in lens design as the equations are somewhat complex and in most cases third-order theory is sufficient for the initial design work which can then be followed by exact ray tracing (section 18.7).

The five Seidel aberrations are:

S_I Spherical aberration
S_{II} Coma
S_{III} Astigmatism
S_{IV} Curvature of Field
S_V Distortion

Each of these will be dealt with in this chapter. They all occur together and the purpose of lens design is to reduce their size to an acceptable level. It is possible to derive equations of varying complexity for the aberrations due to refraction (or reflection) at spherical surfaces, aspheric surfaces and thin lenses. Almost all these equations are approximations, they give simplicity while still providing valid relationships between the parameters of the lens and the image quality. As the amount of aberration is reduced in the design process, the error due to the approximation also decreases.

These equations may be expressions of **longitudinal aberration** where the effect of the error is measured along the axis; **transverse aberration** where the 'miss-distance' of a particular ray is calculated on a given image surface; or a **wavefront aberration** where the optical path difference between the real wavefront and a perfect spherical wavefront is calculated.

This last approach gives a very useful overall view of the situation. The sag of a spherical curve (*Figure 18.1a*) is given by

$$x = R - (R^2 - r^2)^{1/2}$$

$$= R \left[1 - \left(1 - \frac{r^2}{R^2}\right)^{1/2} \right]$$

By the Binomial expansion,

$$x = \frac{r^2}{2R} + \frac{r^4}{8R^3} + \frac{r^6}{16R^5} + \frac{5r^8}{128R^7} + \text{higher terms} \dots$$

... if spherical

$$+ Ar^4 + Br^6 + Cr^8 + \text{higher terms} \dots$$

... if not spherical

These extra 'correction' terms of the polynomial have been added deliberately and indicate what is needed to obtain a completely general expression for a wave surface which has been deformed from a spherical shape. The added terms refer to third-order theory, fifth-order theory and seventh-order theory respectively.

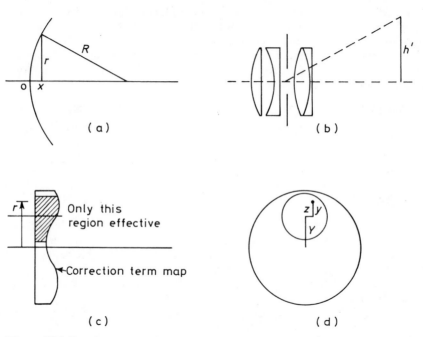

Figure 18.1 Development of primary aberration theory

Most lens systems have some sort of stop which renders only parts of each surface available to the light of a given image point (*Figure 18.1b*). On a map of the correction term's value, the value of r (*Figure 18.1c* and *d*) is given by a combination of position within the stop (y, z) and the displacement, Y of the centre

$$r = [(Y + y)^2 + z^2]^{1/2}$$

If we consider only the third-order correction term, we have

$$Ar^4 = A [(Y + y)^2 + z^2]^2$$

as a perfectly general description of the way changes in Y, y and z can modify x, the axial location of the wavefront to this level of accuracy.

Expanding this, we find that the wavefront aberration, W, is given by an equation of the form:—

$$W = A(y^2 + z^2)^2 + AY[4y(y^2 + z^2)] + AY^2(6y^2 + 2z^2) + AY^3(4y) + AY^4$$

But Y is proportional to h', the image height, so that when we allow the co-efficients to be independent, we have

$$W = \tfrac{1}{8}A_I(y^2 + z^2)^2 + \tfrac{1}{2}A_{II}h'[4y(y^2 + z^2)] + \tfrac{1}{8}A_{III}h'^2(6y^2 + 2z^2) + \tfrac{1}{2}A_V h'^3(y)$$

Spherical Ab. (circular function)	Coma (displaced circular function)	Astigmatism (elliptical function)	Distortion (displacement function)

$$(18.01)$$

The last term has been omitted as it does not change the waveform shape; and dividing through by 8 brings the terms into a better correspondence with the spherical reference wavefront.

The elliptical function degenerates into orthogonal straight lines when $(2y^2 + 2z^2)$ or $(6y^2 + 6z^2)$ are subtracted. The fourth Seidel aberration, curvature of field (S_{IV}) does not show up on this simple analysis as it retains a spherical shape to the wavefront and does not change with stop position. It arises out of our unnatural desire for flat image planes! The term

$$\tfrac{1}{8}A_{IV}h'^2(2y^2 + 2z^2)$$

is found to give the required image shift.

It is usual to calculate the Seidal coefficients for the extreme ray allowed by the stop and the extreme angle of field. This means that the maximum values of y, z and h' are used for S_I, S_{II}, S_{III} etc. Then, when we wish to calculate the wavefront aberration, W, for a particular ray, we use S_I, S_{II}, etc. for the co-efficients and the *fractional* values of y, z, and h' for that ray with respect to their maxima. This wavefront aberration is the actual distance between the aberrated wavefront and the reference sphere centred on the paraxial image. When we include the curvature of field modification, equation (18.01) now becomes

$$W_{(h'yz)} = \tfrac{1}{8}S_I(y^2 + z^2)^2 + \tfrac{1}{2}S_{II}h'[y(y^2 + z^2)] + \tfrac{1}{4}(3S_{III} + S_{IV})h'^2 y^2$$
$$+ \tfrac{1}{4}(S_{III} + S_{IV})h'^2 z^2 + \tfrac{1}{2}S_V h'^3 y \qquad (18.02)$$

The third and fourth terms now give the astigmatic effect with respect to a flat image plane through the paraxial focus. Equation (18.02) serves as a guiding beacon for a subject which, at this level, is a hotch-potch of ill-connected relationships.

From the wavefront aberration, the transverse aberration T.A., may be obtained (in the meridional plane) by

$$\text{T.A.} = l'\,\frac{\partial W}{\partial y} \qquad (18.03a)$$

where l' is the final image distance (or radius of the reference sphere). The longitudinal aberration, L.A., follows from

$$\text{L.A.} = \frac{l'}{y} \text{ T.A.} \tag{18.03b}$$

Both equations (18.03) are correct to a sufficient approximation for third order theory. The great value of the wavefront expression is that the path difference caused by one surface in a system can be added directly to that due to the next surface, and so on, without loss of accuracy and the total value used in (18.03) to find the residual transverse or longitudinal aberration of a system. This can also be done using each element of a system (as a thin lens). This latter approach contains the inaccuracies of the thin lens approximation but gives a quicker route into lens system aberrations. In the following sections we will discuss either the longitudinal or transverse aberration, as appropriate, together with the wavefront aberration expression, and concentrate on the thin lens. No great attention will be paid to signs as this can be confusing rather than instructive at this level.

18.2 CHROMATIC ABERRATION AND CHROMATIC DIFFERENCE OF ABERRATION

The nature of **chromatic aberration** was dealt with in Chapter 13. It was seen there that as the refractive index of optical materials shows a variation according to the wave-length of the light used there is a variation in lens power and principal planes with wave-length. In this chapter we shall assume the light is mono-chromatic light so that chromatic effects do not arise.

However, all these monochromatic aberrations are affected by changes in refractive index, so that, for example, the amount of spherical aberration from a given lens will be different for red light than for blue light. Fortunately, these **chromatic differences of aberration** are much smaller than the aberrations them-selves and need not be considered at this level. However, when the image of a well-corrected lens is examined in detail the effects of this may be seen to be as large as the effect due to the residual chromatic aberration.

18.3 SPHERICAL ABERRATION

This aberration is unique among the five as it is the only one affecting axial images. This is because the value of the other aberrations, whose terms in (18.02) contain h', is zero for on-axis points. Spherical aberration is independent of field angle and so affects all points in the field similarly.

The general nature of spherical aberration has been briefly discussed in section 5.9 which then restricted further discussion to paraxial rays. In section 5.12, working from *Figure 5.14*, the paraxial equation (5.02) was found using *two* approximations. We assumed then that Snell's Law was $ni = n'i'$ and, further, that

$$u = \frac{y}{AB}; \quad u' = \frac{y}{AB'}; \quad a = \frac{y}{AC}$$

Working now from *Figure 18.2* with a marginal ray having non-paraxial values of i and i', we need $n \sin i = n' \sin i'$. In triangle DBC we can apply the sine rule to give

$$\frac{\sin i}{l - r} = \frac{\sin u}{r} \tag{18.05a}$$

or

$$\sin i = l \sin u \left(\frac{1}{r} - \frac{1}{l} \right) \tag{18.05b}$$

The same equations can be obtained for the refracted ray using triangle DB'C. Joining (18.05b) for incident and refracted rays via Snell's law we have

$$n \left(\frac{1}{r} - \frac{1}{l} \right) l \sin u = n' \left(\frac{1}{r} - \frac{1}{l'} \right) l' \sin u' \tag{18.06}$$

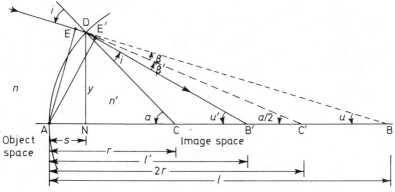

Figure 18.2 Refraction at a spherical surface—all positive diagram

This is the same as the paraxial case only if we assume $l \sin u = l' \sin u'$. As $l \sin u$ is equal to EA, the perpendicular dropped from the incident ray to A; while $l' \sin u'$ equals E'A; the difference between these lengths is the key to spherical aberration. Only when l is infinite or equal to r does EA equal y, the paraxial assumption. At all other times there is a small discrepancy. We can see from *Figure 18.2* that

$$l \sin u = \frac{ly}{[y^2 + (l - s)^2]^{1/2}} = y \left(1 - \frac{2s}{l} + \frac{s^2 + y^2}{l^2} \right)^{-1/2}$$

where

$$s = r - r \left(1 - \frac{y^2}{r^2} \right)^{1/2} = \frac{y^2}{2r} \left\{ 1 + \frac{y^2}{4r^2} + O \left(\frac{y}{r} \right)^4 \right\}$$

and

$$s^2 = \frac{y^2}{4r^2} \left\{ 1 + \frac{y^2}{2r^2} + O \left(\frac{y}{r} \right)^4 \right\}$$

by the Binomial expansion; O indicating higher order terms. Therefore,

$$l \sin u = y \left\{ 1 - \frac{y^2}{2rl} \left[1 + \frac{y^2}{4r^2} + O\left(\frac{y}{r}\right)^4 \right] + \frac{y^2}{l^2} \left[1 + \frac{y^2}{4r^2} + O\left(\frac{y}{r}\right)^4 \right] \right\}^{-1/2}$$

and, again by the Binomial expansion,

$$l \sin u = y + \frac{y^3}{2rl} - \frac{y^3}{2l^2} + \frac{y^5}{8r^3 l} + \frac{y^5}{4r^2 l^2} - \frac{3y^5}{4rl^3} + \frac{3y^5}{8l^4} + O(y)^7$$

A similar equation can be obtained for $l' \sin u'$ and, putting these into (18.06), we have, as far as the third order

$$n\left(\frac{1}{r} - \frac{1}{l}\right) + \frac{ny^2}{2l}\left(\frac{1}{r} - \frac{1}{l}\right)^2 = n'\left(\frac{1}{r} - \frac{1}{l'}\right) + \frac{n'y^2}{2l'}\left(\frac{1}{r} - \frac{1}{l'}\right)^2 \quad (18.07)$$

This shows that the aberration in l' (the longitudinal aberration) varies as y^2. This is to be expected as $-y$ has no different meaning to $+y$ for an axial object. Obviously, (18.07) is the same as the paraxial case when the first-order terms only are taken.

In (18.06) we have equality when $l = l' = 0$ and when $l = l' = r$ for all values of u and u'. Thus, there is no spherical aberration at these points which are cases where EA and E'A = y. There is a further condition of equality when EA = EA'; when they are inclined at equal and opposite angles to DA. This gives (*Figure 18.2*)

$$\beta = \beta' = \frac{a}{2} - u = u' - \frac{a}{2}$$

This is equivalent to

$$\frac{1}{l'} = -\frac{1}{l} + \frac{1}{r}$$

which may be used with equation (5.02) to show that at this position

$$\frac{l}{r} = \frac{n + n'}{n}$$

and

$$\frac{l'}{r} = \frac{n + n'}{n'}$$

Figure 18.3 shows these **aplanatic points**. The absence of spherical aberration at the last of these points can be shown graphically using Young's construction (section 5.8). Aplanation means freedom from spherical aberration and coma. In none of these cases is it possible to have a real image of a real object but a combination of the second and third condition is often used in the design of large aperture lenses to be free from spherical aberration and coma, e.g. microscope objectives, Fizeau interferometers (section 14.14).

When (18.07) is applied to the thin lens case, considerable interaction is found between l and l', the object and final image distances; and r_1 and r_2, the radii of curvature of the first and second surface. Two new factors may be defined, which substantially reduce the complexity of the aberration equations.

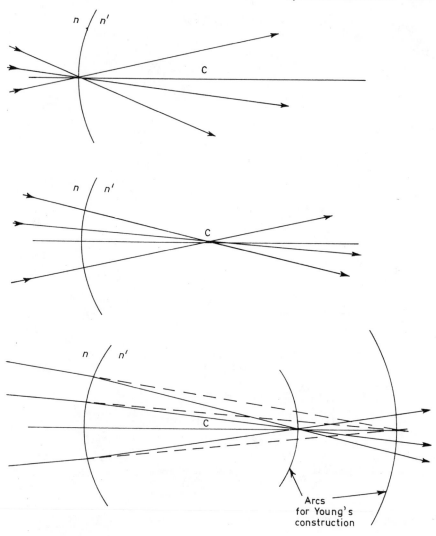

Figure 18.3 Aplanatic points of a spherical refracting surface

These factors are defined as*

$$X = \frac{r_2 + r_1}{r_2 - r_1} = \frac{R_1 + R_2}{R_1 - R_2}$$

(18.08a)

which is dimensionless and known as the **bending factor**, and

$$Y = \frac{l' + l}{l' - l} = \frac{L + L'}{L - L'}$$

(18.08b)

*Some writers give different definitions but those above give all positive signs in equations (18.09) and (18.35).

which is also dimensionless and known as the **conjugates factor.**

Using these it is found that the difference between the paraxial and marginal foci is given by

$$l' - l'_m = \tfrac{1}{2}y^2 l'^2 F^3 (\alpha X^2 + \beta XY + \gamma Y^2 + \delta) \qquad (18.09)$$

where

$$\alpha = \frac{n+2}{4n(n-1)^2}$$

$$\beta = \frac{n+1}{n(n-1)}$$

$$\gamma = \frac{3n+2}{4n}$$

$$\delta = \frac{n^2}{4(n-1)^2} \qquad (18.10)$$

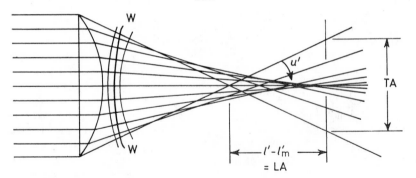

(a) $R_1 = 0$ and $L = 0$; therefore $X = -1$ and $Y = -1$

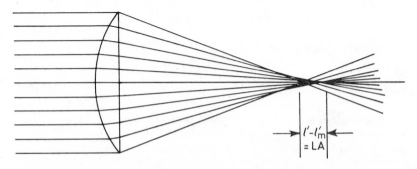

(b) $R_2 = 0$ and $L = 0$; therefore $X = +1$ and $Y = -1$

Figure 18.4 Spherical aberration for single positive lens (to scale)

The values of these coefficients for four values of refractive index are given below (for ϵ and ξ see section 18.5).

n	α	β	γ	δ	ϵ	ξ
1.4	3.79	4.28	1.11	3.06	2.14	1.36
1.5	2.33	3.33	1.08	2.25	1.67	1.33
1.6	1.56	2.71	1.06	1.78	1.35	1.31
1.7	1.11	2.27	1.04	1.48	1.13	1.29

As they all reduce with increasing index, a lens made out of high-index glass will have less spherical aberration than one of equal power made from low-index glass. For a plano-convex lens (n = 1.5) of 80 mm efl the spherical aberration is shown to scale, using (18.09), in *Figure 18.4*. The differences between the two drawings amply demonstrates the effect of the bending factor X. The length LA is Longitudinal Third-Order Spherical Aberration. The distance TA is the Transverse Third-Order Spherical Aberration. Clearly TA = LA tan u' which compares with (18.03b). It is also clear that the best image obtainable is a good deal better than the value shown for TA. The smallest diameter blur occurs $^3/_4$ of the way to the marginal focus and is necessarily TA/4. The shift of the utilised image plane is known as defocussing even though it improves the image.

This shift of focus can be examined from the point of view of the wavefront aberration. The wavefront W–W is drawn in *Figure 18.4a* with each ray normal to it. It is apparent that it is not a sphere and the optical path differences between this wavefront and a reference sphere constitute the wavefront aberration. Different reference spheres result from defocussing and give different values for the aberration. The thin lens contribution to the maximum wavefront aberration due to spherical aberration with respect to the paraxial focus is

$$W_{\text{S.A.}} = \tfrac{1}{8}(S_{\text{I}}) = \tfrac{1}{8}y^4 F^3 \left(\alpha X^2 + \beta XY + \gamma Y^2 + \delta\right) \tag{18.11}$$

which reduces to (18.09) when equations (18.03) are applied. Note that (18.11) does not contain l' directly (only as a component of Y) and straightforward summation over a number of lenses is possible. The wavefront aberration varies with y^4, the transverse with y^3 and the longitudinal with y^2.

The shape of the wavefront aberration can be shown in pictorial form as in *Figure 18.5a* where the hatched areas show the wavefront difference from a paraxial reference sphere. An alternative presentation of the aberration is by means of a **spot diagram** where rays, equally spaced in the aperture of the lens, are traced by computer and points of intersection with the chosen image surface are plotted. The regions of greatest density of dots show the brighter parts of the image. The actual intensity profile of the image may also be computed and forms the **point spread function**. Often the intensity across the image of a line object is calculated, giving the **line spread function** which can be utilised with the M.T.F. concept discussed in sections 18.10 and 18.11.

Yet again calculations may be made to obtain the locations of the **caustic surfaces**. These are three-dimensional surfaces in space which define the intersections of adjacent rays in image space. In general, only adjacent rays along two directions across a lens intersect, thus giving two surfaces. In the case of spherical aberration these caustic 'sheets' comprise a trumpet-shaped surface and a straight line on the axis (as a degenerate surface) shown in *Figure 18.5b*. An

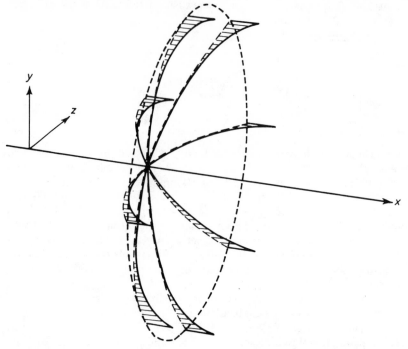

*Figure 18.5a Representation of spherical aberration. The broken lines represent
a spherical reference wavefront advancing along the axis. The solid lines represent
the aberrated wavefront and the hatching the aberration*

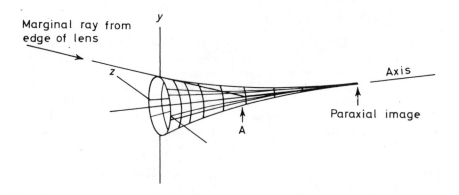

Figure 18.5b Caustic surfaces for spherical aberration

image plane cutting through these sheets will show a bright spot and ring. A ring
only is seen when nearer to the lens than A. This approach can be useful when
the image plane is indeterminate. By far the most common representation, is a
simple graph showing where a ray from a given height on the lens intersects the
axis, as used in *Figure 18.6b*.

The bending factor, X can be changed without altering the power of the lens.

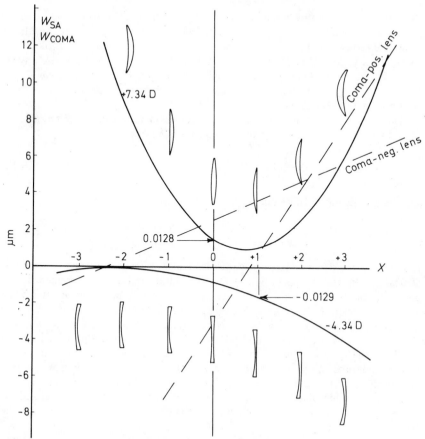

Figure 18.6a Spherical aberration contributions of possible doublet components (see section 18.5 for coma discussion)

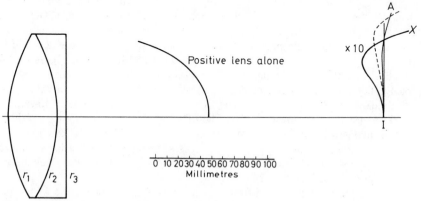

Figure 18.6b Spherical aberration correction in F/2 doublet lens (to scale)
$r_1 = -r_2 = 141$ mm; $n_1 = 1.5175$; $d_1 = 44$ mm
$r_3 = -8333$ mm; $n_2 = 1.6225$; $d_2 = 8$ mm
The broken line shows O.S.C.—see coma, section 18.5

Thus, for any given power and conjugates the spherical aberration may be varied throughout a range by changing X. Since (18.11) is quadratic in X the result is a parabolic variation with a minimum value but with no intersection with the x-axis. Thus, the spherical aberration of a single positive lens is never zero and always in the same direction. This may be termed positive aberration (or negative aberration) according to different conventions. Ambiguity can be avoided by referring to it as undercorrect spherical aberration.

The bending for minimum spherical aberration is found by differentiating (18.11) with respect to X, regarding Y constant. We obtain

$$X_{min} = \frac{-2(n^2 - 1)}{n + 2} \, Y \tag{18.12}$$

and the coefficient at this value is given by

$$S_{I(min)} = y^4 F^3 \left[\frac{n^2}{4(n - 1)^2} - \frac{n}{4(n + 2)} \, Y^2 \right] \tag{18.13}$$

Due to the F^3 dependence, a negative lens always has finite negative spherical aberration. In section 13.11 a $+ 3$ D achromatic doublet was designed having components, $+ 7.34$ D and $- 4.34$ D, which have refractive indices of 1.5175 and 1.6225 respectively. The effect of bending these lenses is calculated [using (18.11)] and plotted in *Figure 18.6a* assuming light from a distant object is incident first on the positive component. Thus $Y = 1$ for the positive lens and $+ 2.4$ for the negative. Equation (18.11) calculates the contribution of each lens to the spherical aberration of the wavefront. For thin lenses in contact y is effectively the same for each lens and could be left out of the calculation, but for realism a value of 10 mm (0.01 m) have been assumed and the actual shift in the wavefront is plotted. The influence of the F^3 term on the magnitude of the two curves is easily seen.

In the suggested design in section 13.11, the positive component was equiconvex while the negative was almost plane-concave. Therefore, the relevant X values are 0 and slightly more than 1. As can be seen from the values indicated in *Figure 18.6a*, the third-order spherical aberration is virtually eliminated for a distant object. Clearly, other combinations can be chosen to do this. The actual choice is usually governed by extra conditions such as the need to cement components together, or the need to correct coma (section 18.5) or the best correction for fifth-order spherical aberration. In the scale drawing of *Figure 18.6b*, which shows the actual longitudinal aberration, curve IA, calculated by ray tracing, the dramatic improvement in quality between the positive lens alone and the doublet can be clearly seen. The thicknesses of the components have been set at 44 mm and 8 mm respectively. The fifth-order contribution gives a residual longitudinal error of about 2 mm within an $f/2$ cone. The curve IX shows this aberration exaggerated by $\times 10$ and emphasises the rapid deterioration in aberration if the aperture is pushed too far. The effect of making the final surface plano is covered in Question 24.

The choice of a slightly different image plane can improve the image but care must be taken in interpreting graphs of this nature because (a) the outer zones of the lens carry more energy per unit change in y; (b) diffraction effects may become large with respect to the residual geometric aberration. When new conjugates are used with the same lens construction, the spherical aberration correction will be less good although still generally better than a singlet lens. If the lens in *Figure 18.6b* is reversed, however, the longitudinal aberration for $y = 40$ mm increases from 1.6 mm to 28 mm while that of a single lens bent for minimum aberration would only be about 7 mm for the same power and y value. Most achromatic doublets are designed for use with their more convex surface facing the incident light from a distant object.

18.4 OBLIQUE ABERRATIONS: 1–CURVATURE OF FIELD AND ASTIGMATISM

In the preceding section the paraxial approximation was shown to be insufficient to describe the image of an axial object when this is formed by a lens of appreciable aperture. Different problems arise when the lens aperture is small but the

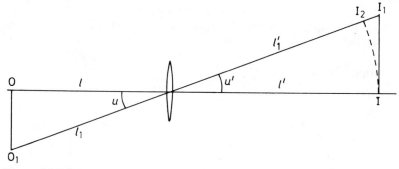

Figure 18.7 Image curvature

object is considerably off-axis. In *Figure 18.7* we have the familiar example of a lens forming an image II_1 of the object OO_1. However, it is clear that the length l_1 is longer than l. If, therefore

$$\frac{1}{l'} - \frac{1}{l} = \frac{1}{f}$$

and

$$\frac{1}{l'_1} - \frac{1}{l_1} = \frac{1}{f}$$

are both to be true then l'_1 must be shorter than l'. The image must be curved to be somehow like II_2. Strictly, it has a paraboloidal shape and is known as the **Petzval Surface**.

Calculations for the single surface case (see *Figure 5.16*) show that this curvature may be expressed approximately as a sphere of radius r'_i given by

$$\frac{1}{n'r'_i} - \frac{1}{n\,r_i} = \frac{n-n'}{n\,n'\,r} \tag{18.14}$$

where r is the actual radius of curvature of the surface and r_i the radius of curvature of the object (which may not be plano for later surfaces in a system). The right-hand side of (18.14) is the *contribution* of each surface to the curvature of the final image. If the contributions of the two surfaces are added together we easily obtain the thin lens contribution to field curvature as

$$C = \frac{-F}{n} \tag{18.15}$$

where n is the index of the lens material. If a lens system is to provide a flat image field from a flat object the sum of these contributions must be zero

$$\Sigma_k \frac{F_k}{n_k} = 0 \tag{18.16}$$

which is known as the Petzval condition. Only by combining lenses of different refractive index can this condition be satisfied and field curvature eliminated as (18.15) is not affected by lens shape or image conjugates.

The curvature referred to here is *not* that of the wavefront. The maximum wavefront aberration is given by:

$$W_{\text{Field Curve}} = \tfrac{1}{4}S_{IV} = \tfrac{1}{4}H^2 \frac{F}{n} \tag{18.17}$$

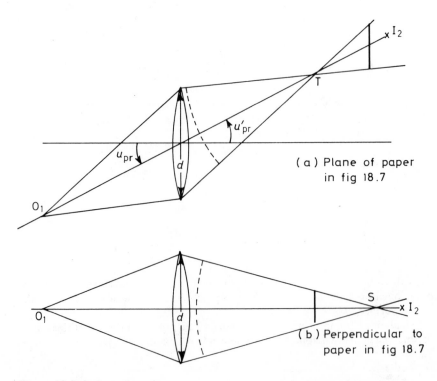

(a) Plane of paper in fig 18.7

(b) Perpendicular to paper in fig 18.7

Figure 18.8 Astigmatism

where I is the Lagrange Invariant, $n'h'u'$, which was developed in section 5.15. For a thin lens in air, this reduces to $h'y/l'$. If (18.03) is applied to (18.17), we find the longitudinal aberration, that is, the distance from the paraxial image plane to the relevant point on the Petzval surface, is $\frac{1}{2}h'^2 F/n$. This is also obtained if the sag. formula is applied to (18.15).

A further consideration of the off-axis image is the difference between the section of the wavefront lying in the plane of the paper (T) and that, containing the ray $O_1 I_2$ (*Figure 18.8*) but perpendicular to the paper (S). The optical path difference due to the lens is zero at its edge if we make its thickness there zero. At the centre, the optical path difference is the same for each section of the wavefront as it is that experienced by the ray $O_1 I_2$ which is called the **principal ray**. However, in the case of the S section this occurs across the whole diameter, d of the lens, while for the T case the diameter is foreshortened by the angle u_{pr} to be $d \cos u_{pr}$ (or $d \cos u'_{pr}$). The radius of curvature of the sections of the emerging wavefront can be obtained from the sag. formula and, as the sags are equal, we have

$$\frac{y^2}{2r_S} = \frac{y^2 \cos^2 u'_{pr}}{2 r_T} \tag{18.18}$$

If the incident beam is from a distant object then these radii become focal lengths. We then have

$$f'_T = f'_S (1 - \sin^2 u'_{pr})$$

and, when the difference is a small fraction of the paraxial focal length;

$$f'_S - f'_T = f' \sin^2 u'_{pr} \tag{18.19}$$

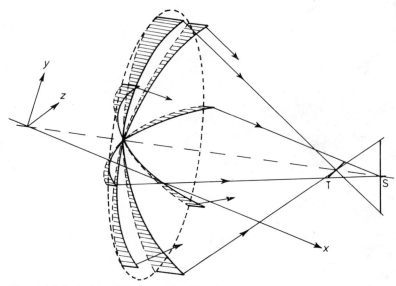

Figure 18.9 Astigmatism—wavefront and rays
Broken line arcs—reference wavefront
Solid line arcs—actual wavefront
Hatched areas—wavefront aberration

which is called the astigmatic difference or **Sturm's Interval**. This longitudinal aberration is *not* reduced by stopping down the lens as it is independent of y.

A pictorial representation of the wavefront aberration is given in *Figure 18.9*. The actual wavefront has a toroidal shape similar to that described in section 8.8. In this case, however, the astigmatism is occurring with a spherical lens receiving the light obliquely. In ophthalmic texts this astigmatism is often referred to as oblique astigmatism to distinguish it from that occurring (on axis) with toric lenses. Oblique astigmatism arises at all spherical surfaces used off-axis, the wavefront aberration giving rise to the fan of rays, as in *Figure 18.9*, which pass through the short line at T and a second perpendicular line at S. There is, therefore, a general similarity in the form of the beam in the two cases.

The length of each of these lines depends on the astigmatic difference and the lens aperture; the transverse aberration does depend on y as predicted by (18.03b). Between the two lines, roughly midway, there is, if the aperture is circular, a point where the blur is circular. This is the circle of least confusion which is also proportional to y.

Because the longitudinal aberration is independent of y it is possible to develop an imitation of paraxial mathematics in a form which applies along narrow *oblique* pencils of light. In *Figure 18.10* ADEGH is a spherical surface between two media. C is the centre of curvature and CA a radius in the plane of the paper. Q is an object point in the plane of the paper and this plane is called the meridional plane or tangent plane. The principal ray, QA, the central ray of

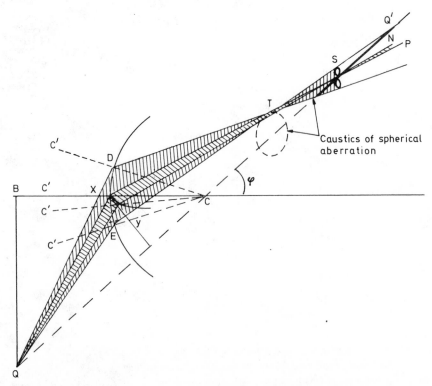

Figure 18.10 Astigmatism at a single spherical surface

the bundle is in this plane and remains in it after refraction to Q'. The plane perpendicular to the meridional plane and containing the principal ray is the equatorial or sagittal plane. This plane is tilted on refraction. The points QAGHTSNP lie in the sagittal planes (approximately).

Consider four limiting rays of the pencil, which is supposed to be very narrow, but is shown with its width exaggerated. Rays QD and QE are in the plane of the paper while rays QG and QH (not drawn) are above and below the plane of the paper. QD and QE remain in the plane of the paper and meet the principal ray at T. The ray QG is refracted back towards the paper and the ray QH forwards towards the paper and they meet the principal ray at S. Other rays are refracted in such a way that they pass through a short line at T and a second perpendicular line at S. These focal lines are really small caustic surfaces, that at T is called the **tangential line focus** and is a short arc of a circle centred on the axis of the surface, while that at S is called the **radial** or **sagittal line focus** and is more exactly a narrow figure of eight. Note that the *actual line* of T lies in the sagittal plane and vice versa!

The reason for the shapes and positions of these lines is clear when their relationship with the basic caustic surfaces of spherical aberration is recognised. In *Figure 18.10*, the axis for which the object and image would be regarded as axial has been shown as a dashed line. All our oblique narrow pencil does is to select a small non-axial aperture on the surface and the relevant parts of the caustics.

When the aperture is very small, the marginal ray paths with respect to the principal ray show very small changes in the angles of incidence and refraction. If we differentiate Snell's Law

$$n \, di \cos i = n' \, di' \cos i' \tag{18.20}$$

In *Figure 18.10* we can use the triangles XDC and XAG to obtain

$$di = i_2 - i_1 = \angle C'DQ - \angle C'AQ = \angle DCA - \angle AQD$$

remembering that $\angle AQD$ as drawn is a negative angle. The last two angles can be approximated to give

$$di = \frac{AD}{r} - \frac{AD \cos i}{t}$$

where r is the radius AC and t is the object distance along the principal ray in the tangential plane. By the same method we find

$$di' = \frac{AD}{r} - \frac{AD \cos i'}{t'}$$

where t' is the image distance in the tangential plane. Putting these expressions into (18.20) we have,

$$\frac{n' \cos^2 i'}{t'} - \frac{n \cos^2 i}{t} = \frac{n' \cos i' - n \cos i}{r} \tag{18.21}$$

where i and i' are the incidence and refracted angles of the principal ray.

In the other plane we know that QAS and QGS intersect at S on the line QCS as this is the axis of symmetry. In the triangle AQC,

$$\frac{QC}{\sin i} = \frac{s}{\sin \angle ACQ}$$

and similarly

$$\frac{SC}{\sin i'} = \frac{s'}{\sin \angle A\hat{C}S}$$

where s and s' are equal to the object and image distances along the principal ray in the sagittal plane.

As $\angle ACS + \angle ACQ = 180°$, we have

$$\frac{SC}{s' \sin i'} = \frac{QC}{s \sin i}$$

which together with Snell's Law gives

$$\frac{n'SC}{s'} = \frac{nQC}{s}$$

If we project SC and QC onto the line C' AC by multiplying both sides by cos φ we have QCcos $\varphi = (s \cos i - r)$ and SC cos $\varphi = (s' \cos i' - r)$ so that

$$\frac{n'}{s'} - \frac{n}{s} = \frac{n' \cos i' - n \cos i}{r} \qquad (18.22)$$

Equations (18.21) and (18.22) are commonly known as the **Coddington Equations** and may be used in a method known as differential ray tracing to obtain the location of astigmatic images of oblique pencils through a lens system by taking distances *along* the principal ray and not the axis. The value

$$\frac{n' \cos i' - n \cos i}{r}$$

is called the oblique power K of the surface. For a thin lens $s'_1 = s_2$ and $t'_1 = t_2$ in the usual way so the oblique power becomes

$$(n' \cos i' - n \cos i)\left(\frac{1}{r_1} - \frac{1}{r_2}\right) = \left(\frac{n' \cos i' - n \cos i}{n - 1}\right) F \qquad (18.23)$$

where n is the refractive index of the lens material and F the paraxial power.

We then have the equations

$$S' = S + \left(\frac{n \cos i' - \cos i}{n - 1}\right) F \quad = S + F_s \qquad (18.24a)$$

and

$$T' = T + \left(\frac{n \cos i' - \cos i}{(n - 1)\cos^2 i}\right) F \quad = T + F_t \qquad (18.24b)$$

where S, T, S' and T' are the object and image vergences along the principal ray in the two planes of regard.

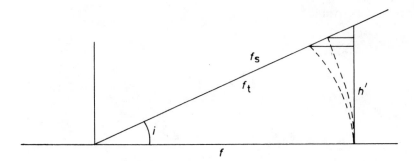

Figure 18.11a Astigmatism curvatures

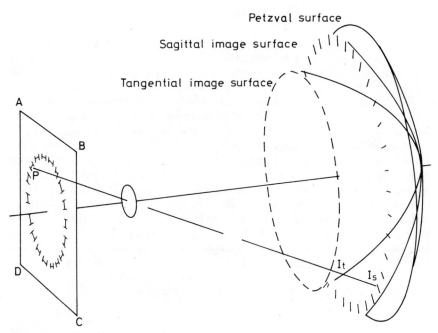

Figure 18.11b Astigmatic surfaces

In *Figure 18.11a*, the sags for the curved images at height h' on the paraxial image plane are,

$$\left(\frac{f}{\cos i} - f_s\right)\cos i \quad \text{and} \quad \left(\frac{f}{\cos i} - f_t\right)\cos i$$

or

$$\frac{F_s - F\cos i}{F^2} \quad \text{and} \quad \frac{F_t - F\cos i}{F^2}, \text{approximately} \tag{18.25}$$

By the Binomial expansion

$$\cos i = 1 - \frac{h'^2}{2f^2}$$

$$\cos i' = 1 - \frac{h'^2}{2n^2 f^2}$$

and

$$\cos^2 i = 1 - \frac{h'^2}{f^2}$$

all to third order. The respective sags be found, using (18.24) and (18.25) to be,

$$\left(1 + \frac{1}{n}\right)\frac{h'^2 F}{2} \quad \text{and} \quad \left(3 + \frac{1}{n}\right)\frac{h'^2 F}{2} \tag{18.26}$$

The general shape of these curves is given in *Figure 18.11b*, the so-called cup and saucers diagram. If ABCD is a plane object, a point p upon it is imaged as a short arc on the tangential image surface and a short radial line on the sagittal image surface. If p is already part of a line then the image of the line may appear to be in focus on either of the surfaces if the object line is orientated in the same direction as the image of the point. Thus the circle on ABCD will appear to be in focus at I_t and the spokes I_s, if the aperture is small. Object lines at other orientations do not form a good image.

The diagram shows a third surface, the Petzval surface. This is not an image surface at all in the presence of astigmatism. If astigmatism is corrected all images of objects on ABCD will be formed on this surface (assuming narrow pencils). Third-order astigmatism is the difference between the image surfaces and the Petzval surface. The tangential image is always three times further from the Petzval surface even when the astigmatism is negative. Astigmatism in a thin lens cannot be controlled except by having a stop some distance from the lens (see section 18.6). The maximum wavefront aberrations for a thin lens is given by

$$W_{\text{Astig S}} = \tfrac{1}{4}\,(S_{III} + S_{IV}) \text{ for the sagittal image} \tag{18.27a}$$

and

$$W_{\text{Astig T}} = \tfrac{1}{4}\,(3S_{III} + S_{IV}) \text{ for the tangential image} \tag{18.27b}$$

where $S_{III} = H^2 F$; and $S_{IV} = H^2 F/n$, as given in (18.17), is the effect of field curvature. Equations (18.27) assume a circular lens. They constitute the third and fourth terms of the general equation (18.02) and include the effect of field curvature. H is the Lagrange Invariant, $n'h'u' = h'y/l'$ in the thin lens case. If equations (18.03) are applied to (18.27) we obtain the longitudinal astigmatic aberration which is the sag values obtained as (18.26). If we apply the sag formula to (18.26) we can obtain approximate curvatures for the image surfaces of *Figure 18.11b* which are the very simple formulae,

$$\text{Sagittal Curvature} = -\left(1 + \frac{1}{n}\right)F \tag{18.28a}$$

and

$$\text{Tangential Curvature} = -\left(3 + \frac{1}{n}\right)F \qquad (18.28b)$$

These oblique aberrations with narrow pencils do not depend on the lens bending even though *Figure 18.10* suggested that astigmatism could be regarded as the spherical aberration of the principal ray. This is because the y shown in that diagram depends on r in such a way that the effects for the first and second surface of the lens have a cancelling effect if $1/r_1 - 1/r_2$ is a constant which is the case when bending a thin lens of constant power.

The difference between the two expressions of (18.26) may be shown to be approximately equal to (18.19), where $f'_s - f'_t$ was obtained by much simpler reasoning, if u_{pr} of the latter is set equal to i, and $\sin i = h'/f \cos i$ is used.

18.5 OBLIQUE ABERRATIONS: II–COMA

Although coma is usually categorised as the second monochromatic aberration it is here dealt with after astigmatism and curvature of field because it requires both an off-axis object *and* an appreciable pupil. Its form gives it an unsymmetrical appearance somewhat like the tail of a comet—hence its name. It will be shown later that it depends on the first power of the off-axis angle making it the first oblique aberration to appear as the field is increased. In early optical instruments it was therefore a very troublesome aberration particularly when accurate measurements were required. Fortunately, it is not very difficult to correct in a single thin lens which may have zero coma.

For a single refracting surface it is possible to regard coma as part of the spherical aberration. However, this is not very rewarding as the coma effect for moderate field angles is only a small part of the spherical aberration (and we cannot, as in the case of astigmatism, reduce the spherical aberration by considering a small aperture). In order, therefore, to demonstrate coma we consider a case where the spherical aberration is zero. In *Figure 18.12a* parallel light from a distant axial object is assumed to be refracted at the second principal plane of a thick lens in air. Furthermore, the image formed is assumed to be free from spherical aberration.

If now the object is moved to cause a small change in the angle of the incident rays (shown with broken lines) it is possible to construct the new refracted rays by shifting them through the same angle. This is valid for small angles up to say $5°$. When these new rays are produced towards their intersection they are found to be aberrated. This is coma.

In *Figure 18.12b* the same procedure has been applied but the second principal plane is assumed to be curved! When the centre of its curvature is coincident with the image we find the off-axis image to be unaberrated. Before discussing the importance of this we will consider the form of the aberration illustrated in *Figure 18.12a*. It can be seen from this diagram that the image size for the outer zones of the lens is greater than for the central zone. Whereas spherical aberration is a difference in axial location of the image for different zones of the lens, coma is a difference in magnification.

The diagram (*Figure 18.12a*) shows only the tangential coma. When various

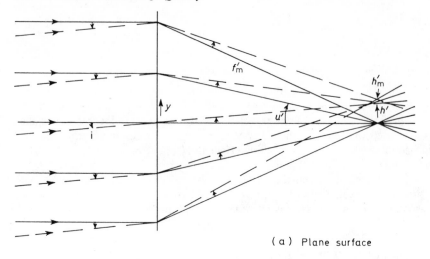

(a) Plane surface

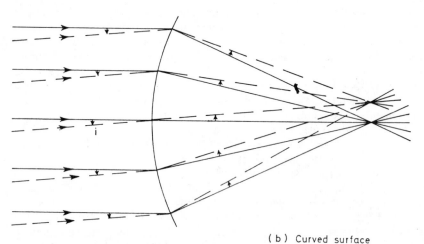

(b) Curved surface

Figure 18.12 The dependence of coma on form of equivalent surface

parts of the lens are designated as in *Figure 18.13a*, the related parts of the image are as in *Figure 18.13b*. The region A at the top of the lens *and* the region A at the bottom combine to produce the image at A. The line CcPcC is in the sagittal plane of the lens and so the distance PC on the image is called the sagittal coma. As with astigmatism the tangential coma, PA, is three times the sagittal coma, PC. Because the light from the annulus ABCD is spread over a considerable area, the actual intensity of the image is mainly in the triangle Pdb. Thus the sagittal coma value is a much better indication of the image blur. With such a complicated ray pattern the wavefront representation, *Figure 18.13c*, gives a clearer description of the aberration.

In section 5.15, it was shown that for objects and images close to the axis the magnification expression gave rise to the Lagrange Invariant,

$$nhu = n'h'u' = H \quad (5.09)$$

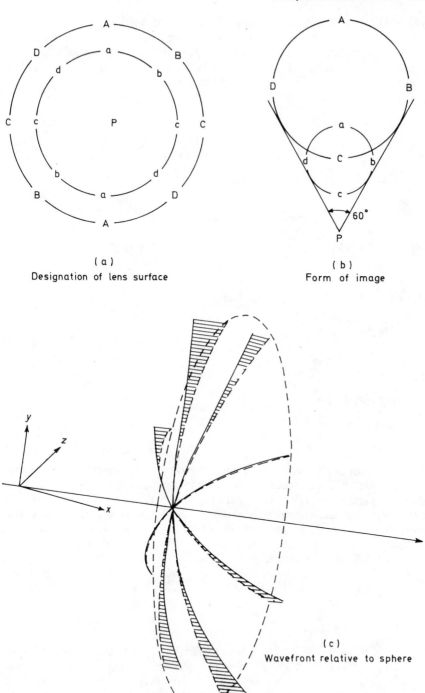

(a)
Designation of lens surface

(b)
Form of image

60°

(c)
Wavefront relative to sphere

Figure 18.13 Representations of coma

within the paraxial region. We may extend this by remembering that the ray through the centre of curvature of a spherical surface is undeviated. Therefore, correct magnifications may be obtained from it, for the *sagittal* images of *Figure 18.10* which lie along it. In *Figure 18.14* this ray has been drawn together with refracted rays to the axial image. By similar triangles

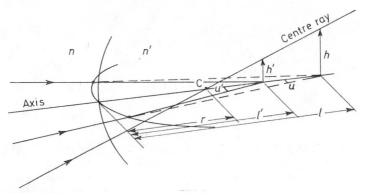

Figure 18.14 Construction for sine rule

$$\frac{h}{l - r} = \frac{h'}{l' - r} \tag{18.29}$$

We also have from (18.06)

$$n(l - r) \sin u = n'(l' - r) \sin u'$$

Putting this with (18.29), we obtain

$$n\, h \sin u = n'h' \sin u' \tag{18.30}$$

This is known as the **Sine Condition**, discovered independently by Abbe and Helmholtz in 1873. Although applying only to sagittal rays, this famous relationship is of great value in optics as it links together the axial and off-axis images. It does not matter that the ray through the centre of curvature does not exist. The important feature is the invariant nature of the condition allowing it to be applied successively to a number of surfaces. Thus, it certainly applies to thin lenses and also to lens systems.

In applying it to the distant object case of *Figure 18.12*, we have $\sin u = y/l$ for each sagittal ray, even though u is very small. Thus

$$h' = \frac{h}{l} \left(\frac{n}{n'}\right) \frac{y}{\sin u'}$$

and for h' to be constant over different values of y, we must have $y/\sin u'$ equal to a constant as u' is the only other variable. This term is equal to the slant refracted ray lengths, and so it is a constant only when the equivalent surface is a sphere. In *Figure 18.12a*, where coma is present, we can see that $y/\sin u'$ is not a constant because the slant lengths vary.

For a distant object we may call the extreme slant length the marginal focal length, f'_m. If we define the extent of the mismatch as $(h' - h'_m)/h'$, we have

$$\frac{h' - h'_m}{h'} = 1 - \frac{h'_m}{h'} = 1 - \frac{y_m \sin u'}{y \sin u'_m} = 1 - \frac{f'_m}{f'} \tag{18.31}$$

This expression, equal to zero for zero coma, is called the **Offence against the Sine Condition** or O.S.C. The transverse sagittal coma is then given by,

$$\text{Transverse Coma}_s = h' \text{ (OSC)} \tag{18.32a}$$

In the tangential plane, as shown in the figure, $\cos i$ enters in a similar manner to astigmatism and, to a third-order approximation,

$$\text{Transverse Coma}_t = 3h' \text{ (OSC)} \tag{18.32b}$$

For a thin lens the wavefront aberration is given by

$$W_{\text{Coma}} = \tfrac{1}{2}(S_{\text{II}}) = \tfrac{1}{2} y^2 IF^2 (\epsilon X + \xi Y) \tag{18.33}$$

where I is the Langrange Invariant ($h'y/l'$ for a thin lens) and X and Y are defined as in (18.08) (giving a positive sign inside the brackets). The coefficients ϵ and ξ are given by

$$\epsilon = \frac{n+1}{2n(n-1)} \quad \text{and} \quad \xi = \frac{2n+1}{2n} \tag{18.34}$$

These are tabulated, for four refractive indices, under equations (18.10). When (18.33) is treated by (18.03a) we obtain the equation

$$\text{Transverse Coma}_t = \tfrac{3}{2}y^2 h'F^2 (\epsilon X + \xi Y) \tag{18.35}$$

and the Transverse Coma$_s$ is $\tfrac{1}{3}$ of this (but this simplified treatment does not readily give this).

This coma is often referred to as Central Coma being that exhibited when the stop is adjacent to the lens. The simple expression in the brackets of (18.35) allows bending of the lens (X) to give zero coma for a given conjugates factor (Y). Coma is not normally given directly for a lens. Following the lead of (18.31), it is common practice to represent the residual coma by plotting, on the same graph as the spherical aberration, a graph of the variation of f'_m with y. It can be shown that, in the presence of spherical aberration, the amount of coma varies with the difference between the two curves. In the doublet analysis shown in *Figure 18.6a*, the straight dashed lines give the coma wavefront aberration for $5°$ off-axis. It can be seen that this is not as well corrected as spherical aberration. The actual size of the coma can be checked by calculating the O.S.C. and using (18.32). In *Figure 18.6b*, the broken line shows the variation of f'_m with y-value (using

$$f'_m = \frac{y}{-\sin \alpha'_m}$$

with the ray trace programs of Appendix 6). This also indicates, by the difference between the broken and full lines that coma is not well corrected.

At $y = 70$ mm, the difference between these lines is about 9 mm. Thus, the O.S.C. [from (18.31)] is about 0.027, as the equivalent focal length is 330 mm. The image height for $5°$, given by (6.12), is 29 mm so we find from (18.32a) that the sagittal coma size is about 0.8 mm.

As the axial blur due to spherical aberration is only 0.4 mm at full aperture it is clear that the field of view at full aperture is rather less than ± 5° depending on the application. If the lens is stopped down the coma blur varies only with the square of the aperture (18.35) while the spherical blur reduces as the cube of the aperture. The importance of good coma correction even for small fields is thus demonstrated. The programs of Appendix 6 may be used to trace oblique rays and obtain the tangential coma directly. This will be found to be a little more than 3× the sagittal coma as calculated above due to the limitations of third order theory.

18.6 ABERRATIONS AND STOP POSITION–DISTORTION

The previous sections have dealt with monochromatic aberrations in an order most suited to learning and have assumed that in the thin lens case the controlling aperture was at the lens. As a thin lens under these conditions exhibits no distortion ($S_V = 0$) this aberration has been left until now.

The order of the aberrations given by Seidel and indicated in (18.01) is very important when considering the effect of relocating the aperture stop. Coma, astigmatism and distortion are all affected by stop position if one of the preceding aberrations is present. Curvature of field, as suggested by the development of (18.01) is not affected by stop position. Spherical aberration also is not affected by stop position (except that the amount of the lens used may alter, with finite conjugates) but coma is affected provided there is some spherical aberration present. Likewise astigmatism is affected if there is spherical aberration or coma. Distortion changes with stop position if there is spherical aberration, coma, astigmatism or curvature of field.

In *Figure 18.15* the principal ray, by definition, passes through the centre of the entrance pupil, stop and exit pupil. When the stop is at the lens, $\bar{y}$, is zero. For other stop locations the eccentricity of the principal ray may be defined as

$$Q = \frac{\bar{y}}{y} \qquad\qquad (18.36)$$

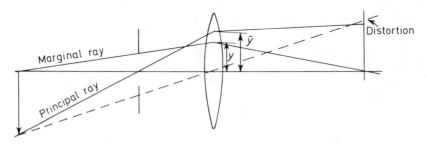

Figure 18.15 Definition of eccentricity, $\bar{y}/y$

Following the notation of Hopkins† we can define new Seidel coefficients for a displaced stop as follows:

$$S^*_I = S_I$$
$$S^*_{II} = S_{II} + QS_I$$

$$S^*_{III} = S_{III} + 2QS_{II} + Q^2 S_I$$

$$S^*_{IV} = S_{IV}$$

$$S^*_V = S_V + Q(3S_{III} + S_{IV}) + 3Q^2 S_{II} + Q^3 S_I \qquad (18.37)$$

where the new coefficients are indicated by the asterisk and the 'old' coefficients, for the stop at the lens, are given by equations (18.11), (18.17), (18.27), (18.33) and $S_V = 0$, for a thin lens. The new values S^*_{II} S^*_{III} and S^*_V can be used in these equations [and (18.33)] in place of the old values to give the wavefront aberration under the new conditions. The relatively simple expressions in (18.37) arise because of the use of maximum values for h', y, (and z) given by the two rays in *Figure 18.15*. The aberrations at other field points and other parts of the pupil are obtained by using the fractional values of h', y, (and z) in (18.02). The simple expression for the coma coefficient in (18.37) is apparent when the similarity between *Figure 18.12a* and *Figure 18.4b* is recognised. Provided the coma is not as large as that shown it is possible, in the presence of spherical aberration, to restrict the area of the lens used so as to centralise the caustic sheets about the image by choosing a new principal ray. This is what is achieved by relocating the stop and the extent of this is measured by Q.

Distortion is evident in *Figure 18.15* but only if there is aberration between the ray that would have passed through the centre of the lens and the actual principal ray. Because the principal ray crossed the axis at the centre of the stop and emerges as if from the centre of the exit pupil, it is sometimes convenient to regard distortion as spherical aberration of the principal ray.

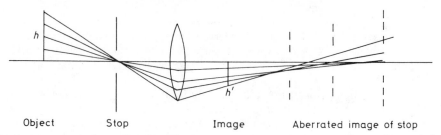

Figure 18.16 Distortion with thin lens

Figure 18.16 shows undercorrect spherical aberration of the stop image, the exit pupil, which gives rise to a reduced magnification at the outer parts of the image. If the object in this case were a square as in *Figure 18.17a*, the image would appear as in *Figure 18.17b*, having a characteristic barrel shape. For a stop on the other side of the lens the image shape of *Figure 18.17c* would occur.

The terms **barrel distortion** and **pincushion distortion** are commonly used to describe these two types. For centred systems these are the only distortions which can occur although, for some corrected lenses, the central parts of the field may suffer from one type and the outer parts from the other.

Obviously the term for S^*_V is rather complex but a few general points can be obtained directly from (18.02) whence,

$$W_{DIST} = \tfrac{1}{2} S^*_V h'^3 y \qquad (18.38)$$

†Hopkins, H. H. *Wave Theory of Aberrations*, Oxford University Press (1950)

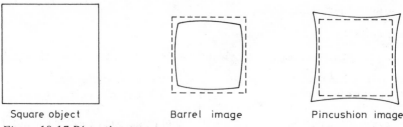

Square object   Barrel image   Pincushion image

Figure 18.17 Distortion types

The direct dependence on y shows that this is a simple tilt of the wavefront. When this expression is reduced to the transverse aberration, by (18.03a) we have

$$\text{Distortion}\ddagger = \text{}^1\!/_2 \left(\frac{S^*_v}{y} \right) l'h'^3 = h'_d - h'$$
(18.39a)

where h' is the expected image height ($h \times$ mag) and h'_d the aberrated height of the principal ray. Thus the actual distortion varies as the cube of the image height. A common measure of distortion is as a percentage error,

$$\text{Percent Distortion} = \frac{h'_d - h'}{h'} \times 100\% = \text{}^1\!/_2 \left(\frac{S^*_v}{y} \right) l'h'^2 \times 100\%$$
(18.39b)

This is quoted for the maximum field angle and values for intermediate field points can be obtained by noting that this measure of distortion varies with the square of the image height.

When looking through a single lens, the eye acts as a distant stop so that distortion is seen. This means that all spectacle lenses suffer from distortion as S_I, S_{II}, S_{III} and S_{IV} cannot be zero for a single lens. When the lens is placed very close to the eye the distortion is very much less even though the image may be blurred.

The correction of distortion arises mainly from its change of sign as the stop is moved from one side to the other side of the lens. If we consider the case where a lens system has to work at a magnification of − 1, the use of two identical lenses symmetrically placed each side of the stop offers some advantages. In *Figure 18.18* the eccentricity factors for lens A and B are clearly subject to:

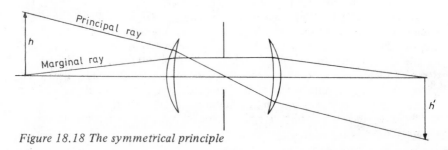

Figure 18.18 The symmetrical principle

‡The division by y occurs here because the S* term contains y. That in (18.38) is the *fractional* value of y.

$$Q_A = -Q_B$$

Similarly $\quad X_A = -X_B \quad$ Bending factors

$$Y_A = -Y_B \quad \text{Conjugates factors}$$

From these

$$S^A_I = S^B_I$$
$$S^A_{II} = -S^B_{II}$$
$$S^A_{III} = S^B_{III}$$
$$S^A_{IV} = S^B_{IV}$$

When these values are used in (18.37) we find that

$$S^{*A}_V = -S^{*B}_V$$

so that the residual distortion is zero. We also obtain zero coma, but spherical aberration, astigmatism and field curvature remain uncorrected. The use of doublets in this symmetrical layout is discussed in section 18.9.

With a single lens, the spherical aberration never goes to zero. The coma is, therefore, always susceptible to correction by the use of a relocated stop. Furthermore, the equation for astigmatism may be examined for the effect of stop position. S^*_{III} has a maximum or minimum value when

$$\frac{\partial S^*_{III}}{\partial Q} = 0.$$

From (18.37)

$$\frac{\partial S^*_{III}}{\partial Q} = 2S_{II} + 2QS_I = 2S^*_{II}$$

Thus the correction of coma occurs at the same time as astigmatism is a maximum or minimum. With undercorrect spherical aberration, the coma-free stop position gives the most backward curving astigmatism. In the design of single lens eyepieces and magnifiers it is common practice to choose a bending which gives zero coma at the eye position as this also gives minimum astigmatism (see also section 18.9).

18.7 RAY TRACING: 1—MERIDIONAL RAYS

Previously we have seen how Snell's Law governs the refraction of a ray of light at a surface

$$n' \sin i' = n \sin i$$

The formula also applies to reflection in air or inside an optical medium if we set n' equal to $-n$. Thus, this equation is exact and applicable at all the interfaces likely to be encountered in an optical system. In section 5.8, it was seen that rays could be traced through a surface using a graphical construction to an accuracy limited only by the accuracy of the drawing. The same process may be done mathematically by taking sines and arcsines in sequence to evaluate (18.06). Ray tracing in itself does not give an analytical description of a lens system but it is possible to develop a sequence of ray traces which assess the performance of a given lens system and then by changing the system parameters slightly it is possible to do the sequence of ray traces again to find out if the second system is better than the first. If it is, the changes can be made again in the same direction and the whole process repeated until the best system is found.

This process of iteration is called an **optimisation program** and considerable work has gone into the design of such programs so that they reach the best design in the shortest amount of computing time. As a lens of 5 elements will have up to 10 surfaces, 9 thicknesses and spacings and 5 refractive indices, rather complex mathematics is required to compute which changes are most likely to improve the performance. This is a technology in its own right and well beyond the scope of this book.

However, ray tracing can be done on any pocket calculator which is programmable with more than about 40 steps. *DO NOT* try it on a non-programmable calculator—it is very difficult to avoid errors and doubtful if much can be learned from the process. Because different calculators use different languages and a wide variety of features it is not possible to give detailed programs for all types. The intention is to provide the reader with the necessary mathematics and advice with some examples, which are given in Appendix 6.

A considerable amount of information can be obtained about a simple lens by ray tracing in two dimensions. Almost every ray diagram in this book has used **meridional** rays which lie in the plane of the page before and after refraction. Thus, the normal to the surface at the point where the ray strikes it must also lie in the plane of the page. In optometry, with toric surfaces abounding and the eye not at the centre of the lens, most rays lie in different planes and a full three-dimensional skew ray trace is needed. The basic procedure is the same for both (but skew rays need a much larger number of program steps), and may be set out as follows:

(1) Set up incident ray
(2)
    (a) Define surface
    (b) Calculate intersection point with surface
    (c) Calculate normal to surface at intersection
    (d) Calculate angle of incidence and, hence, refraction
    (e) Calculate details of refracted ray ready for next surface.
(3) Repeat 2 for following surfaces
(4) Repeat 1, 2, 3 for other rays
(5) Use information on all rays to calculate image position and extent of aberrations.

In the two-dimensional meridional case parts of stage two in the procedure above may be condensed in the mathematics so that it fits into a small calculator, but overall the procedure is followed.

In *Figure 18.19*, the familiar all positive diagram is used with the object ray, coming from an off-axis object. The first stage of the procedure is to define the incident ray. For simplicity we assume the object of height y is at $x = 0$. The ray leaves it at gradient angle α to the axis travelling in a medium of refractive index n. Note that the value of u is positive for the all-positive diagram when the

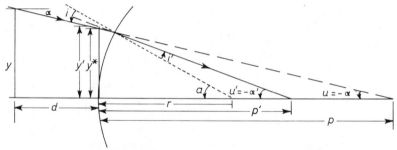

Figure 18.19 All positive diagram for meridional ray trace

gradient angle α of the ray is negative. Although slopes as defined in section 5.4 assist the derivation of formulae, in ray tracing gradients (and direction cosines) are positive when xyz are all increasing.

The values of y, α, n define the incident ray (Stage 1). The definition of the surface is contained in the parameters r, d, n', which are the radius of curvature, the axial distance from the object (or previous surface) and the subsequent refractive index. These will need changing as the calculation progresses through the lens system.

The simples refraction mathematics comes from

$$i + u = a = i' + u' \tag{18.40a}$$

and

$$n \sin i = n' \sin i' \tag{18.40b}$$

for every refraction and

$$\frac{p - r}{\sin i} = \frac{r}{\sin u} = \frac{-r}{\sin \alpha} \tag{18.40c}$$

for every ray [from (18.05a)]
  In the diagram

$$y^* = y + d \tan \alpha \tag{18.41a}$$

$$p = \frac{y^*}{-\tan \alpha} \tag{18.41b}$$

From (18.40c, 18.41)

$$\sin i = (y + d \tan \alpha + r \tan \alpha)\frac{\cos \alpha}{r} \tag{18.42a}$$

Also

$$\sin i' = \frac{n}{n'}(y + d \tan \alpha + r \tan \alpha)\frac{\cos \alpha}{r} = (y' + r \tan \alpha')\frac{\cos \alpha'}{r} \quad (d' = 0) \tag{18.42b}$$

The first expression gives i' so that α' is known from (18.40a). The second expression then gives

$$y' = r \left(\frac{\sin i' - \sin \alpha'}{\cos \alpha'} \right)$$ (18.43)

This approach includes stages 2b and 2c only by implication but we have calculated the angles of incidence and refraction (2d) and if we put the values y', α', and n' into the same calculator locations which had $y\alpha$ and n(2e) then we may proceed to the next surface (3) and repeat the process. If a plano surface is encountered it is a simple matter to insert a very large radius of curvature but then inaccuracies arise. Most computer programs use curvatures rather than radii and angles with intercept heights rather than lengths as this avoids large numbers. The main inaccuracy in the scheme above is that when r becomes large, $\sin i'$ − $\sin \alpha'$ becomes very small indeed and pocket calculators may not be consistent at these values. To avoid this, each of the programs given in the appendix has a conditional jump so that values of r larger than the check number inserted (say 10^5) are assumed to be plano and the calculation truncated by putting y' equal to b. Plano r values should then be inserted as 10^{50}.

It is useful if a program can include a short extra calculation to leave, in the display, the value of p', the image ray intercept with the axis (as measured from the pole of the last surface). This is not really needed but serves as a check and allows spherical aberration to be easily calculated for axial objects (as in *Figure 18.16b*).

Other images can be found by ray tracing two rays and storing α' and y' of the previous ray in the unused storage registers. The location (x_i, y_i), where they intersect, can be found by calculating with respect to the pole of the last surface using:

$$x_i = -\left(\frac{y'_1 - y'_2}{\tan \alpha'_1 - \tan \alpha'_2} \right)$$

and y_i from $y_i = x_i \tan \alpha'_1 + y'_1$.

Such calculations will yield values for tangential coma, tangential astigmatism and distortion.

18.8 RAY TRACING: 2—SKEW RAYS

As with the previous section, the mathematics described here is not necessarily that used in a full-sized computer for ray tracing. The aim here is to provide an introduction to the procedures required, together with equations which best represent the mathematics involved. These can be programmed onto the larger programmable calculators such as the HP-97 or TI-59 or a home computer, where the purpose is that of getting actual figures for educational use rather than speed and efficiency.

The procedure given in the previous section must be followed but we start this exposition with the actual refraction. Snell's Law applies as usual.

$$n \sin i = n' \sin i'$$ (18.44)

It is necessary to find i and i' in terms of the incident ray, the normal to the surface and the refracted ray. This is obtained from the very simple equations

$$\cos i = \alpha_1 \alpha_0 + \beta_1 \beta_0 + \gamma_1 \gamma_0$$ (18.45)

where $\alpha_1 \beta_1 \gamma_1$ and $\alpha_0 \beta_0 \gamma_0$ are the direction cosines of the incident ray and the normal, and similarly

$$\cos i' = \alpha'_1 \alpha_0 + \beta'_1 \beta_0 + \gamma'_1 \gamma_0 \qquad (18.46)$$

where $\alpha'_1 \beta'_1 \gamma'_1$ are the direction cosines of the refracted ray.

We require the last three values independently and using (18.44) (18.45) and (18.46) plus the property that the incident ray, the normal and the refracted ray lie in the same plane it can be shown that

$$n(\alpha_1 - \alpha_0 \cos i) = n'(\alpha'_1 - \alpha_0 \cos i')$$
$$n(\beta_1 - \beta_0 \cos i) = n'(\beta'_1 - \beta_0 \cos i')$$
$$n(\gamma_1 - \gamma_0 \cos i) = n'(\gamma'_1 - \gamma_0 \cos i') \qquad (18.47)$$

Alternatively we can set these out as

$$n'\alpha'_1 = n\alpha_1 + (n \cos i - n' \cos i') \alpha_0$$
$$n'\beta'_1 = n\beta_1 + (n \cos i - n' \cos i') \beta_0$$
$$n'\gamma'_1 = n\gamma_1 + (n \cos i - n' \cos i') \gamma_0 \qquad (18.48)$$

Thus the refracted ray can be calculated if we know the direction cosines of the incident ray, the point of interception of the incident ray with the surface and the direction cosines of the normal to the surface at that point.

The direction cosines of the incident ray may be defined by a knowledge of two points on that ray—usually the object itself and some point in the entrance pupil. Defining these (*Figure 18.20*) as $x_1 y_1 z_1$ and $x_2 y_2 z_2$ we have

$$(x_2 - x_1)^2 + (y_2 - y_1)^2 + (z_2 - z_1)^2 = T^2 \qquad (18.49)$$

where T is the length from $x_1 y_1 z_1$ to $x_2 y_2 z_2$ and then

$$\alpha_1 = \frac{x_2 - x_1}{T} \text{ or } \left(\frac{x_0 - x_1}{T_0} \right)$$

$$\beta_1 = \frac{y_2 - y_1}{T} \text{ or } \left(\frac{y_0 - y_1}{T_0} \right)$$

$$\gamma_1 = \frac{z_2 - z_1}{T} \text{ or } \left(\frac{z_0 - z_1}{T_0} \right) \qquad (18.50)$$

However, we need the intercept point $(x_0 y_0 z_0)$ with the surface. If this is a spherical surface, radius r, we have

$$x_0^2 + y_0^2 + z_0^2 = r^2 \qquad (18.51a)$$

if the centre of the surface is at the origin of the co-ordinate system.

Or $\quad (x_0 - r)^2 + y_0^2 + z_0^2 = r^2 \qquad (18.51b)$

if the pole of the surface is at the origin.

Or $\quad (x_0 - c)^2 + y_0^2 + z_0^2 = r^2 \qquad (18.51c)$

if the centre of the sphere is at $x = c$, which may be calculated from the actual thicknesses, separations and radii of the system. From (18.50) and (18.51c) we find that the distance T_0 from $x_1 y_1 z_1$ to the point of interception is given by

$$T_0^2 + 2[(x_1 - c)\alpha_1 + y_1 \beta_1 + z_1 \gamma_1] T_0 + [x_1^2 + y_1^2 + z_1^2 - 2x_1 c + c^2 - r^2] = 0 \qquad (18.52)$$

from which T_0 can be calculated and $x_0 y_0 z_0$ found by using T_0 in (18.50) (α and T are positive for rays left to right).

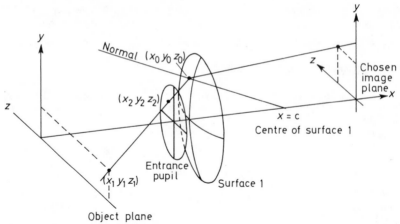

Figure 18.20 Skew ray tracing

From here it is a simple matter, with a sphere, to obtain $\alpha_0 \beta_0 \gamma_0$ as

$$\alpha_0 = \frac{x_0 - c}{r}; \quad \beta_0 = \frac{y_0}{r}; \quad \gamma_0 = \frac{z_0}{r} \qquad (18.53)$$

These are all the equations needed to trace a skew ray through a spherical surface centred on the x-axis. If equations (18.53) are put straight into (18.48), the last two terms become

$$\frac{n \cos i - n' \cos i'}{r}$$

which is the oblique power, K, developed in (18.21) and (18.22) for the Coddington formulae.

Aspheric and elliptical or rolling toric surfaces are naturally more complicated, but not greatly so. Reflecting systems can be accommodated by putting $n' = -n$, but take care with signs! Spot diagrams can be easily generated by using (18.50) on the last ray of the system. When x_2 is the axial location of the image surface y_2 and z_2 give the penetration point.

To recapitulate, the procedure of section 18.7 is followed (1) by letting the ray find its own direction cosines from a given object point and one of an array of points in the entrance pupil. Then, with the first surface defined (2a), the interception point is found via (18.52) (2b). Stages 2c, 2d and 2e follow using equations (18.53), (18.45), (18.48) and (18.46). Rather than the sine version of (18.44), $\sin i = \sqrt{1 - \cos^2 i}$ can be used.

When a number of rays have been traced a spot diagram can be built up. Retaining the direction cosines of the final rays allows this to be converted to longitudinal aberration.

For further reading on skew rays see

A Guide to Instrument Design SIRA–Taylor and Francis, London (1963)
Chapter 4 Optical Design–W. T. Welford.

Ray Tracing Procedures Glen A. Fry (1970) College of Optometry, Ohio State University.

18.9 LENS DESIGN

In the previous sections some indication of the science and art of lens design has been given. The interrelationship between the aberrations and with the stop location is an inexact analytical tool which the designer uses to assess the potential of a lens system, but it is to ray tracing and computer aided auto-design that he turns to finalise the design and accurately predict its performance.

These auto-design methods are entirely outside the scope of this book but some indication of the aberration balancing act can be given. The relative importance of each aberration depends very much on the purpose of the lens being designed—particularly the field of view required. In the case of a vary narrow field of view system for use as a collimator or telescope objective, the aberrations of spherical aberration and coma together with chromatic aberration are the most important and the work of *Figure 18.6a* and *b* would be carried out for different wave-lengths of light using different glass types in combination. This would be followed with optimisation by ray tracing to balance the high-order aberrations.

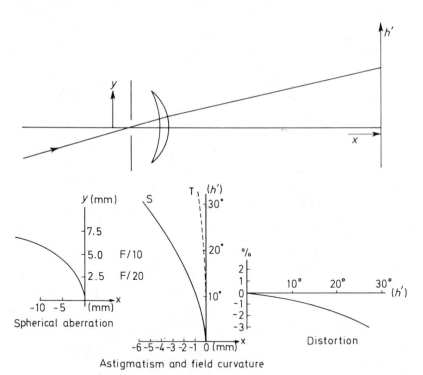

Figure 18.21 Landscape lens (elf 100 mm)

For a photographic lens a larger field of view is usually required. The earliest solution to this problem was the so-called Landscape lens. By choosing a lens bending and stop location that gives low coma and negative astigmatism it is possible to obtain moderate performance over fairly wide fields. The negative astigmatism counteracts the effect of field curvature but nothing can be done about spherical or chromatic aberration without resorting to a doublet design (or aspheric curves). *Figure 18.21* shows the shape and performance of a representative singlet design. Notice the considerable amount of distortion, which also varies with wave-length. Although a nearly flat tangential field is obtained, considerable curvature remains with the sagittal image. The alarming amount of spherical aberration severely limits the F/no.

The same balance of aberrations can be made with the stop behind the lens but in this case the distortion is pincushion and the residual coma and lateral chromatic aberration is of the opposite sign. The symmetrical principle arises from this and, when two similar single lenses are arranged each side of the stop, good correction of all aberrations is possible except spherical and (longitudinal) chromatic aberration. These can be corrected with compound lenses and the Protar Lens, designed by Rudolf for the German company Zeiss, in 1890, is an

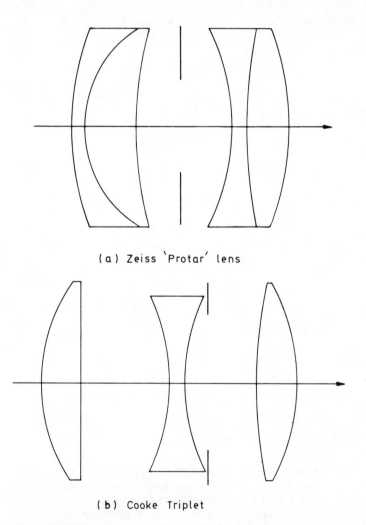

(a) Zeiss ʻProtarʼ lens

(b) Cooke Triplet

Figure 18.22 Basic photographic lenses

example of this. *Figure 18.22a* show the general symmetry, modified to some extent because of the unequal conjugates and different glass types. This was the first lens to be called an anastigmat. An alternative solution was designed by H. Denis Taylor for the British Company, Cooke and Sons in 1893. Known as the Cooke Triplet it has the very minimum number of variables to achieve correction of all the third-order aberrations. *Figure 18.22b* shows its general form.

The symmetrical principle and the triplet solution have been the starting points of two families of lenses developed largely during the years of this century. Generally, these later lenses give improved performance at the cost of greater complexity in the form of more elements. Very approximately, the F/no. achievable by a camera lens of N elements is given by

$$F/no. = \frac{\theta}{3N}$$

where θ is the total field coverage in degrees.

In the Protar Lens, five of its six surfaces are concave to the stop, indicating the correction of field aberrations. Symmetrical lenses of five or six elements can cover fields of $60°$ to $90°$, provided the aperture is not greatly faster than F/4. On the other hand derivatives of the triplet are not much used above $60°$. At lower fields, $(30°)$ the basic three element design is sometimes used at apertures as fast as F/2.5, particularly for projection lenses. For wider fields an early modification was to change the last element to a cemented doublet as in the well-known Tessar lens.

Basic eyepiece designs are described in section 10.13. The eyering constitutes the stop of the system and, as this is where the eye is located, it is not possible to locate lens elements on both sides of the stop as with symmetrical systems. Distortion is therefore very difficult to correct. Furthermore, the chromatic difference of distortion (and astigmatism) becomes significant. The condition for zero lateral chromatic aberration developed in section 13.12 is often not fully met so that this residual lateral colour partially compensates for the chromatic difference of magnification.

Although the stop of the eyepiece is usually the image which it forms of the objective in a telescope system, the rays are precisely limited by it. This is useful to the designer as it ensures that badly aberrated edge rays do not reach the eye. When an eyepiece is used to examine a real object rather than an image formed by an objective it is customary to call it a magnifier. The triplet system of *Figure 10.8a* may be used as an eyepiece or a magnifier. The magnifier case is not subject to equation (10.05) where a reduction in the focal length of the eyepiece reduces the stop size.

The eye may adopt any position limited only by the diameter of the lenses of the magnifier. In recent years **biocular magnifiers** have been developed with useful magnifications of $\times 5$ and above. These are lens systems of very low F/no. (large numerical aperture) so that an equivalent focal length of about 40 mm is obtained with an overall diameter of 80 mm which permits an observer to view the magnified image with both eyes. Although the magnifier may be F/0.5 the small pupils of the eyes make the *used* apertures about F/6.0. The fact that both these small apertures are situated remotely from the lens axis changes the third-order aberrations so that, for instance, spherical aberration in the lens shows as astigmatism across the eye pupils, and coma generates an anamorphic distortion.

18.10 IMAGE QUALITY ASSESSMENT–POINTS AND PATTERNS

The previous sections have discussed the aberrations of lenses from the stand-point of geometrical optics, without considering diffraction effects. This is a valid procedure for low quality lenses but the effects of diffraction are of increasing importance as the aberrations are reduced. Often these effects can be predicted and allowed for during the design of a lens. The diffraction effects associated with the image normally constitute an extra ripple in the pattern predicted by geometrical ray tracing.

When the aberrations become very small the image of a point appears as the Airy disc given in Plate 8 but with some of the energy taken from the central maxima and appearing in the rings, thus enlarging the image size. The ratio between the reduced central intensity and that for zero aberrations is called the **Strehl Intensity Ratio** and has proved to be a useful yardstick for high quality lens design.

In section 15.12 the concept of Fourier Transforms was introduced in a very simplified way. This was developed further in section 15.15 to apply to spatial frequencies as well as temporal changes. The idea of spatial frequencies having different contrasts or modulations was developed in section 11.14. We now bring together all three concepts.

A single point traced across a photograph transparency, for example, will see changes in light intensity due to the light and dark regions. Whatever shape of waveform this is, it can be represented by a host of different spatial frequencies having different modulations; its Fourier Transform. If now this same trans-parency is placed in an enlarger we can find how well the enlarging lens reproduces the image on the photographic paper by asking how well it reproduces each of the spatial frequencies. Due to the effects of diffraction and aberrations the higher spatial frequencies will be smeared out and lose contrast in comparison with the lower (coarser) spatial frequencies.

A real object, such as an outdoor scene or a microscope sample, usually con-tains a very wide range of spatial frequencies. An image, on the other hand, is restricted to that which is transmitted by the lens. In the case of a photograph it is also limited by the ability of the photographic material to respond to the image variations, and in the case of a television system, by the response of the camera tube and electronics, transmission medium and receiver display. The value of considering the response of each stage to specific spatial frequencies is seen in (18.56). When a single frequency waveform, a sine curve of SHM, is modified in a *linear* way, each point on it is blurred to the same amount and the effect is to reduce its amplitude and possibly change its phase but the result is a sine wave *of the same frequency*.

In the optical case (because intensities can never be negative) we can express the intensity variation of Plate 1, for example, as

$$I = I_0 \left(1 + C_m \sin 2\pi s x\right) \tag{18.54}$$

[see also (14.05a)] where I_0 is the mean intensity and C_m the modulation as defined in section 11.14 (C_m is unity only when the intensity falls to zero at the dark lines), s is the spatial frequency and x, the location across the pattern.

The **principle of linearity** means that $I' = f(I)$, that is, the right-hand side is an expression containing only the first power of I, as in the straight line equation, and not I^2, I^3 etc. Each point on the image is made up of so much of the corresponding point of the object sine wave plus strictly defined amounts of the neighbouring points according to the smearing. This summation follows the principle of superposition and the addition amounts to the addition of sine waves of differing phases and amplitudes but identical frequencies. Two such waves were added in section 12.4 *Figure 12.4a* and in the general case it can be shown that

$$I' = I'_0 \, [1 + C'_m \sin (2\pi s'x' + \epsilon)] \qquad (18.55)$$

The primed terms have the same meanings for the image as for the object. The important point is that only one spatial frequency, s' is involved, linked to s by the magnification. The image is another single frequency sine wave! We find that I'_0/I_0 is determined by the light gathering power and transmission of the lens and C'_m/C_m by the quality of the lens as determined by its aberrations. The phase shift, ϵ is caused by the so-called odd aberrations of coma and distortion which have an asymmetric action on the image.

The **Modulation Transfer Function** (MTF) introduced in section 15.15 can now be defined simply as C'_m/C_m, the ratio of the input and output modulation. In imitation of electronic amplifiers, where this approach was first developed, it is also known as the **Spatial Frequency Response** of the lens or imaging system. When an imaging system has more than one linear component, the modulation of a given frequency in the image can be found by multiplying the object modulation at the related frequency by the MTF values of the component parts.

$$C'_m = C_m \, \text{MTF}_1 \, \text{MTF}_2 \, \text{MTF}_3 \text{ etc.} \qquad (18.56)$$

This simple approach cannot be used when component parts interact coherently as with the individual elements in a lens or two lenses used in a relay system. Then the aberrations of the overall system must be determined to give a system MTF value. It can be used when the several parts of an imaging system act independently of each other as in the case of a distant object of contrast, C_m; the atmospheric degradation, MTF_1; the camera lens, MTF_2; a photographic plate exposed and developed, MTF_3; a microscope used to study it, MTF_4; and the visual system of the observer, MTF_5. Two of the above have non-linear properties: the photographic process and the eye. Quite often, however, they are assumed to be linear and normally the inaccuracy introduced is not significant.

Given that the modulation of the object is unity for all spatial frequencies, then the modulation of the image is equal to the modulation transfer function of the lens and vice versa. This is utilised in lens testing as described in the next section. The use of a point object as described in the sections on the Seidel aberrations is not excluded from the MTF concept. The image of a point gives the **point spread function** and of a narrow line, the **line spread function**. Because the spatial frequency content of a point or line is known, the MTF can be calculated from the blurred image although this is obviously a more complex process than for a single sine wave. The more restricted the MTF, the wider the spread function and vice versa. A similar approach can be adopted by looking at the image of a sharp boundary between light and dark. This is known as **edge gradient analysis**.

18.11 LENS TESTING

The purpose of almost every lens or lens system is to form an image, whether real or virtual, at some location other than that of the object. The quality of this image is determined by the type and size of the residual aberrations of the lens system, the accuracy of its focussing and ultimately by the diffraction effects due to the image forming light being restricted to that which gets through the lens aperture. The Airy disc described in section 15.15 is normally the best image which can be achieved. When lens systems are being considered it is common for off-axis images to receive less light through the lens due to **vignetting**. This is the action whereby a series of holes (the lenses) spaced along an axis combine to give a reduced lozenge-shaped aperture when viewed at an angle. Reducing the aperture increases the size of the Airy disc. Thus, the basic diffraction image quality of lenses is generally worse at the edge of the image format than at the centre.

With modern lens design it is often possible to make the residual aberrations so small that the diffraction image is only slightly modified by the aberrations. This is particularly the case when a high quality photographic lens is used at a small aperture, say F/16. It is customary to measure very small residual aberrations by the number of wavelengths deviation they cause: 3λ of coma etc.

When the aberrations of the system are large considerable blurring of the image is found. This does not necessarily mean the system is 'poor' in the widest sense of the word. It may be that the need for it to be low cost, or lightweight or very wide aperture means that the optimum design exhibits large residual aberrations. The specification of lenses with larger residuals can be more difficult than those which are nearly diffraction limited as there are more ways in which they can be right!

As an alternative to examining the image itself, Foucault allowed it to form its own pinhole camera. By placing a screen behind the image of a point source a patch of light is found similar in shape to the aperture of the lens system. At the image location a sharp edge or **knife-edge** is moved across so that it progressively cuts into the beam. If this is done at the image and it is a point image, then the patch of light will darken evenly and quickly. When the image is aberrated the patch takes on typical patterns associated with the aberration type and where the edge is located. The intensity at a particular point in the patch is given by the slope of the wavefront at the other end of that ray. **Foucault knife-edge testing**, as this is called, is very sensitive and an identical system is used, with high quality lenses or mirrors, when we wish to measure the deformation of a wavefront caused by inhomogeneities in optical materials or by changes in refractive index of air or liquid flowing past aerofoils or obstructions. In this so called **Schlieren test** the optical material or fluid flow is placed close to the imaging system while the edge is placed at the image of the small source.

This test recognises that a point image is produced by a spherical wavefront while an aberrated image must have a deformed wavefront. These deformations can be measured by interference, as described in sections 14.12 and 14.13, when they are not larger than a few wave-lengths. However, the methods of spatial frequency response described in the previous section have a much wider application and are nowadays routinely used on medium to high quality lenses. The test method comprises the presentation to the lens of a pattern (grating) of a given spatial frequency and unity contrast modulation. In the specified image location a narrow slit is moved across this and the intensity variations measured using a photo-multiplier tube. It is usual to plot these on a graph with spatial frequency on the *x*-axis and relative contrast on the *y*-axis, that is, the modulation at spatial frequency as compared to the modulation at the very lowest frequency. This means that the graphs *Figure 18.23* always start at $y = 1$. This so-called normalization is a useful simplification but does remove from the graph information on the veiling glare of the lens caused by poor surface quality or dirty surfaces because this reduces the image contrast at all spatial frequencies.

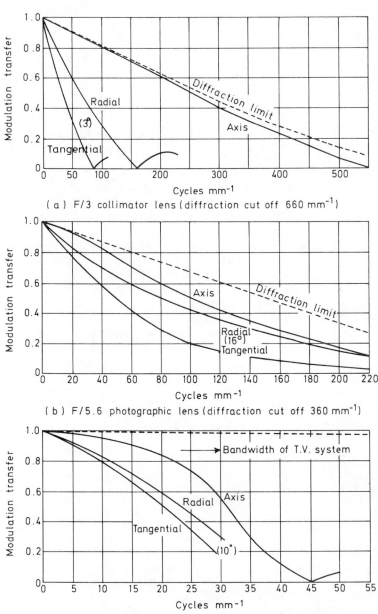

(a) F/3 collimator lens (diffraction cut off 660 mm⁻¹)

(b) F/5.6 photographic lens (diffraction cut off 360 mm⁻¹)

(c) F/0.9 low-light TV lens (diffraction cut off 2240 mm⁻¹)

Figure 18.23 MTF curves of various lenses

For ordinary lenses under ordinary conditions the relative contrast never rises above unity. Some photographic processes can give curves above unity due to a contrast enhancing action of the developer. The eye also is less responsive at low and high spatial frequencies (section 20.7).

The curves at *Figure 18.23* show the response curves for different types of lenses. While a lens may be made diffraction limited for axial objects it is very difficult to maintain this over an appreciable field. *Figure 18.23a* shows a narrow field star tracker lens which does not need high performance off-axis. The residual astigmatism means that the grating image has a different modulation in the meridional plane to that in the tangential plane. The frequency, s_0, at cut-off is given by

$$s_0 = \frac{2\,\text{NA}}{\lambda} \tag{18.57}$$

and in this case has been caclulated for F/3 (NA = 0.166) and λ equal to 0.5 μm. The diffraction limited curve may be used for other lenses by merely changing the scale on the x-axis so that s_0 satisfies equation (18.57).

In the case of the photographic lens the performance at the centre and edges of the field are much more alike although astigmatism still shows. (*Figure 18.23b.*) For a low-light TV camera lens the need is for maximum light rather than high frequency response as the electronic systems have a sharp cut-off. The design therefore maximises the response at those spatial frequencies which are of use (*Figure 18.23c*).

If the lenses in *Figure 18.23* were measured for other image surfaces, that is, defocussed, it is possible that one or other of the off-axis curves would show an improved response but generally they become poorer. If the photographic lens of *Figure 18.23b* were stopped down to a smaller aperture the response curves would normally show an improvement. However, the diffraction limit curve would show a degradation and once the actual lens performance approached the diffraction limit further aperture reduction would give a worsening response.

CHAPTER 18   *EXERCISES*

1. Explain what is meant by the spherical aberration of a lens. An object is viewed over the edge of a card placed near to the eye. When the card is moved up and down, the object appears to move, the motion being *with* the card when the object is distant, and against the card when the object is near. What optical defects of the eye does this indicate? Illustrate your answer with diagram.

2. (*a*) Using the ray plotter, find graphically the paths of the refracted rays conjugate to rays parallel to the axis, incident in one plane at heights (y) of 1, 2, . . ., 9 cm on a spherical surface of radius + 10 cm. Medium of object space is air and for image space n = 1.5.

(*b*) Find by calculation the Gauss focal point F′ and hence the distances along the axis from this point to each of the points where the refracted rays cut the axis.

(*c*) Tabulate these quantities and plot them against y^3.

(*d*) Draw the spherical aberration curve.

(*e*) Plot the log of the longitudinal aberration for each ray height against log y and so derive the relationship between L.A. and y.

3. Determine by means of the ray plotter the axial spherical aberration of a spherical surface of radius − 10 cm separating air from glass of index 1.5, with the incident ray parallel to the axis and the incidence height 7 cm.

4. A parallel pencil of light is incident on a spherical surface of radius + 10 cm separating air from glass (n = 1.5). Draw accurately the refracted rays, show the caustic curve and

the curve of aberration. What will be the positions of the pair of conjugate points at which there is no spherical aberration and what will be the magnification at those points?

5. Explain and prove Young's construction for finding the exact path of a ray of light refracted at a spherical surface.

Find the positions of the conjugate aberrationless (aplanatic) points of a positive spherical refracting surface of radius 2 inches separating air from glass of index 1.6.

6. Prove that the distances (from the vertex) of the aplanatic points of a spherical refracting surface are given by

$$r\left(1 + \frac{n'}{n}\right) \text{and } r\left(1 + \frac{n}{n'}\right)$$

What is the magnification at these points

7. Explain from first principles (using the concepts of wave curvature, or otherwise) why a thin positive lens traversed obliquely by a pencil of light from a star does not yield a point image. Describe the appearances which could be found on a focusing screen held at different distances in the convergent beam if the lens were a + 5 D 'sphere', and the light incident at 45° to the axis.

8. Apply Young's construction to determine the spherical aberration of the cornea. Assume the latter to be spherical and of radius 7.8 mm; refractive index of aqueous humour 1.33. Take a ray parallel to the visual axis at an incident height of 4 mm.

9. Give a general description, with the aid of a diagram of the defect of lenses called coma. How would you show this experimentally?

10. By means of Young's construction, or by using the ray plotter, find the astigmatism produced when a narrow pencil from an infinitely distant object is incident at an angle of 20° to the axis on a spherical refracting surface of radius + 5 cm separating air from glass of refractive index 1.5.

(*a*) When the beam is central.
(*b*) When the stop is placed 9 cm in front of the surface.
(*c*) When the apparent stop is placed 9 cm behind the surface.

11. A narrow pencil of light from an infinitely distant object is incident on a spherical refracting surface of +5 cm radius separating air from glass of $n = 1.5$. The angle of incidence is 50°. Calculate the distance from the point of incidence to the primary and secondary focal lines. Check by graphical construction.

12. A spherical surface of radius + 8 inches separates two media of refractive indices 1.36 and 1.70 respectively.

Light from the left is converging in the first medium, towards an axial point 18 inches to the right of the surface vertex. Find the position of the image point and show that there will be no spherical aberration.

13. A narrow pencil is incident centrally on a + 5 D spherical lens in a direction making 30° with the optical axis of the lens. Explain briefly the nature of the refracted pencil.

14. A narrow pencil of parallel rays in air is refracted at a spherical surface of radius r into a medium of refractive index $\sqrt{3}$. If the angle of incidence is 60°, show that the distance of the primary and secondary image points from the point of incidence are given by $\frac{3}{4} r \sqrt{3}$, and $r \sqrt{3}$ respectively.

15. Explain with the help of a diagram the following terms associated with the refraction by a lens of an oblique pencil: meridian plane, sagittal plane, astigmatic difference, place of least confusion.

16. What is meant by radial astigmatism in lenses? Give examples of the effect produced by it in obtaining an image of a star covered field.

17. A spectacle lens of power + 5 D is made to form the image of a distant axial source of light on a ground glass screen; the lens is then slightly tilted about one diameter. Explain with numerical details the changes thus produced in the image and the further changes observed as the screen is moved nearer the lens. N.B. If F is the axial power, and F_S and F_T are respectively the sagittal and tangential powers, while θ is the angle of obliquity, then

$$F_S = F \left(1 + \frac{\sin^2 \theta}{3} \right) \text{ approximately}$$

$F_T = F_S \sec^2 \theta$ approximately.

18. Write a short account of the aberrations in the image produced of a flat object by an uncorrected optical system. Give diagrams.

Explain, giving reason, which aberrations are of most importance in the case of (*a*) a telescope objective, (*b*) a spectacle lens.

19. Explain the terms barrel and pincushion distortion and show how they are produced by a single convex lens and stop.

20. Draw the all-positive diagram for refraction at a spherical surface and from it deduce the five simple equations used in the trigonometrical computation of rays through such a surface.

Write down also the corresponding equations for computing a paraxial ray and from these latter equations derive the fundamental paraxial equation:

$$\frac{n'}{l'} - \frac{n}{l} = \frac{n' - n}{r}$$

21. Write down the equations used in computing the path of a ray of light after refraction at a spherical surface. Apply them to compute the refracted ray in the following case:

$r_1 = + 377.85$ mm, $n = 1.0$, $n' = 1.5119$.

The axial object point is situated 487 cm in front of the surface; the incidence height of the ray is 12.7 cm. Show how to arrange the work in tabular form and calculate the first six quantities.

22. Trace by trigonometrical computation the course, through the lens the constants of which are given below, of a ray that crosses the axis at a point B 30 cm in front of the first surface and is incident on the lens at an incident height of 2 cm. Find also the point where a paraxial ray incident through B crosses the axis after refraction.

$r_1 = + 10$ cm, $r_2 = - 20$ cm, $d = 3$ cm, $n = 1.624$.

23. In general, the image formed by a simple spherical lens suffers from certain defects or aberrations. Enumerate these aberrations.

If the object consists of a point source of white light situated on the optical axis of the lens, with which aberration or aberrations will the image be affected when the aperture of the lens is (*a*) small; (*b*) large?

If the point source is moved away from the axis, what aberration or aberrations will be manifest when the aperture of the lens is small?

Give clear diagrams.

24. The example doublet of section 18.3 has image locations (for a distant axial object) as given in the table below, where *y* is the height of the incident ray from the axis.

y	1	20	40	50	60	64	66	68	72	mm
l'(*p'*)	286.20	285.68	284.52	284.17	284.67	285.36	285.85	286.51	288.36	mm

Use a programmable calculator with one of the ray tracing programs in appendix 6 and calculate, for the same rays, the image locations for an identical doublet except that the last surface is plano. Ignore the shift in F' and plot the spherical aberration curves for each lens with respect to its paraxial focus. Which lens has the most spherical aberration? If the maximum permitted longitudinal aberration is ± 2 mm, which lens would allow the greater aperture for an image plane located at the paraxial focus?

25. List the primary monochromatic aberrations which can affect the quality of an image of an object produced by a lens system.

Explain one of these aberrations, and consider how it may be reduced or eliminated.

26. Describe the image ray patterns characteristic of spherical aberration and coma when a lens with these aberrations images a distant point object.

Explain methods by which these two aberrations can be minimised.

27. A plano-convex lens of index 1.5 is 50 mm in diameter, 2.5 cm thick, and the convex surface has a radius of curvature of 40 mm. By construction find the amount of *longitudinal* and *lateral* spherical aberration when collimated light is incident on the plane surface.

28. List the aberrations from which lens systems may suffer, and explain how each occurs, using diagrams where applicable.

Which of the aberrations you describe are of importance in
(*a*) A slide projector.
(*b*) A high power microscope?

29. List and describe the aberrations from which optical systems may suffer.

State which of these are of importance in
(*a*) Telescopes.
(*b*) Film Projectors.

30. A plano-convex lens of index 1.6 is 50 mm in diameter, 2 cm thick, and the radius of curvature of the convex surface is 48 mm. By construction find the amount of longitudinal spherical aberration when collimated light is normally incident on the plano surface. Also find the amount of lateral spherical aberration.

31. Derive the equations used in computing the path of a ray of light after refraction at a spherical surface. Apply these equations to compute the refracted ray in the following case:

$$r_1 = 37.79 \text{ cm}; \quad n = 1.00; \quad n' = 1.512$$

The axial object point is situated 487 cm in front of the surface. The incident height of the ray is 12.7 cm. Show how to arrange the work in tabular form and use four figure logarithims.

32. Explain briefly what is meant by:
(*a*) Disc of least confusion.
(*b*) Astigmatism.
(*c*) Field curvature.
(*d*) Distortion.
Explain qualitatively how the primary field curvature of a simple camera lens can be compensated by the introduction of overcorrected astigmatism in order to obtain a flat image field.

33. Trace by trigonometrical computation the path through a spherical surface of a ray that crosses the axis at a point B 500 cm in front of the surface and is incident on the surface at a height of 15 cm.

$$r_1 = +400 \text{ mm}; \quad n = 1.000; \quad n' = 1.512$$

Use four figure logarithms.

34. A thin + 10 D lens 40 mm in diameter exhibits 0.50 D of positive spherical aberration for marginal rays.
(*a*) For a distant object, what is the amount of lateral spherical aberration (in mm) for this lens?
(*b*) If the lens is stopped down to only a 20 mm aperture how much longitudinal spherical aberration (in mm) does it then have for distant objects?

35. Explain what is meant by spherical aberration.

A parallel beam of light strikes the plane surface of a plano-convex lens normally. The radius of curvature of the convex surface of the lens is 20 cm and its refractive index is 1.5. Calculate the position of the second focal point for the paraxial rays, and, by using the laws of refraction, the longitudinal spherical aberration for a ray which strikes the lens 5 cm from the axis.

Chapter 19
Lens systems — general

19.1 CENTRED SYSTEM OF SPHERICAL SURFACES. PARAXIAL REGION

In dealing with the thin lens in Chapter 6 and with lens systems and thick lenses in Chapter 9, we have considered only the case in which both object and image spaces are air. We may, however, have systems in which the object and image spaces have different refractive indices, the eye being an important example of this kind, and the general case of an optical system will now be considered.

Taking a system of centred surfaces as shown in *Figure 19.1* we can trace the path of the rays through the system as in Chapter 18, or for pencils within the paraxial region we may apply the results, previously obtained for a single refracting surface (Chapter 5), to each of the surfaces in turn.

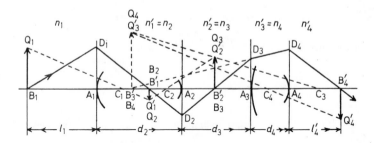

Figure 19.1 Centred system of spherical surfaces. Showing path of rays from B_1 and additional construction for determining image sizes

Within the paraxial region, pencils from all points in the object plane remain homocentric after refraction at the first surface, and an image of the object is formed. This image is the object for the next surface, and as the pencils remain homocentric after refraction at the second surface, a second image is formed, which becomes the object for the third surface, and so on through the system.

An object perpendicular to the axis in the object space is imaged in a plane perpendicular to the axis by each surface in turn, and the final image in the image space is perpendicular to the axis and geometrically similar to the object.

In *Figure 19.1* the path of a paraxial ray $B_1 D_1$ from an object point B_1 on the axis is shown. Images are formed by each surface in turn at the points B'_1,

B'_2, B'_3 and B'_4 respectively, where the ray crosses the axis, either actually or when produced back, in which case the image is virtual. If $B_1 Q_1$ is a small object perpendicular to the axis, then the images formed by each surface will be perpendicular to the axis through B'_1, B'_2, etc. The image of Q_1 formed by the first surface is found by taking a ray from Q_1 through C_1 meeting the first image plane at Q'_1. A line from Q'_1 through C_2 fixes the position of Q'_2 and so on for each surface in turn.

19.2 CHANGE OF VERGENCE OF ADVANCING WAVEFRONT

Figure 19.2 shows a wavefront at A immediately after passing a refracting surface. The wavefront is converging in a medium of refractive index n' towards a focus B' distant l' from A.

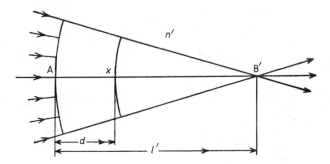

Figure 19.2 Change of vergence of advancing wavefront

The *actual* vergence at A is therefore $1/l'$ and the reduced vergence is $n'/l' = L'$.

At some place X the actual vergence has increased to $1/(l' - d)$. The reduced vergence is therefore

$$\frac{1}{\dfrac{l'}{n'} - \dfrac{d}{n'}} = \frac{\dfrac{n'}{l'}}{1 - \dfrac{d}{n'}\dfrac{n'}{l'}}$$

$$= \frac{L'}{1 - \dfrac{d}{n'} L'} = L'_x \tag{19.01}$$

If a second refracting surface exists at X, L'_x would be the value of L for the second surface.

19.3 CALCULATION FOR PARAXIAL PENCILS THROUGH A CENTRED SYSTEM

The focal power of each surface is given by

$$
\left.
\begin{array}{r}
\dfrac{n'_1 - n}{r_1} = F_1 \\[2em]
\dfrac{n'_2 - n_2}{r_2} = F_2 \\[2em]
\dfrac{n'_k - n_k}{r_k} = F_k
\end{array}
\right\}
\qquad (19.02)
$$

Each of these powers is added to the reduced vergence of the oncoming wavefront in turn so that we have

$$
\left.
\begin{array}{ll}
\text{First surface} & L'_1 = L_1 + F_1 \\[2em]
& L_2 = \dfrac{L'_1}{1 - \dfrac{d_2}{n_2} L'_1} \\[3em]
\text{Second surface} & L'_2 = L_2 + F_2 \\[2em]
& L_3 = \dfrac{L'_2}{1 - \dfrac{d_3}{n_3} L'_2} \\[3em]
k\text{th surface} & L'_k = L_k + F_k \\[2em]
& l'_k = \dfrac{n'_k}{L'_k}
\end{array}
\right\}
\qquad (19.03)
$$

Thus giving the position of the final image.

As the image formed by one surface is the object for the next, the size of each successive image and thence the size of the final image can be found as follows:

$$
m_1 = \frac{h'_1}{h_1} = \frac{L_1}{L'_1}; \quad m_2 = \frac{h'_2}{h_2} = \frac{L_2}{L'_2}; \text{etc.}
$$

and $h'_1 = h_2$; $h'_2 = h_3$; etc.

Hence the total magnification

$$
m = \frac{h'}{h} = \frac{L_1 \times L_2 \times \ldots L_k}{L'_1 \times L'_2 \times \ldots L'_k} = \Pi_1^k \frac{L}{L'}
\qquad (19.04)
$$

where $h = h_1$, the height of the original object, and $h' = h'_k$, the height of the final image.

If the original object is at infinity and subtends an angle w, $L_1 = 0$ and $h'_1 = f_1 \tan w$; then final $h' = h'_1 \times$ product of subsequent magnifications

$$h' = f_1 \tan w \times \Pi_2^k \frac{L}{L'} \qquad (19.05)$$

Considering the system as a whole, and making use of our definition of focal length in terms of the apparent size of the object [equation (5.11)]

$$h' = f \tan w$$

where f = first focal length of the system.

Hence

$$f = f_1 \times \Pi_2^k \frac{L}{L'} \qquad (19.06)$$

By tracing the light through the system in the opposite direction we have

$$f' = f'_k \times \Pi_{k-1}^1 \frac{L}{L'} \qquad (19.06a)$$

Later it will be shown that

$$\frac{f'}{f} = -\frac{n'}{n}$$

where n and n' are the refractive indices of the original object and the final image spaces respectively.

For each surface

$$n_1 h_1 u_1 = n_2 h_2 u_2 = n_3 h_3 u_3 = \ldots$$

so for the whole system

$$n\,h\,u = n'h'u' \qquad (19.07)$$

the Lagrange relation, $u = u_1$ being the angle that a ray from the axial object point makes with the axis in the object space and $u' = u'_k$, the angle the ray makes with the axis in the final image space.

19.4 CARDINAL POINTS OF A CENTRED SYSTEM

Any system as a whole forms, of an infinitely distant object, either a real inverted image or a virtual erect image and is therefore either

> Positive, convergent
> or Negative, divergent

As we have previously seen with the thin lens combination or the thick lens in air, the result is as if the refraction took place completely at a plane distant f' back from F' for light in one direction and at a plane distant f back from F for light in the other direction. These two planes are the principal planes of the system and the points at which they intersect the axis, the principal points, P' and P. The distances f and f' can be found from equations (19.06) and (19.06a), and by taking L' and L in turn equal to zero we can find the distance f'_V of F′ from the last surface.

We can therefore represent any lens system by its two focal planes through the principal foci F and F′ and its two principal planes through the principal points P and P′ where PF = f and P′F′ = f'. As was shown in Chapter 9, the principal planes are planes of unit magnification, and therefore any ray in the object space incident towards a point at a given distance above the axis in the first principal plane travels in the image space as though from a point in the second principal plane at the same distance from the axis.

19.5 NUMERICAL EXAMPLE—SCHEMATIC EYE

A useful numerical example of the above work is provided by the **schematic eye**. This is an optical system based on average values for the optical parameters of the human eye and is intended to closely represent the optical effects of the eye. The working given in the table is also used in sections 19.10 and 19.11. The values used are those of the **Emsley Schematic Eye** which consists of three refracting surfaces, the first representing the cornea and the second and third, the two surfaces of the crystalline lens. Other schematic eyes are considered in Chapter 20.

In the table the full working is given so that the student can check the figures on his calculator and also obtain a comparison with other schematic eyes. All dimensions are given in metres (except focal lengths as indicated). This gives the reciprocals directly in dioptres. From part A of the table, L'_3 = + 79.88 D, and therefore

$$l'_3 = \frac{1.3333}{79.88} \text{ m} = + 16.69 \text{ mm}$$

Hence the distance of F′ from the cornea (compare *Figure 19.1*).

$$= 16.69 + d_2 + d_3 = 16.69 + 3.6 + 3.6 = 23.89 \text{ mm}$$

$$f = f_1 \, \Pi_2^3 \, \frac{L}{L'} = -23.40 \times 0.8539 \times 0.8276 = -16.53 \text{ mm}$$

$$f' = \frac{-n'}{n} \, f = 1.3333 \times 16.53 = 22.04 \text{ mm}$$

The position of the second principal point = 16.69 − 22.04 = − 5.35 mm from the third surface = + 1.85 mm from the cornea.

The focal power of the eye is

$$F = \frac{1}{0.01653} = \frac{1.3333}{0.02204} = + 60.48 \text{ D}$$

		First surface	Second surface	Third surface
Given data	n	1.0000	1.3333	1.4160
	r	0.0078	0.0100	$-$ 0.0060
	n'	1.3333	1.4160	1.3333
	d	0.0036	0.0036	
	d/n	0.0027	0.00254	
	$n' - n$	0.3333	0.0827	$-$ 0.0827
	$1/r$	128.2	100.0	-166.64
	F	42.73	8.27	13.78
	f (mm)	-23.40		
	f (mm)	31.20		
	L	0	48.31	66.10
	L'	42.75	56.58	79.88
	$1/L'$	0.0234	0.01767	0.01253
	$1/L' - d/n$	0.0207	0.01513	$-$
	$L'/[1 - (d/n)\,L']$	48.31	66.10	$-$
PART A	L/L'	$-$	0.8539	0.8276

PART B

Combining second and third surfaces

$F_1 + F_2$	22.05
d/n	0.00254
$(d/n)\,F_1 F_2$	0.2897
F	21.76
f (mm)	-61.26
f' (mm)	61.26
e	0.002146
e'	-0.001288

Complete system

$F_1 + F_2$	64.49
d/n	0.00431
$(d/n)\,F_1 F_2$	4.007
F	60.48
f (mm)	-16.53
f' (mm)	22.04
e	0.001551 or 1.551 mm
e'	$-$ 0.004060 or -4.060 mm

19.6 SPECIAL CASE OF TWO SURFACES

Focal length and focal power

The focal length of a combination of two surfaces is given by

$$f = f_1 \; \Pi_2^2 \; \frac{L}{L'} = f_1 \times \frac{L_2}{L'_2}$$

$$\frac{1}{f} = \frac{1}{f_1} \times \frac{L_2 + F_2}{L_2} = \frac{1}{f_1}\left[1 + F_2 \frac{(f'_1 - d_2)}{n_2}\right] = -\frac{F_1}{n_1}\left[1 + \frac{F_2}{F_1} - \frac{d_2}{n_2}F_2\right]$$

or

$$-\frac{n_1}{f} = F_1 + F_2 - \frac{d_2}{n_2}F_1 F_2 = F \qquad (19.08)$$

giving the general expression for the power of a combination of two surfaces forming a lens, the media in front and behind having refractive indices n_1 and n_3 respectively.

By tracing the light in the opposite direction we obtain

$$\frac{n_3}{f'} = F_1 + F_2 - \frac{d_2}{n_2}F_1 F_2 = F \qquad (19.08a)$$

19.7 BACK VERTEX POWER AND POSITIONS OF PRINCIPAL POINTS

When the light is incident from an infinitely distant object the reduced vergence immediately on leaving the last surface is the **back vertex power**, B.V.P. of the system. This value is of particular importance in the case of the spectacle lens.

$$\left. \begin{array}{l} \text{B.V.P.} = \dfrac{F_1}{1 - \dfrac{d_2}{n_2}F_1} + F_2 = \dfrac{F_1 + F_2 - \dfrac{d_2}{n_2}F_1 F_2}{1 - \dfrac{d_2}{n_2}F_1} \\[4em] \qquad = \dfrac{F}{1 - \dfrac{d_2}{n_2}F_1} = \dfrac{n_3}{f'_V} \end{array} \right\} \qquad (19.09)$$

The distance from the second surface to the second principal point is

$$f'_V - f' = \frac{n_3}{F_1 + F_2 - \dfrac{d_2}{n_2}F_1 F_2}\left(1 - \frac{d_2}{n_2}F_1 - 1\right) = -n_3 \frac{d_2}{n_2}\frac{F_1}{F}$$

Calling this distance e' and the distance of the first principal point from the first surface, e, we have

$$e' = -n_3 \frac{d_2}{n_2}\frac{F_1}{F}$$

or

$$\frac{e'}{n_3} = -\frac{d_2}{n_2}\frac{F_1}{F} \qquad (19.10)$$

and similarly

$$e = n_1 \frac{d_2}{n_2} \frac{F_2}{F}$$

or $\quad \dfrac{e}{n_1} = \dfrac{d_2}{n_2} \dfrac{F_2}{F}$ (19.10a)

It will be seen that all these expressions are similar to those obtained in Chapter 9 for a thick lens in air, in which case n_1 and $n_3 = 1$.

19.8 CONJUGATE RELATIONS OF CENTRED SYSTEM

Having once determined the cardinal points of a given lens system we may make use of their properties to find the position and size of the image of *any* object formed by that system. In this way the rather tedious process of tracing the light through each surface of the system is avoided, and any further calculations with the given system are greatly simplified. It was for this purpose that the theory, originally due to Gauss, was developed, and with a knowledge of the cardinal points, any further work with the system is concerned only with the object and image spaces.

Thus if the positions of the cardinal points are known, we can easily find the position of the image of a given object by construction. In *Figure 19.3* F and F' are the principal foci and P and P' the principal points of a system. Then a ray QD becomes D'F' after refraction, and a ray QF after refraction becomes E'Q' parallel to the axis. These refracted rays intersect at Q', the image of Q, and as we are dealing with the paraxial region, a perpendicular to the axis through Q' gives the point B', the image of B.

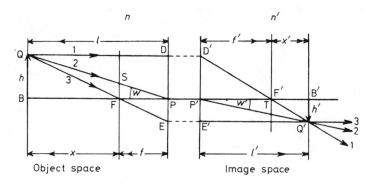

Object space   Image space

Figure 19.3 Cardinal points of centred system

Relations connecting the object and image, similar to those for a single surface or thin lens, can easily be found from the geometry of the figure.

Since the image of a point S in the first focal plane is at infinity and a ray from S parallel to the axis when refracted must pass through F', then this refracted ray and P'T must be parallel and hence $F'T = -FS$. Incidentally this provides a construction for finding the path of the refracted ray corresponding to any incident ray, similarly to that already described for a thin lens.

$$FS = f \tan w = fw$$

and $FT = f' \tan w' = f'w'$

Therefore

$$\frac{f'}{f} = -\frac{w}{w'}$$

since $F'T = -FS$

For any system $n\,h\,u = n'h'u'$, and therefore for the principal points, since $h' = h$, we have $nu = n'u'$, i.e., $nw = n'w'$, or an incident ray QP leaves the system as P'Q' as if it had suffered one refraction at a surface separating n and n'

hence $\dfrac{f'}{f} = -\dfrac{n'}{n}$ $\qquad\qquad$ (19.11)

$$m = \frac{h'}{h} = \frac{B'Q'}{BQ} = \frac{B'Q'}{P'D'} = \frac{F'B'}{F'P'} = -\frac{x'}{f'} \left.\vphantom{\begin{array}{c}1\\1\\1\\1\end{array}}\right\}$$

also

$$m = \frac{B'Q'}{BQ} = \frac{PE}{BQ} = \frac{FP}{FB} = -\frac{f}{x}$$

$\qquad\qquad$ (19.12)

and $xx = ff'$ (Newton's relation) $\qquad\qquad$ (19.13)

$$x = -f + PB \quad\text{and}\quad x' = -f' + P'B'$$

$$(-f + PB)(-f' + P'B') = ff'$$

$$PB \cdot P'B' - PBf' - P'B'f = 0$$

or $\dfrac{f'}{P'B'} + \dfrac{f}{PB} = 1$

$$\frac{n'}{P'B'} \cdot \frac{f'}{n'} - \frac{n}{PB}\left(-\frac{f}{n}\right) = 1$$

$$\frac{n'}{P'B'} - \frac{n}{PB} = \frac{n'}{f'} = -\frac{n}{f} = F$$

Thus it will be seen that if the distances of object and image are measured from the respective principal points, these distances are connected with the power of the system by the same simple expression as was previously obtained for a surface or thin lens, i.e.,

$$\frac{n'}{l'} - \frac{n}{l} = F \quad\text{or}\quad L' = L + F$$

$\qquad\qquad$ (19.14)

Magnification (lateral)

$$m = \frac{h'}{h} = \frac{B'Q'}{BQ} = \frac{l'w'}{lw}$$

$$= \frac{l'}{n'} \bigg/ \frac{l}{n} = \frac{L}{L'} \tag{19.15}$$

Magnification is given by the relation between the *reduced* distances of object and image in the same way as with a single surface.

19.9 NODAL POINTS

Two further points, as was mentioned in Chapter 9, are usually added to the focal and principal points under the heading of Cardinal Points; these are the **nodal points**. Of all the rays from Q (*Figure 19.4*) which reunite in a conjugate image point Q' there is one pair of conjugate rays for which $u = u'$, that is, the refracted ray is unchanged in direction although displaced. These conjugate rays cut the axis at two special conjugate points, the nodal points.

Figure 19.5 shows a parallel pencil from a distant object point Q incident on a system. The ray FD is refracted parallel to the axis and, since the image is

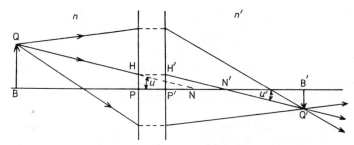

Figure 19.4 Nodal points

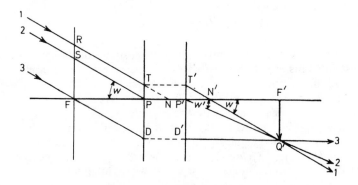

Figure 19.5 Centred system. Positions of nodal points

formed in the second focal plane, it is situated at Q'. The second nodal point N' is located by drawing the line $Q'N'T'$ through Q' parallel to FD, and as this is a ray that leaves the second principal plane at a distance $P'T'$ from the axis, the conjugate incident ray is a ray parallel to FD meeting the first principal plane at a distance $PT = P'T'$ from the axis. This incident ray RT locates the first nodal point N.

$$NN' = TT' = PP'$$

or the nodal points have the same separation as the principal points

$$T'D' = TD = RF$$

Therefore

$$\left. \begin{array}{c} FN = D'Q' = P'F' = f' \\ \text{similarly} \\ N'F' = FP = -f \end{array} \right\} \qquad (19.16)$$

Thus the distance between the *first* focal and nodal points is equal to the *second* focal length, and the distance between the *second* focal and nodal points is equal to the *first* focal length, the distance being measured *from* the focal *to* the nodal point, that is, in the opposite directions to the focal lengths.

Relative to the principal points the nodal points are displaced towards the focal point corresponding to the higher refractive index, i.e. if $n' > n$ the nodal points are displaced towards F'.

In a system for which $n' = n$ (such as a thick lens in air) the focal lengths are numerically equal and it is easily seen that the principal and nodal points then coincide; the combined points are often called the **equivalent points**.

Nodal points are of considerable importance in connection with image size, as this is determined by the ray emerging from the second nodal point. Frequent use of this fact is made in work on the optical system of the eye, where the size of the retinal image of any object is quickly found when the distance of the retina from the second nodal point is known.

In a single refracting surface the nodal points coincide at the centre of curvature C, while the principal points coincide at the vertex of the surface A.

19.10 COMBINATION OF TWO CENTRED SYSTEMS

From the above expressions we can find analytically the positions of the cardinal points of any system consisting of two surfaces. It is now necessary to consider how we can deal with the combination of two such systems.

Figure 19.6 represents two systems 1 and 2, the first having its principal points at P_1 and P'_1 and its focal points at F_1 and F'_1; the second system has its principal points at P_2 and P'_2 and its focal points at F_2 and F'_2. Two rays, in opposite directions, are traced graphically through the combination, using a similar construction to that in *Figure 6.14*. it is clear that F and F' are the principal foci of the compound system, and as the two rays intersect at H and H' at equal distances PH and P'H' above the axis, PH and P'H' are the principal planes. Then $PF = f$ and $P'F' = f'$, the focal lengths of the combined system.

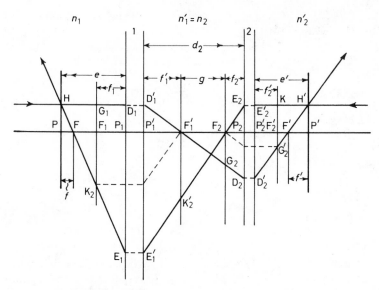

Figure 19.6 Combination of two systems

As F and F_2 are conjugate points with respect to system 1, from expression (19.13)

$$F_1 F \times F'_1 F_2 = f_1 f'_1$$

Similarly F'_1 and F' are conjugate points with respect to system 2 and

$$F_2 F'_1 \times F'_2 F' = f_2 f'_2$$

then

$$F_1 F = \frac{f_1 f'_1}{g}$$

and

$$F'_2 F' = - \frac{f_2 f'_2}{g}$$

where

$$F'_1 F_2 = g$$

In the similar triangles FPH and $FP_1 E_1$

$$\frac{FP}{FP_1} = \frac{PH}{P_1 E_1}$$

and as

$$PH = P_2 E_2 \text{ and } P_1 E_1 = P'_1 E'_1$$

$$\frac{FP}{FP_1} = \frac{P_2 E_2}{P'_1 E'_1}$$

In the similar triangles $F_2 P_2 E_2$ and $F_2 P'_1 E'_1$

$$\frac{P_2 E_2}{P'_1 E'_1} = \frac{F_2 P_2}{F_2 P'_1}$$

and hence

$$FP = \frac{F_2 P_2}{F_2 P'_1} \cdot FP_1$$

$$FP_1 = FF_1 + F_1 P_1 = \frac{-f_1 f'_1}{g} - f_1 = \frac{-f_1}{g}(f'_1 + g)$$

and

$$F_2 P'_1 = F_2 F'_1 + F'_1 P'_1 = -g - f'_1$$

Putting

$$FP = -f \text{ and } F_2 P_2 = -f_2$$

we obtain

$$f = \frac{f_1 f_2}{g}$$

Similarly we find

$$f' = -\frac{f'_1 f'_2}{g}$$

From the last two equations

$$F = -\frac{n_1}{f} = \frac{n'_2}{f'} = -\frac{n'_2 g}{f'_1 f'_2}$$

$$= -\frac{g}{n_2} \cdot \frac{n_2}{f'_1} \cdot \frac{n'_2}{f'_2} \tag{19.17}$$

$$= -\frac{g}{n_2} \cdot F_1 . F_2 \tag{19.17a}$$

For two systems separated by air, $n_2 = 1$ and $F = -g F_1 F_2$. The microscope is a system of this type where g is the optical tube-length.

When $g = 0$, $F = 0$ and we have a system of zero power—an afocal or telescope system (compare section 10.8).

$$P'_1 P_2 = d_2 = f'_1 + g - f_2$$

$$g = d_2 - f'_1 + f_2$$

substituting this value of g in expression (19.17a) gives us

$$F = F_1 + F_2 - \frac{d_2}{n_2} F_1 F_2 \tag{19.18}$$

From the geometry of the figure we also obtain

$$P_1 P = e = n_1 \frac{d_2}{n_2} \frac{F_2}{F} \tag{19.19}$$

and

$$P'_2 P' = e' = -n'_2 \frac{d_2}{n_2} \frac{F_1}{F} \tag{19.19a}$$

It will be seen that the above expressions for the combination of two systems are the same as those for the combination of two surfaces.

The focal power and the positions of the cardinal points of *any* system of centred surfaces may be found by the above method. Each pair of adjacent surfaces may be combined to form a system, the systems so formed are then combined again in pairs, and so on until the cardinal points of the complete system are found. Thus, with a system of six surfaces, the cardinal points are first found of each of the three systems formed by combining surfaces 1 and 2, 3 and 4, and 5 and 6. The cardinal points of the system formed by the combination of the first two of these systems are found, and this system is then combined with the third.

As an example we will consider the schematic eye which we have previously calculated. Here it will be convenient to find the power and principal points of the system formed by the second and third surfaces, representing the crystalline lens, and combine this with the first surface. The work is shown in part B of the table in section 19.5.

It will be seen from the table that the first principal point is + 1.551 mm from the cornea. The second principal point is − 4.060 mm from the second principal point of the crystalline lens and this is − 1.288 mm from the third surface; the second principal point is therefore 7.2 − 1.288 − 4.060 = + 1.85 mm from the cornea. The first principal focus is − 16.53 + 1.55 = − 14.98 mm from the cornea, and the second principal focus 1.85 + 22.04 = + 23.89 mm from the cornea. The distances from the cornea of the first and second nodal points respectively are − 14.98 + 22.04 = + 7.06 mm and 23.89 − 16.53 = + 7.36 mm.

19.11 LENS MIRROR OR THICK MIRROR

These terms are applied to any system of surfaces or lenses, the last surface of which acts as a mirror, the light being thereby reflected back through the system. Such systems occur in the Mangin mirror frequently used as the reflector in various projector lamps, in the shaving glass formed by silvering the back surface of a double convex lens and in the various surfaces of the eye media, the reflections from which give rise to the Purkinje images.

In *Figure 19.7* P_1 and P'_1 are the principal points of a system in front of a convex spherical reflecting surface at A_2 with radius of curvature $A_2 C_2$. A ray TD_1 incident parallel to the axis will, after refraction and reflection, travel along the path $D_2 E'_1 E_1 S$, and the total effect of the complete system will be the same as if the light had undergone a simple reflection at a mirror of suitable curvature at a position D'_2. $A'_2 D'_2$ represents therefore the *equivalent mirror* which can replace the complete lens-mirror system.

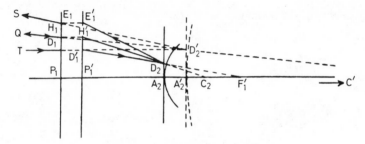

Figure 19.7 Lens mirror or thick mirror

Position of equivalent mirror

Light from any point D_2 on the mirror surface will, after passing through the system in front of the mirror, emerge as though it had come from D'_2; that is, D'_2 is the image of D_2 as formed by the system. But D'_2 is a point on the equivalent mirror corresponding to any point D_2 on the actual mirror, hence the equivalent mirror occupies the position of the image of the actual mirror as formed by the system before the mirror.

Position of the centre of curvature of equivalent mirror

A ray QH_1 incident in such a direction that after refraction at the lens system it meets the mirror normally at D_2 will be reflected and refracted back along its original path. Such a ray must be incident towards the centre of curvature C' of the equivalent mirror. Hence this ray is, after reflection coming apparently from C_2 the centre of the mirror, and after emergence from the system is coming apparently from C', then C', the centre of curvature of the equivalent mirror, is the image of C_2 as formed by the system in front of the mirror.

Example: To find the position and radius of curvature of the equivalent mirror corresponding to the anterior surface of the crystalline lens in the simplified schematic eye (section 19.5). The reflection from this surface gives rise to the third image of Purkinje.

Position of equivalent mirror

Taking anterior surface of crystalline lens as object, the light travelling from crystalline lens to cornea,

$$l = -0.0036 \text{ m} \quad L = -\frac{1.333}{0.0036} = -370.4D$$

$$n = 1.333$$

$$F = +42.73D \text{ (power of cornea)}$$

$$n' = 1$$

$$l' = -3.052 \text{ mm}$$

Therefore the equivalent mirror is situated 3.05 mm behind the cornea.

Position of centre of curvature of equivalent mirror

Taking the centre of curvature of the anterior surface of the crystalline lens as object,

$$l = -0.0136 \text{ m} \quad L = -\frac{1.333}{0.0136} = -98.06\text{D}$$

$$L' = -98.06 + 42.73 = -55.33\text{D}$$

$$l = -18.08 \text{ mm}$$

Therefore the centre of curvature of the equivalent mirror is 18.08 mm behind the cornea, and the radius of curvature is $18.08 - 3.05 = +15.03$ mm.

If a lamp is placed, say 300 mm in front of the cornea, the position and magnification of the third image of Purkinje may now be found as follows:
Object distance from equivalent mirror,

$$l = -303 \text{ mm}$$

Radius of equivalent mirror $r = +15.03$ mm

$$L' = L + F = L - 2R$$

$$= \frac{1000}{-303} - \frac{2000}{15.03} = -136.4\text{D}$$

Therefore $l' = +7.33$ mm.

Hence, the image is situated $3.05 + 7.33 = 10.38$ mm behind the cornea. The magnification of the image

$$m = \frac{7.33}{303} = 0.0242 \text{ or practically } \frac{1}{40}$$

CHAPTER 19   *EXERCISES*

1. An equi-convex lens having curves of 10 cm radius, a central thickness of 2 cm and refractive index 1.61 closes one end of a long tube filled with water. An object 5 cm long is placed 60 cm in front of the lens; find the position and size of the image.

2. Explain the terms: equivalent focal length, principal points and nodal points, and illustrate each by a diagram of the eye.

3. A $+ 10$D equi-convex lens of glass of $n = 1.62$ is mounted in the centre of a tank 20 cm long and having thin plane glass ends; the tank is filled with water ($n = 1.33$). Find the power of the system and the position of the focus when parallel light is incident on one end of the tank.

4. What are the properties of the principal and nodal points of a lens?
A double convex lens of glass of refractive index 1.53 having curves of 5D and 8D respectively and thickness of 4 cm has its second surface immersed in water ($n = 1.33$). Find the focal lengths and positions of the principal and nodal points.

5. A thin $+ 10$D equi-convex lens of refractive index 1.53 forms one end of a long tube filled with water of refractive index 1.33; find the positions of the principal foci and the equivalent focal length of the system.

6. Find the focal power and positions of the nodal and principal points of an aphakic eye, radius of cornea = 8 mm; refractive index = 1.336. What correcting lens is necessary to focus distant objects on the retina, the lens being placed 15 mm from the cornea and the axial length of the eye being 24 mm?

7. A long tank filled with water has an equi-convex lens of n = 1.58, thickness $^3/_4$ inch and radii of curvature of 2 inches fitted in one end. Find the position of the principal, nodal and focal points.

8. Define, with diagrams, the cardinal points of a thick lens system.
A parallel beam of light is directed along the axis of a thick, biconvex lens. After refraction at the first surface the beam is reflected by the second surface, which is silvered. After a further refraction at the first surface, the beam comes to a point focus. Where is this focal point? The radius of the curvature of both lens surfaces is 20 cm, the refractive index of the lens is 1.5 and its axial thickness is 5 cm.

9. Show that in a concentric lens, i.e., one in which the two surfaces have a common centre of curvature, the principal points coincide at the centre of curvature.

10. In Gullstrand's Schematic Eye, the cornea and the aqueous humour have the following constants:

Thickness of cornea	0.5 mm
Radius of anterior surface	+7.7 mm
Radius of posterior surface	+6.8 mm
Refractive index of cornea	1.376
Refractive index of aqueous humour	1.336

Find the two focal lengths and the power of the cornea.

11. A hollow globe of glass (n = 1.52) has an external diameter of 10 inches and an internal diameter of $9^1/_2$ inches; it is filled with water. Find the position at which a narrow parallel pencil of light incident on the globe will be focused. What will be the positions of the principal points of the system?

12. Calculate the equivalent focal length of a lens system consisting of three thin positive lenses of 4 inches, 5 inches and 6 inches focal length respectively, each separation being 1 inch.

13. An object 50 mm high is placed 250 mm in front of the cornea of the schematic eye (section 19.5). Find, by tracing through the various surfaces in turn, the position and size of the image formed by the system. Compare the result with that obtained by making use of the equivalent focal power and the positions of the principal and nodal points (section 19.5 Table A and B).

14. Find the equivalent focal length and the positions of the principal points of the Rapid Landscape photographic lens having the following constants:

r_1 = −120.8 mm		n_1 = 1
r_2 = − 34.6 mm	d_2 = 6 mm	n_2 = 1.521
r_3 = − 96.2 mm	d_3 = 2 mm	n_3 = 1.581
r_4 = − 51.2 mm	d_4 = 3 mm	n_4 = 1.514
		n_5 = 1

15. Find from the following data (all distances in mm) the equivalent focal length and the positions of the principal points of the Cooke Series IV photographic objective:

r_1 = + 19.44		n_1 = 1.000
r_2 = −128.3	d_2 = 4.29	n_2 = 1.6110
r_3 = − 57.85	d_3 = 1.63	n_3 = 1.000
r_4 = + 18.19	d_4 = 0.73	n_4 = 1.5754
r_5 = +311.3	d_5 = 12.9	n_5 = 1.000
r_6 = − 66.4	d_6 = 3.03	n_6 = 1.6110
		n_7 = 1.000

16. Having given the following data of Helmholtz's schematic eye at rest, find:
(*a*) Power and focal lengths of crystalline lens.
(*b*) Positions of principal points of crystalline lens.

(c) Power and focal lengths of cornea.
(d) Power of whole schematic eye.

Data:

Radius of cornea at vertex	+7.829 mm
Radius of anterior surface of crystalline lens	+10.0 mm
Radius of posterior surface of crystalline lens	−6.0 mm
Depth of anterior chamber	3.6 mm
Thickness of crystalline lens	3.6 mm
Refractive index of aqueous humour	1.3365
Refractive index of crystalline lens	1.4371
Refractive index of vitreous humour	1.3365

17. An equi-convex lens has surfaces of 15 cm radius of curvature and a central thickness of 2 cm and $n = 1.51$. Part of the light which enters the lens from a distant luminous point is reflected at the second surface and emerges again from the first. Where will this reflected light focus?

18. What will be the focal power and position of the equivalent thin mirror corresponding to a meniscus lens having curvatures $r_1 = -15$ cm, $r_2 = -25$ cm, central thickness 2 cm, $n = 1.53$, the second surface of the lens being silvered?

19. Using the data for the schematic eye given in section 19.5, find the position and size of the image formed by reflection at the second surface of the crystalline lens (the fourth image of Purkinje), the object being 5 cm diameter and situated 30 cm from the cornea.

20. Discuss the significance of the nodal points of an optical system, and indicate how they are made use of in the design of a panoramic camera.

A lens of refractive index 1.5, thickness 5 cm and having two convex surfaces of radius of curvature 10 cm is used in air. Determine the position of the nodal points.

21. Two thin lenses A and B 2.5 cm apart make up a telephoto lens. Their focal lengths are: A + 6 cm, B − 4 cm (i.e. $F_A = +16^2/_3$ D, $F_B = -25$D).

Find (a) the position of the film for a distant object on the side of lens A to give a sharp image, and (b) the positions of objects in sharp focus and the magnification, when the film is moved away from the lens from the position given by:
(i) 1 cm   (ii) 2 cm   (iii) 3 cm   (iv) 4 cm

22. A schematic relaxed eye consisting of a refracting cornea and a lens has the following constants:

$r_1 = 7.8$ mm $n_{aqueous} = {}^4/_3$

$r_2 = 10$ mm $n_{lens} = 1.416$

$r_3 = -6$ mm

Position		
	Cornea	0
	Lens anterior surface	+ 3.6 mm
	Lens posterior surface	+ 7.2 mm

A distant object subtends an angle of 2° at the eye. Determine the size and position of the image obtained by reflection from the surface r_3.

Chapter 20

The eye as an optical instrument

20.1 INTRODUCTION

In Chapter 10 a simplified description of the eye was given in terms of the refraction of light as developed in the earlier chapters. In Chapter 19 some further calculations were carried out to establish the first-order parameters of the eye as a lens system. Now with the help of the topics covered in Chapter 11 onwards we are able to study the eye as an optical instrument. Even though the emmetropic eye is found to have considerable variation from subject to subject, it is possible to obtain an overall picture of its performance as a detector of those variations of light which we call objects.

The treatment in this chapter is not meant to be comprehensive nor should it be as many of these topics are treated in books on Physiological Optics. Increasingly, however, the eye has been evaluated in terms of instrumental and physical optics. In particular, the interaction between the eye and optical display instruments has been found to need a knowledge of vision in terms which can be incorporated into the design process of the equipment. This chapter lays the basic ground work for this, but there is no discussion of binocular vision phenomena such as stereopsis and convergence. Furthermore, all time varying factors have been omitted as have colour vision phenomena.

Thus, the sections which follow constitute working descriptions of the eye which when applied to real situations will give valid results provided the real situation is adequately described and that sufficient allowance is made for the very sizeable differences between subjects.

20.2 MORE SCHEMATIC EYES

The Emsley Schematic eye used in section 19.5 is an example of a three-surface or **simplified schematic eye**. Many earlier examples exist due to Listing (1851) Helmholtz (1866), Tscherning (1898) and Gullstrand (1909). Details of Helmholtz, Gullstrand (no. 2) and Emsley simplified schematic eyes are given in *Figure 20.1*. The differences between them, while not large, were significant steps in understanding the parameters of the eye although the modifications suggested by Emsley were mainly to simplify computation.

Reduced schematic eyes have also been suggested using only one refractive surface and these are often adequate for calculating straightforward values like retinal image sizes. The **Emsley 60-Dioptre eye** and the **Ogle 17 mm eye** are very nearly the same. *Figure 20.2* gives their salient values.

444

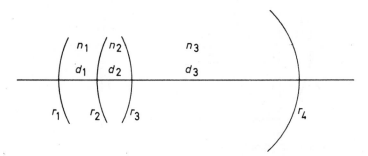

Radii of Curvature	Helmholtz (1866)	Gullstrand (1909)	Emsley (1955)
r_1	7.83	7.80	7.80
r_2	10.00	10.00	10.00
r_3	−6.00	−6.00	−6.00
Axial Distances			
d_1	3.60	3.60	3.60
d_2	3.60	3.60	3.60
d_3	15.03	16.97	16.70
Refractive Indices			
n_1	1.3365	1.336	1.333
n_2	1.4371	1.413	1.416
n_3	1.3365	1.336	1.333

Figure 20.1 Simplified schematic eyes—all dimensions in mm

There have been numerous schematic eyes using four refracting surfaces. *Figure 20.3* shows the Gullstrand-Le Grand eye and the Schematic Eye given in the US Military Handbook 141. The cornea is represented by a lens of finite thickness having power in these cases of about 43.5D. The variation across the population in actual corneal powers is from about 38D to 48D. The precision of these schematic eyes is therefore that of representing the *average* case. Comparison between these eyes and the simplified versions may be made using the procedure of section 19.5.

The representation may be improved firstly by describing the cornea not by spherical surfaces but by elliptical surfaces. The crystalline lens is also more complex than *Figure 20.3* indicates having an increasing refractive index towards the centre in the form of a shell-like structure and also having surfaces which are aspheric. These changes affect to the aberrations of the eye which are dealt with in section 20.5.

In terms of the first-order parameters the schematic eyes suggested differ mainly in their locations for the cardinal points. The eye given in *Figure 20.3* for example has principal points and nodal points which differ in their distances from the cornea by up to 0.3 mm compared with the Emsley eye of section 19.5. These give rise to slightly different locations for the entrance and exit pupils of the eye. The real pupil is located at the anterior surface of the lens 3.6 mm from the cornea but the power of the cornea generates an entrance pupil only 3 mm from the cornea and magnified by ×1.12. When pupil sizes are referred to in

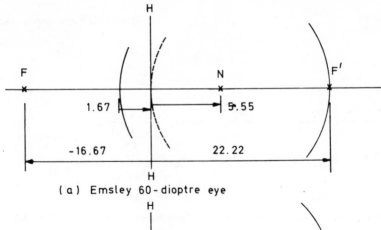

(a) Emsley 60-dioptre eye

(b) Ogle 17-mm eye

Figure 20.2 Reduced schematic eyes—all dimensions in mm

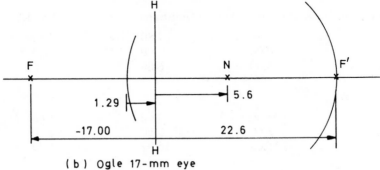

Radii of Curvature	Gullstrand-Le Grand	MIL Handbook—141
r_1	7.8	7.98
r_2	6.5	6.22
r_3	10.2	10.20
r_4	−6.0	−6.17
r_5	−12.3	
Axial Distances		
d_1	0.55	1.15
d_2	3.05	2.39
d_3	4.00	4.06
d_4	16.60	17.15
Refractive Index		
n_1	1.3771	1.376
n_2	1.3374	1.336
n_3	1.420	1.420
n_4	1.336	1.337

Figure 20.3 Schematic eyes—all dimensions in mm

ophthalmic work it is always the entrance pupil which is meant as this is the measured pupil as seen from the outside world.

As the centre of rotation lies 13 mm behind the cornea, the entrance pupil moves as the eye fixates on different objects, by a lateral distance given in millimetres by 10tan θ where θ is the angle of rotation. In real eyes there is some variation in this centre to cornea distance and also a wandering of the centre at different directions of gaze.

20.3 CHROMATIC ABERRATION OF THE EYE

The various media of the human eye exhibit dispersion and in the absence of negative correcting elements, its optical system suffers from longitudinal chromatic aberration. Thus the image of a blue object is formed closer to the lens than the image of a red object. It is estimated that the average dispersion is only slightly greater than that of distilled water. For the 60-dioptre eye of Emsley the power error against wave-length can be calculated as shown in *Figure 20.4* using the known refractive indices of distilled water.

In spite of about 2D chromatic aberration over the visible spectrum the eye

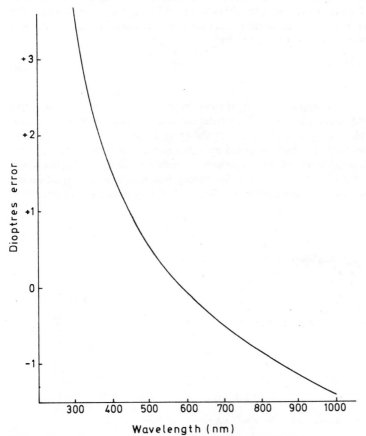

Figure 20.4 Chromatic aberration of the eye

continues to operate without being conscious of the error. It is known that for some people a polychromatic image is needed for good accommodation. If the chromatic error is eliminated by compensating optics it is found that the ability to see low-contrast large objects is improved but there is little effect on fine detail acuity.

The extent of the chromatic aberration over the visible spectrum corresponds to an image shift of about 0.5 mm. This would be partially compensated if the receptor layers for blue light are displaced forward in the retina to be in front of those receiving red light, but this is not thought to be the case. When light is received by the eye, some of it is reflected in the regions of the retina. The sharpness of the image formed on the retina can therefore be studied. It does appear that red light penetrates deeper into the retina than blue light.

However, very careful subjective measurements of best focus for different wave-lengths show a curve closely following that expected from the longitudinal chromatic aberration calculated above. The locations at which light is reflected are therefore not necessarily coincident with those at which the light is absorbed.

20.4 DIFFRACTION AND THE EYE

In Chapter 15 it was seen that the effects of diffraction constituted a lower limit to the size of image which an optical system could form of a point object. The eye is no exception to this. In section 15.11 we saw that the minimum resolveable angle, w, between two images is given by

$$w = \frac{1.22\lambda}{b} \qquad \text{[from (15.09)]}$$

where λ is the wave-length of the light used (in object space) and b the diameter of the limiting circular aperture. This criterion is known as the **Rayleigh Limit** and can be plotted against pupil size as shown in *Figure 20.5*. This graph also contains experimental values obtained by a number of workers. It can be seen that for small pupil diameters the experimental points, while following the shape of the curve, are slightly better than the theory which is, after all, based on the arbitrary figure of $\lambda/4$ path difference. Although this pessimism has been noted

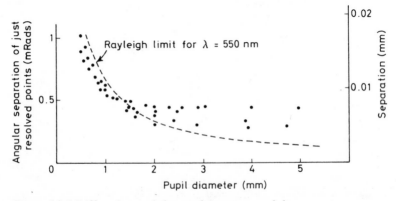

Figure 20.5 Diffraction and the resolving power of the eye

for other instruments the Rayleigh Limit continues to be the criterion in general use.

For pupil diameters above 2 mm the eye is considerably worse than that predicted by diffraction alone. This is due to the aberration of the eye which increases with increasing aperture. In general terms the eye is commonly assumed to be diffraction limited for pupil diameters of 2 mm or less. When the separation distance on the retina is calculated (using the 17 mm reduced eye) for the 2 mm pupil it is found to be commensurate with the retinal receptor spacing at the fovea.

20.5 ABERRATIONS OF THE LENS AND CORNEA

The Seidel aberrations treated in sections 18.3–18.6 of Chapter 18 are applicable to the eye only in general terms as the refracting surfaces of the eye are not regular or spherical. Because the eye rotates to fixate each object of interest so that its image falls on the foveal part of the retina it might be supposed that only spherical aberration of the optical surfaces is of interest. This aberration is certainly the major interest but the common axis of the optical surfaces does not exactly align with the fovea (section 10.1). This means that the eye is used with its optical system off-axis by about 5°–10°. Off-axis aberrations are of interest when light must be accurately focussed through the optics at a large field angle for photo-coagulation work on the retina (section 12.16).

Spherical aberration of the eye may be seen by holding an opaque edge across the pupil of the eye while observing an object. When the edge is more than half-way across the eye the object will appear to move as the remaining rays are

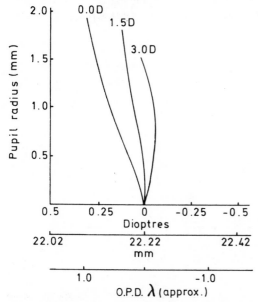

Figure 20.6 Spherical aberration of eye across horizontal meridian (after Ivanoff, Les Aberrations de L'oeil (1953))

restricted to those nearer and nearer to the periphery. Care must be taken to accommodate properly for the object distance.

A more accurate method is to observe a straight line near a positive lens. By using screens one part of the line is illuminated by one small source and the other part by another small source. The distances of these sources is arranged so that they are both imaged by the lens onto the pupil of an observer's eye. Lateral movement of the sources allows one image to be central while the other is some distance off-axis. In the presence of spherical aberration the parts of the line appear displaced. An adjustment can be provided to realign them.

By this method Ivanoff measured the spherical aberration at different viewing distances as shown in *Figure 20.6*. These are the average values over a number of subjects and defined about the 'achromatic axis of the eye' which passes through the fovea and a non-central point of the pupil such that there is no chromatic dispersion. It is seen that the aberration is undercorrect for distant viewing (periphery more powerful than centre) while it is overcorrect for near viewing. The spherical aberration of any of the schematic eyes given in this chapter may be readily calculated by using the meridional ray tracing procedures of Chapter 18. All of them overestimate the undercorrect aberration. This is due to their spherical surfaces. The aspheric surfaces of the real eye tend to reduce spherical aberration.

However, these surfaces are far from regular. The knife-edge test of section 18.11 may be used objectively if light is reflected off the retina. The variations in luminance associated with knife-edge testing can be analysed by computer for different orientations of the edge. The results obtained by **Berny and Slansky** using such a method show a very irregular pattern. *Figure 20.7 for a specific subject* has very little which can be called spherical aberration.

In 1866 Helmholtz noted that measurements across the pupil 'do not form a continuous series at all, so that the conception of spherical aberration does not apply'. This is certainly the case if only one subject is considered, but the aber-

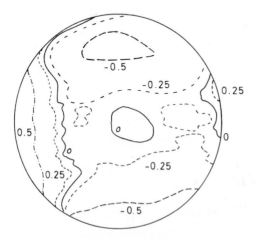

Pupil diameter 7 mm

Figure 20.7 Irregular aberration of eye. Figures indicate OPD in wave-lengths (after Berny and Slansky, Optical Instruments and Techniques. Ed. J. Hume Dickson (1969))

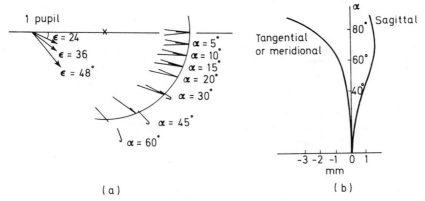

Figure 20.8 Astigmation with schematic eyes (a) Caustic analysis MIL−HDBK−141 Eye (after Shealy and Rosenblum (1975) Opt. Eng., 14/3, 237-240 (b) Seidel analysis of Gullstrand-Le Grand Aspherics Eye (after Lotmar (1971), J.O.S.A., 61/11, 1522-1529)

rations of the generalised eye as a mean over several subjects are definable in terms of the Seidel theory.

For the off axis case, **Shealy** has calculated the caustic surfaces for the MIL−HDBK−141 schematic eye. This is shown in *Figure 20.8a*. The usual shape for spherical aberration occurs on axis but as the angle is increased the two sheets separate showing the presence of considerable astigmatism. The same action can be shown by tracing rays at different field angles. *Figure 20.8b* shows the results obtained with the Gullstrand-Le Grand eye with aspheric surfaces. The x-axis shows distances along each ray with respect to the spherical retinal radius r_s. These values are found to be less than those obtained from spherical surfaces showing that the natural aspherics of the eye tend to reduce astigmatism. In the graphs of *Figure 20.8* the angle α is the visual angle, that is the off-axis angle of the object in front of the eye. Inside the eye this angle is reduced to the angle ϵ as shown in *Figure 20.8a*.

20.6 OPTICAL PERFORMANCE OF THE EYE

Even though the aberrational analysis of the eye is rendered difficult by its non-uniformity, it is still possible to obtain a measure of its optical quality using the concept of the modulation transfer function developed in section 18.11. The visual perception of modulated contrast in a scene may be considered in two parts. Firstly, the optics of the eye must form an image on the retina; secondly, the neural processes must respond to the image. In this section we are concerned only with the first of these—the M.T.F. of the optics of the eye.

Methods of measurement range from imaging a fine point or line onto the retina and analysing its sharpness (or line spread function) to imaging a full sinusoidal grating onto the retina and measuring the contrast modulation of it. All these so-called objective methods (which do not need a response from the subject) use light reflected from the retina and therefore use the optics of the eye twice. There is also some loss of quality in the reflection process at the retina. The double-pass effect can be allowed for by taking the square-root of the

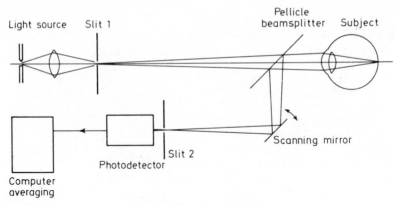

Figure 20.9 Measurement of optical performance of the eye

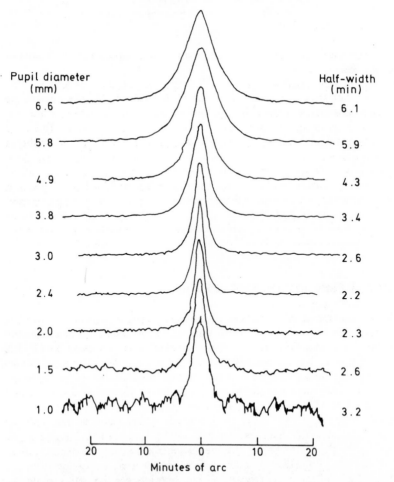

*Figure 20.10 External line spread function of the human eye (white light)
(after Campbell and Gubisch)*

measured modulation (assuming that the retinal reflection is diffuse). The second effect is more difficult. Clearly, the light which is reflected has not become physiologically effective since it has not been absorbed by the receptors. However, the objective approach does allow a determination of the minimum quality of the optics of the eye. Earlier measurements were limited by the apparatus and techniques used and gave somewhat poorer values than were found subsequently.

The more recent determinations have used the simplest possible apparatus. *Figure 20.9* shows an illuminated slit viewed by the observer via a pellicle beam-splitter. This beamsplitter directs light reflected by the retina and refocussed by the optics of the eye onto a second slit via a front reflection mirror which can scan the image across the second slit. A photodetector placed behind the second slit records the profile of the so-called external line spread function attributable to the double passage of light through the eye. The slit widths are small compared with the spread function.

Stray light must be carefully excluded from the system. A very sensitive low noise photodetector must be used with the highest light intensity supportable on the first slit. Even so, most experimenters use a computer to average up to a thousand passes of the retinal image by the scanning mirror so that extraneous effects are reduced. The eye of the subject must be held at optimum focus during this time, as must pupil size and fixation. A beautiful series of results using white light was obtained by Campbell and Gubisch* in 1966 and are shown in *Figure 20.10*, where the curves have been shifted vertically for easy comparison.

The 'half-width' of the spread (that is, the full width at half the peak value) is given for the different sizes of the artificial pupil used (see section 20.8). Notice how the noise increases as the smaller pupil returns less light. Further studies have shown that chromatic aberration and (irregular) spherical aberration contribute in roughly equal parts. Using numerical methods, these line spread functions can be converted to M.T.F. values allowing for the double pass effect and the finite width of the slits. *Figure 20.11* shows these values and, for comparison, a diffraction limited curve for an eye of 1.5 mm pupil (for light of 550 nm). The assumption is made that the retina is a perfect diffuse reflector of the image, which is less than true at the extremes of the visible spectrum. However, these curves constitute real Modulation Transfer Functions being the ratio of output over input as defined in section 18.10. Thus the Modulation Transfer is a dimensionless quantity.

The curves of *Figure 20.11* are obviously related to each other and an empirical mathematical expression has been sought so that its coefficients might be quoted and used rather than all the data of the curves themselves. Jennings and Charman† have shown that a good fit can be obtained using

$$T(s) = e^{-(s/s_c)^n}$$

where s_c is a constant spatial frequency and n is an 'MTF index'. *Figure 20.12* shows the curves of *Figure 20.11* plotted with a log.scale and log.reciprocal ln scale for the x and y axes. These give straight lines to the MTF values if the equation above is a good description. As can be seen, a good fit is obtained with

*Campbell and Gubisch (1966) *J. Physiol.*, 186, p. 558
†Jennings and Charman (1974) *Brit. J. Physiol: Optics*, **29**, p. 64–72

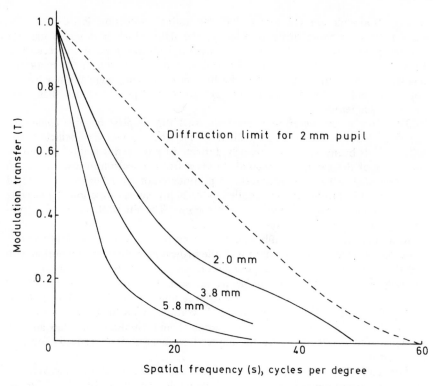

Figure 20.11 MTF values for the optics of the eye (after Campbell and Gubisch)

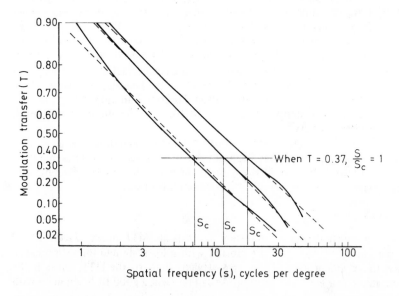

Figure 20.12 Mathematical approximation to MTF of the optics of the eye (after Jennings and Charman)

s_c values shown and n (the slope) nearly equal to one in each case.

Thus, for example, the MTF of the optics of the eye may be given by

$$T(s) = e^{-(s/11)}$$

where s is in cycles per degree and the pupil size is 3.8 mm.

20.7 TOTAL PERFORMANCE OF THE EYE

When the response of the retinal receptors in the eye are included in an assessment of 'performance' we can no longer invoke a strictly defined modulation transfer function, as it is not possible to obtain an output value in a quantified form (although use has been made of the evoked potentials which are measured when electrodes are attached to the scalp near the visual cortex). A further limiting feature concerns the extent to which the eye-brain system may be treated as linear. Although linearity is essential to the concepts of frequency response methods, small departures from linearity do not generate large inaccuracies and the methods may be applied, with care, to the visual response.

The description of visual performance most closely related to frequency response methods is that of the **contrast sensitivity** or **threshold contrast**. This involves determining the minimum contrast, Cm, of a sine-wave grating of spatial frequency, s, for which the subject can detect its presence or orientation. It is found that the eye has an optimum spatial frequency at which the minimum contrast which can be seen is lower than that at higher or lower spatial frequencies. It is also found that other parameters such as luminance, overall size of grating, viewing time, non-uniform backgrounds, etc., all affect the curve. Visual parameters such as pupil-size, accommodation, peripheral or foveal viewing, monocular or binocular viewing etc. also influence the curve. In some studies non-sinusoidal objects have been studied, leading to important advances in the understanding of the visual system. Threshold contrast curves have been obtained for almost every conceivable condition of test and subject.

Such curves are not true M.T.F. curves for the reasons stated above. The 'output' in this method is a 'seen' or 'not seen' response from the human subject. The curve given is therefore the boundary between combinations of contrast and frequency which are seen and combinations which are not seen. When contrast modulation is plotted on the y-axis and spatial frequency on the x-axis, using logarithmic scales on both, a J-shaped curve is found, for most conditions and subjects, similar to that shown in *Figure 20.13*. There is no point in normalising this curve as it is not a ratio and cannot be used to calculate anything about points in contrast frequency space other than those on the curve. The points on the curve have no mathematical significance (as if the M.T.F. was zero or unity at these points) other than specifying the input condition for which the signal rises sufficiently above the noise of the system, to be detected. Sometimes the log reciprocal of contrast is plotted on the y-axis. This gives the curve of greater similarity to M.T.F. curves and to avoid confusion this practice will not be followed here. Given a family of contrast sensitivity curves such as *Figure 20.14*, for different average scene luminances, it is permissible to interpolate between the curves to find threshold contrast values for other scene luminances, but it is notoriously difficult to extrapolate from such curves to different experimental conditions. *Figure 20.14* shows that increasing scene luminance improves the

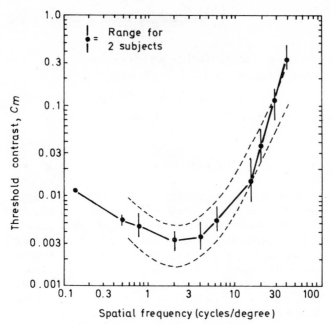

Figure 20.13* Contrast sensitivity under optimum conditions (after Van Meeteren and Vos (1972) Vision Research, 12, 825–833)

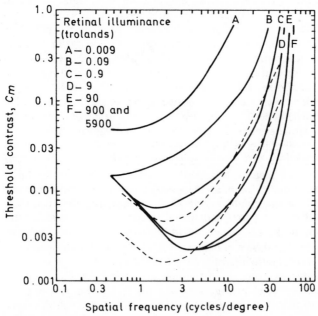

Figure 20.14* Contrast sensitivity variation with light level (after Van Ness and Bouman (1966) J. Opt. Soc. Am., 56, 689–694)

*Figures 20.13, 20.14 and 20.15 are reproduced from Design Handbook for Imagery Interpretation Equipment, Farrel and Booth, The Boeing Aerospace Company whose permission for this use is gratefully acknowledged. The dashed lines on these figures indicate an estimate of contrast sensitivity for the best and worst of 90% of people under optimum viewing conditions. Contrast is defined as modulation as in section 11.14.

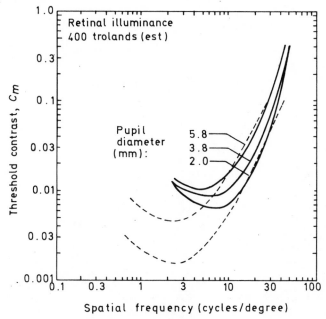

Figure 20.15* Contrast sensitivity variation with pupil size (after Campbell and Green (1965) J. Physiol., 181, 576–593)

performance of the eye as evidenced by the smaller threshold contrast values at each frequency.

Another family of curves may be obtained when the apparent scene luminance is kept constant but the effective pupil size adjusted. *Figure 20.15* shows how a reduction in the pupil size gives a better visual performance when the retinal illumination is kept constant, as might be expected from the improving optical quality. The values given, however, show a poorer performance at lower spatial frequences than those in *Figure 20.14*. Each study used a different subject and different test method, the main influence being the more restricted overall field used for the latter results. Comparison between *Figures 20.11* and *20.15* shows that the general range of useable frequencies is similarly bounded for high frequencies but the reduction in visual performance for the lower spatial frequencies is a retinal rather than optical effect.

20.8 VARIATION IN VISUAL PERFORMANCE WITH FOCUS

The ability of the eye to accommodate sets it apart from most optical instruments. However, it does not necessarily achieve a perfect in-focus condition at all times. Indeed, for the aberrated eye a single perfect in-focus position does not exist. In the first place, the considerable chromatic aberration of the eye means that an incorrect focus for one colour may be a good focus for another. It does appear that the mechanism of colour vision can mitigate the effect of poor focus by concentrating on the in-focus colour.

In the second place, the presence of spherical aberration means that the

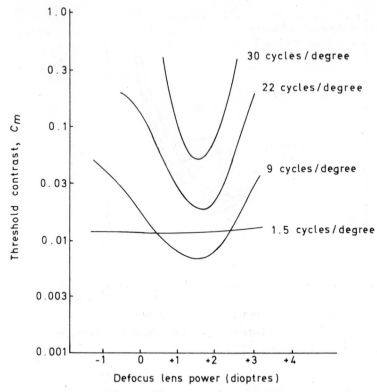

Figure 20.16 Variation of contrast sensitivity with defocus (2 mm pupil) (after Campbell and Green (1965) J. Physiol., 181, 576–593)

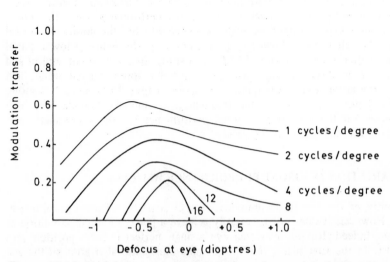

Figure 20.17 Modulation transfer of optics of eye with various amounts of defocus (5 mm pupil) (after Charman and Jennings (1976) Brit. J. Physiol. Optics, 31, 119–134)

'correct' focus is not at the paraxial focus. Indeed, it may be shown that, at intermediate pupil sizes, the best response for higher spatial frequencies may be obtained at a different focus condition than for lower spatial frequencies. The presence of irregular aberrations serves to blur things even more!

Figure 20.16 shows how various amounts of artificial defocus, introduced by lenses in front of a drugged eye, raise the threshold contrast for the spatial frequencies indicated. The effect is generally symmetrical about a value of + 1.5D due to the near object and the subjects' refractive error. With a larger pupil it is possible to see an unsymmetrical effect. The curves of *Figure 20.17* are for an eye with a 5 mm pupil and are derived from double-pass line spread functions on the retina. This is therefore a true M.T.F. representation. Here again the subject shows refractive error, partly due to the use of red light. However, the peak modulation transfer value occurs at a different defocus depending on the spatial frequency considered. The shift shown is approximately commensurate with about 1D of spherical aberration.

20.9 THE MEASUREMENT OF CONTRAST SENSITIVITY

A graphic demonstration of the threshold effect of contrast is provided by Plate 11. This picture interacts with the eye to give an analogue of *Figure 20.13*. The vertical lines have a sinusoidal profile the spatial frequency of which increases from left to right in the same way as the x-axis of *Figure 20.13*. For reasons of space and photography it is not possible to provide a range much bigger than log 1.5 which means that the lines on the right are about 30 times closer than those on the left. Clearly we can make this range cover the left half of the x-axis by viewing the plate from a short distance or the right half by moving it 30 times further away.

The effect of the y-axis is achieved by making the sinusoidal profile increase in modulation, again in a logarithmic manner, with increasing value of y. It is difficult to be precise about the actual contrast modulation values following photographic processing and printing. However, the range is sufficient for the J-shape, as the locus of the points where each line becomes invisible, to show clearly under normal viewing conditions. As the viewing distance is varied the J-shape moves laterally without significantly changing its shape although the contrast threshold of the lower spatial frequencies is affected by the apparent size of the plate.

Although such a photograph is useful in a textbook it is not the way to obtain accurate experimental results. For this purpose a sinusoidal grating is generated on an oscilloscope. A very rapid vertical scan is applied to the spot giving a vertical line, while a much slower scan moves this line horizontally to effectively fill the screen with light. This luminance level can then be modulated by applying a sine wave to the brightness grid at a frequency between the two scan rates. As this frequency is varied, vertical sine wave gratings of different spatial frequencies appear on the tube face. The apparent contrast of these gratings may be changed by adjusting the voltage amplitude of the brightness modulating sine wave. This adjustment may be controlled by the subject so that he can determine when the grating is just visible. The change in this voltage with spatial frequencies will give the threshold response curve for the given viewing conditions.

A better method is to increase the modulation from zero and the subject

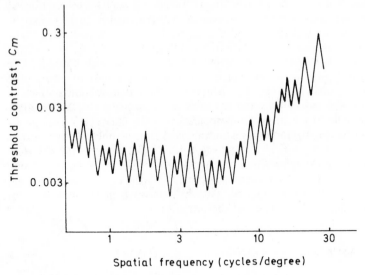

Figure 20.18 Contrast sensitivity curve obtained with automatic equipment (after Sekuler and Tynan)

responds when it reaches threshold. A variation in this method has been proposed for clinical assessment of contrast sensitivity. This approach was originally used to test our other frequency-dependent sense, hearing. In the method, proposed by Sekuler and Tynan*, the observer sees, on pressing a button, a sinusoidal grating of high contrast and low spatial frequency which immediately begins to reduce its contrast at the rate of 4 dB/sec (about a factor of 0.4). As soon as the observer can no longer see the grating he releases the button. As well as recording the contrast and spatial frequency, the equipment then begins to increase the contrast and the observer presses the button as soon as he can see the grating again whereupon the contrast begins to reduce. At the same time the spatial frequency of the gratings is increasing logarithmically (by doubling every 50 secs) to a maximum of 32 cycles per degree). The values of contrast and spatial frequency are continuously recorded so that an output as in *Figure 20.18* is obtained within about six minutes. For clinical screening a photographic method similar to Plate 11 has been used but with only one frequency on each photograph. The photograph is slowly uncovered from the bottom until the patient sees the grating.

20.10 THE MEASUREMENT OF ACCOMMODATION—THE LASER OPTOMETER

Of the three optical factors affecting contrast sensitivity, luminance and pupil size are measurable by direct or nearly direct methods, while accommodation presents some difficulties. While the action of accommodation is stimulated by a number of cues, an important part is played by the blur of the retinal image. However, the reduction of image blur is the purpose of accommodation and so the accommodative function is a closed-loop servo-mechanism. The loop is not

*Sekuler and Tynen (1977) *Am. J. Optometry*, **54**, p. 573–575

perfectly closed and in many situations the state of accommodation is not optimum even when there is no appreciable refractive error.

The measurement of accommodation may be *subjective* where some response from the subject is required or *objective* where the method is independent of subject response. Generally the latter is to be preferred although many of these use light reflected from the retina which, as noted in section 20.3, is not in focus at the same time as physiologically effective light. It is usually important with such methods, that the light enters the eye through particular parts of the pupil. Precise location of the head is therefore necessary. In general the principle of these methods may be understood from the treatment of optics covered earlier in this book.

In recent years a new method of measuring accommodation has become widely used. This uses the apparent speckle pattern produced by coherent light and while this is a subjective method in that a response is required from the subject it is not a 'judgement of sharpness' response and does not, therefore, interfere with the accommodation response. The principle, first suggested by Knoll, is based on the speckle pattern perceived by the eye when observing a diffusing surface illuminated with substantially coherent light from a laser. It was noted that when the observer moved his head the speckles appeared to move also, either with or against the direction of the head movement depending on the accommodation or refractive error of the observer.

In the **laser optometer**, the subject remains still and the diffusing surface is moved, usually by making this in the form of a cylinder and slowly rotating it. The speckle effect arises with the out of focus eye because light from two separate facets on the diffusing surface can illuminate the same point on the retina via different light paths through the lens and cornea. However, because the original source is coherent, there is interference when they combine on the retina. This may result in a dark point or bright point or some intensity between depending on the phase and amplitudes. In the real case many beams are interfering having random phase and amplitude arising from an area on the cylindrical drum determined by the size of the pupil and the extent of the incorrect focus.

The diffraction limit of the pupil determines the smallest size of the speckle

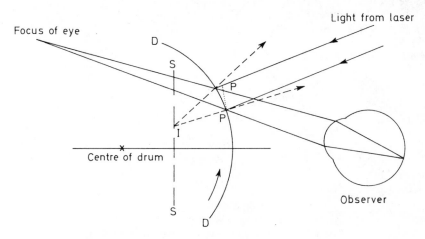

Figure 20.19 Phase change of light scattered from moving surface

and although the statistical nature of the interference allows a range of speckle sizes, they are predominantly near the minimum size. As it turns out that the contrast of the speckles (as with multiple-beam interference) is very high they are easily seen by the eye. If the pupil of the eye is artificially reduced the speckle size increases but too large a speckle makes the movement of the pattern less clear, because the drum movement is constantly bringing a completely new part of its surface into the area causing each speckle. There must be time for the pattern to move before the whole pattern changes.

This movement may be explained in simple terms by considering the Michelson interferometer (section 14.8) which gives an interference pattern of straight fringes when there is a tilt between the two mirrors. If either of the mirrors is moved backwards or forwards the fringes appear to 'run' across the field of view giving an impression of motion. The same action occurs with the speckles. In *Figure 20.19* the drum surface is shown by the full line, DD. The dotted line, PP, is the locus of a supposed tiny facet mirror reflecting light from the laser onto a particular point of the retina for which *the phase at the retina is constant.* The curvature of this line is different from that of the drum and so the movement of the facet mirrors on the drum surface provides a changing path difference which causes the speckles to 'run' across the retina.

The direction in which they run depends on where the eye is focussed. If, for the moment, the drum is assumed to be a specular reflecting cylindrical mirror, the surface of it would cause light to be in phase at the retina when the eye is focussed at the image, I, of the laser source formed by the cylindrical mirror (do not attempt this!). Under these conditions a facet in the rough surface will not change phase at the retina as the drum rotates. In the rough surface case it is found to be sufficient for the eye to be focussed in the same plane as the specular image—the phase changes which then occur remain very small during the life time of each speckle. This plane is called the **plane of stationarity** and under these conditions the growth and decay of each speckle gives a 'boiling' impression but no apparent movement. When the eye is focussed beyond this plane the speckle movement is against the drum motion and vice versa.

The usual arrangement of the optometer is shown in *Figure 20.20*. The laser is diverged onto the rotating drum in an arrangement which may be moved until its plane of stationarity is known to be coincident with the plane of focus of the eye because the subject reports no movement. A shutter is commonly used so

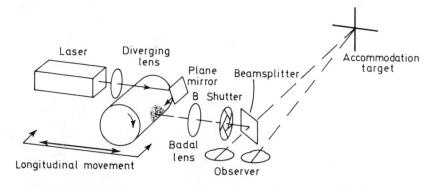

Figure 20.20 Laser speckle optometer

that the speckle pattern is only displayed to the subject for a second or two, while he is viewing the accommodation target and then a new position of the drum system can be adopted according to his response.

Often to save space the drum surface is viewed via a Badal lens as shown. The principle of this is readily understood using Newton's relation. The eye of the subject is placed at the second principal focus of the lens B in *Figure 20.20*. This means that the extra-focal distance, x', of (6.10) (section 6.10) is the distance of the image from the eye. For an object located x metres from the first focal point, the vergence of the light reaching the eye is $1/x' = x/f'^2$ where f' is the focal length of the lens, B, measured in metres. Thus linear changes in object position gives linear changes in image vergence at the eye. Furthermore the apparent size of the image is constant. The image cannot be brought closer to the eye than the focal length of the lens but infinite and positive vergences can be readily obtained for measuring hypermetropia. If the axis of the drum is changed from horizontal through to vertical the astigmatism of the subject may be measured as the plane of stationarity found is valid only in the plane through the eye and containing the drum motion.

CHAPTER 20   EXERCISES

1. List and describe the optical aberrations affecting vision. The fovea of the eye is 5° to 10° away from the axis of the lens and cornea. Which of the oblique aberrations is the most important for foveal vision? How clearly can this be seen in a normal emmetropic eye? Explain the difference between oblique astigmatism and axial astigmatism.

2. What is meant by the external line spread function of the eye? Why can the M.T.F. of the visual system not be calculated from contrast threshold curves?

3. At pupil diameters below 2 mm the eye is diffraction limited. At pupil diameters above 2 mm the eye is aberration limited. Comment and explain.

4. The normal eye has a pupil diameter of 3.8 mm at a scene luminance of 80 Cd/m². An object at 200 metres approximates to a sinusoidal grating of spatial frequency 20 cycles/degree with a contrast modulation of 0.02. Will it be seen? At what distance will it just be seen assuming no change of contrast with distance. (Use *Figure 20.15*).

5. A schematic eye formulated by Ivanoff has the following particulars:

Refractive indices
    Aqueous humour   1.3354
    Crystalline lens   1.44
    Vitreous humour   1.334
Radii of curvature (mm)
    Cornea (single-surface) +8
    Front surface of crystalline lens +10.2
    Back surface of crystalline lens −6
    Depth of anterior chamber (mm) 3.6
    Axial thickness of crystalline lens (mm) 4.0
Determine the equivalent power of the eye and the axial length needed for emmetropia.

6. Discuss the aberrations of the eye, considering in particular how they differ from those of common man-made optical systems and the role of the pupil.

7. What are longitudinal and transverse chromatic aberration? Give details of a possible method for measuring the longitudinal chromatic aberration of the human eye.
A reduced eye consists of a single spherical refracting surface of radius of curvature 5.55 mm and has an axial length of 22.22 mm. It is emmetropic for yellow light. In red and blue light the eye is found to be 0.5D hypermetropic and 0.5D myopic respectively. What is the refractive index of the eye medium in red, yellow and blue light and what is its dispersive power?

8. Explain what is meant by threshold contrast. A sinusoidal grating is just visible to the eye. Which of the following small changes is likely to render it invisible?

(*a*) An increase in the spatial frequency of the grating.

(*b*) An increase in the contrast modulation of the grating.

(*c*) An increase in the illumination of the grating.

(*d*) An increase in the pupil size of the observer.

(*e*) An increase in the focus error of the observer.

(*f*) An increase in the viewing distance.

Comment on the 'don't know' cases.

9. A photo interpreter is examining an aerial reconnaisance photograph. Some very faint detail on it approximates to a sinusoidal grating of spatial frequency, 60 cycles per mm. What microscope magnification will give him optimum vision? Why is the reduction of veiling glare in the instrument so important?

10. By considering the refraction of a spherical wavefront by a spherical refracting surface of radius r, separating medii of refractive indices n and n', derive a formula relating the object and image positions for such a surface.

An eye has a cornea of radius of curvature 7.5 mm and contains a single medium of refractive index 1.33. If its length is 25 mm, how many dioptres of ametropia will the eye suffer from? What corneal radius would render the same eye emmetropic?

If the ametropia was to be corrected by a spectacle lens placed 15 mm in front of the cornea, what power of lens would be required?

11. The Gullstrand–Emsley schematic eye has the following specification:

Radii of curvature

Cornea	+ 7.8 mm
Front surface of crystalline lens	+10.0 mm
Back surface of crystalline lens	− 6.0 mm

Axial separations

Depth of anterior chamber:	3.6 mm
Thickness of crystalline lens:	3.6 mm

Refractive indices

Humours	1.3333(4/3)
Crystalline lens	1.4160

What overall axial length would render this eye myopic with its far point 200 mm in front of the corneal vertex?

12. Describe how the external line spread function of the human eye may be measured. How does it change as the effective pupil of the eye is reduced from 7.0 mm to 0.5 mm?

13. (*a*) Describe the construction and explain the principles of operation of a laser optometer.

(*b*) Describe the characteristics of speckle and explain what is meant by the plane of stationarity.

Appendices

APPENDIX 1 REFLECTION COEFFICIENTS

Material	Percentage of incident light reflected at $30°$ to normal
Magnesium Oxide	96
Plaster of Paris	91
Matt White Celluloid	80–85
White Blotting Paper	80–85
Ground Opal Glass	76
Foolscap Paper	70
Black Cloth	12
Black Velvet	4
3M's Nextel Paint	2
Martin Marietta Black	0.5

APPENDIX 2 PRINCIPAL FRAUNHOFER LINES

Line	Position in spectrum	Element to which line corresponds	Wave-length (nm)
A	Extreme red	O	759.4
B	Red	O	686.7
C	Red	H	656.3
d	Yellow	He	587.6
D_1	Yellow	Na	589.6
D_2	Yellow	Na	589.0
E	Green	Fe and Ca	527.0
b	Green	Mg	518.4
F	Blue	H	486.1
G	Violet	H	434.0
G'	Violet	Fe and Ca	430.8
H	Violet	Ca	396.9
K	Violet	Ca	393.4

APPENDIX 3 OPTICAL GLASS TYPES (CHANCE-PILKINGTON)

Glass type	Mean index n_d	Mean dispersion $n_F - n_C$	V-value V_d	Indices at other wave-lengths 404.7 n_h	435.8 n_g	486.1 n_F	546.1 n_e	656.3 n_C	706.5 n_r	1014.0 n_t	Transmission (25 mm) (550 nm) (%)	Reflection (one surface) (%)	Density (g cm^{-3})
BSC 517642	1.51680	0.00805	64.17	1.53024	1.52668	1.52238	1.51872	1.51432	1.51289	1.50734	98.3	4.3	2.51
*HC 524592	1.52400	0.00885	59.21	1.53896	1.53496	1.53015	1.52611	1.52130	1.51977	1.51403	99.1	4.4	2.55
ZC 508612	1.50759	0.00830	61.16	1.52149	1.51780	1.51334	1.50957	1.50504	1.50358	1.49790	99.4	4.1	2.49
MBC 572577	1.57220	0.00991	57.74	1.58901	1.58451	1.57910	1.57456	1.56919	1.56748	1.56111	99.6	4.9	3.14
DBC 620603	1.62041	0.01028	60.33	1.63774	1.63313	1.62756	1.62286	1.61727	1.61548	1.60876	99.2	5.6	3.60
LF 581409	1.58144	0.01423	40.85	1.60662	1.59961	1.59146	1.58482	1.57723	1.57489	1.56670	99.2	5.1	3.23
EDF 648338	1.64831	0.01916	33.84	1.68307	1.67314	1.66187	1.65285	1.64271	1.63963	1.62918	98.5	6.0	3.74
†EDF 706300	1.70585	0.02353	30.00	1.74938	1.73673	1.72256	1.71140	1.69903	1.69530	1.68273	98.0	6.9	2.99
DEDF 755276	1.75520	0.02738	27.58	1.80589	1.79121	1.77468	1.76167	1.74730	1.74300	1.72889	99.0	7.6	4.79
LAC 713538	1.71300	0.01325	53.83	1.73543	1.72943	1.72222	1.71616	1.70897	1.70668	1.69803	99.6	7.0	3.81
LAF 850322	1.85026	0.02638	32.23	1.89811	1.88450	1.86894	1.85650	1.84256	1.83834	1.82418	97.9	9.0	5.14

*HC 524592 is fine annealed Ophthalmic Crown (Spectacle White) which has indices 0.001 below those given above.

†EDF 706300 is an ophthalmic glass (Highlite) giving high index without high density.

APPENDIX 4 CHIEF SPECTRAL LINES USED IN OPTICS

Designation	Colour	Element	Wave-length
i	U.V.	Mercury	365.02
h	Violet	Mercury	404.66
g	Blue	Mercury	435.84
F'	Blue	Cadmium	479.99
F	Blue	Hydrogen	486.13
e	Green	Mercury	546.07
	Yellow	Mercury	576.9⎱
	Yellow	Mercury	579.1⎰
d	Yellow	Helium	587.56
D_2	Yellow	Sodium	589.0⎱
D_1	Yellow	Sodium	589.6⎰
C'	Red	Cadmium	643.85
C	Red	Hydrogen	656.28
r	Deep red	Helium	706.52
s	IR	Cesium	852.11
t	IR	Mercury	1014.00

APPENDIX 5 REFRACTIVE INDICES OF SOME SOLIDS AND LIQUIDS

	n_D	V
Diamond	2.4173	55.4
Iceland Spar (ordinary ray)	1.6584	48.9
Iceland Spar (extraordinary ray)	1.4864	79.8
Quartz (ordinary ray)	1.5442	69.95
Quartz (extraordinary ray)	1.5533	69.0
Quartz (fused)	1.4585	67.5
Rock Salt	1.5443	42.8
Canada Balsam	1.526	41.5
Fluorite	1.4338	97.3
Ice (ordinary ray)	1.308	
Ice (extraordinary ray)	1.313	
Polymethyl methacrylate (Perspex)	1.4900	53.7
Polystyrene	1.5900	31.0
Monobromnaphthaline (20°C)	1.6582	20.3
Carbon Disulphide (15°C)	1.6303	18.3
Oil of Cassia (10°C)	1.6104	15.9
Cinnamic Ether	1.5607	11.0
Methyl Salicylate (21°C)	1.5319	24.7
Cedar Wood Oil	1.510	
Cedar Wood Oil (hardened)	1.520	
Glycerine (20°C)	1.4730	60.6
Turpentine (11°C)	1.4744	46.5
Water (19°C)	1.3336	54.7

APPENDIX 6 MERIDIONAL RAY TRACING PROGRAMS

These programs are intended for educational use and cover three popular calculators. Their speed of execution varies from 9 seconds to less than 6 seconds per ray surface. The mathematical equations are given in section 18.7 and *Figure 19.4*.

Two ray traces through the system below yielded answers with all three programs which were identical to ± 10⁻⁶ of each of the values given here for use as a check.

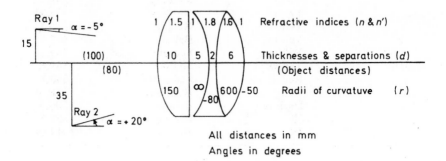

All distances in mm
Angles in degrees

Surface	Ray 1			Ray 2		
	y'	α'	p'	y'	α'	p'
1st	6.249144	−4.127699	86.59303	−5.869276	13.970207	23.592636
2nd	5.527476	−6.198276	50.89554	−3.381518	21.230818	8.704186
3rd	4.996444	−1.832134	156.19908	−1.441554	11.117108	7.336067
4th	4.932529	−2.002319	141.08545	−1.048573	12.514917	4.723987
5th	4.704993	−6.514097	41.20506	+0.283349	20.071025	−0.775501

†*NOTE:* No warranty is implied with these programs. No liability will be accepted for any loss arising from their use.

†Meridional ray trace   ——   Casio fx 201 p

```
ENT    1 : 2 : 3 :
ST     1 :
ENT    4 : 5 : 6 :
GOTO   3 :
1   =  0 ÷ 9√ x 5 + 1 :
7   =  9√ x 1 ÷ 4 + 0 :
8   =  3 x 7 ÷ 6 :
2   =  8 arcsin − 7 arcsin +2 :
3   =  6 :
IF 4 = K1 EXP 5 : 2 : 2 : 1 :
ST 2 :
```

ENTER y, α, n

ENTER r, d, n'

$y^* = y + d \tan \alpha$

$\sin i = \sin \alpha + \dfrac{y^*}{r} \cos \alpha$

$\sin i' = n/n' \sin i$

$\alpha' = i' - i + \alpha$

n' into register 3

$r \geqslant 10^5$?

GOTO 3 :
1 = 8 − 0 × 4 ÷ 9√
GOTO 1 :
MJ
ANS 1:2:
GOTO 1 :

$$y' = \frac{r}{\cos \alpha'} (\sin \alpha' - \sin i')$$

SUB 3 :   Obtains sin α; cos α (to save steps)
0 = 2 sin :   sin α(α′) into register 0
9 = 0 × 0 +/− + K1 :   cos² α(α′) into register 9

OPERATION:
Key in initial conditions : y, α, n;
Key in first surface   : r, d, n'
Key in next surface, etc.
To obtain final y, α key MJ, ANS. Can also check intermediate y, α this way and
then continue ray tracing by keying ANS again.
To obtain y, α on image plane at distance d, key 10^{50}, d, n, where n is the same
as previous n (generally 1).

REGISTER USAGE:

1	Initial ray height above axis, y.
2	Initial ray angle to axis (positive for positive gradient), α
3	Initial refractive index, n
4	Radius of curvature of surface (positive if convex from left) r
5	Distance along axis to surface, d
6	Refractive index beyond surface, n'
7	$\sin i$
8	$\sin i'$
9	$\cos^2 \alpha$; $\cos^2 \alpha'$ (latter in register at close)
0	$\sin \alpha$; $\sin \alpha'$ ” ” ”

†Meridional ray trace — Hewlett Packard HP-25

		X		Y	Z	T	Input: Stack; Registers
00							
01	RCL 2	n		n'	r	d	T − | STO 0 y
02	X⇄Y	n'		n	r	d	
03	STO 2						Z d | STO 1 α
04	÷	n/n'		r	d		Y r | STO 2 n
05	STO 5						
06	R↓	r		d			X n' |
07	STO 4						
08	R↓	d					
09	RCL 1	α		d			*Next surface:*
10	1	1		α	d		Put new d, r, n' into
11	→R	$\cos \alpha$		$\sin \alpha$	d		stack as above.

Step	Instruction			
12	STO 3			
13	÷	$\tan\alpha$	d	
14	×	$d\tan\alpha$		
15	STO + 0			
16	LAST X	$\tan\alpha$		
17	RCL 0	y^*	$\tan\alpha$	
18	RCL 4	r	y^*	$\tan\alpha$
19	÷	y^*/r	$\tan\alpha$	
20	+	$y^*/r + \tan\alpha$		
21	STO × 3			
22	RCL 3	$\sin i$		
23	STO × 5			
24	arc sin	i		
25	STO −1			
26	RCL 5	$\sin i'$		
27	arc sin	i'		
28	STO + 1			
29	RCL 1	α'		
30	1	1	α'	
31	→R	$\cos\alpha'$	$\sin\alpha'$	
32	STO ÷ 4			
33	X ⇄ Y	$\sin\alpha'$	$\cos\alpha'$	
34	STO − 5			
35	÷	$\cot\alpha'$		
36	5	5	$\cot\alpha'$	
37	10^x	10^5	$\cot\alpha'$	
38	RCL 4	$r/\cos\alpha'$	10^5	$\cot\alpha'$
39	X > Y			
40	GTO 44	$\dfrac{r}{\cos\alpha'}$	$10^5\cot\alpha'$	
41	RCL 5	$(\sin i' - \sin\alpha')$	10^5	$\cot\alpha'$
42	×	$\dfrac{r(\sin i' - \sin\alpha')}{\cos\alpha'}$	10^5	$\cot\alpha'$
43	STO 0			
44	R↓	10^5	$\cot\alpha'$	
45	R↓	$\cot\alpha'$		
46	RCL 0	y'	$\cot\alpha'$	
47	×	$y'\cot\alpha'$		
48	CHS	$-y'\cot\alpha'$		
49	R/S	$(=p')$		

Register usage:

0	1	2	3	4	5
y	α	n	$\cos\alpha$	r	n/n'
$y+d\cos\alpha$	$\alpha-i$		$\sin i$		$\sin i'$
	$\alpha-i+i'$			r	$\sin i'-\sin$
	$(=\alpha')$				$\overline{\cos\alpha'}$
y'					

Note:
Use of →R to obtain cosine and sine, saves about 1.2 secs per ray surface.

Note:
This program will not calculate when $\alpha=0$ and $r = \infty$, that is, greater than 10^{12} as the →R calculation at Step 31 gives $\sin\alpha' = 0$ whereupon an inadmissable operation is attempted at Step 35. At $r = 10^{12}$ the error in α' is approx 10^{-11} compared to a flat surface.

†Meridional ray trace — Texas Instruments TI-57

00	RCL 1	d
01	×	
02	RCL 4	α

†No warranty is implied with this program. No liability will be accepted for any loss arising from its use.

Input: Registers

	STO 0 n'	index after surface
03 tan	STO 1 d	distance to surface
04 STO 1	STO 2 r	radius of surface
05 =	STO 3 y	height of ray
06 SUM 3	STO 4 α	slope of ray (degrees)
07 RCL 3	STO 5 n	index in front of surface
08 ÷	t 10^5	test value

Program listing:

```
03  tan          tan α
04  STO  1
05  =            d tan α
06  SUM 3
07  RCL 3        y*
08  ÷
09  RCL 2        r
10  =            y*/r
11  SUM1
12  RCL 4        α
13  cos          cos α
14  PRD 1
15  RCL 1        sin i
16  arc sin      i
17  INV SUM 4
18  RCL 5        n
19  PRD 1
20  RCL 0        n'
21  INV PRD 1
22  STO 5
23  RCL 1        sin i'
24  arc sin      i'
25  SUM 4
26  RCL 4        α'
27  tan          tan α'
28  STO 0
29  RCL 2        r
30  x > t
31  GTO 1
32  ×
33  (
34  RCL 1        sin i'
35  ÷
36  RCL 4        α'
37  cos          cos α'
38  −
39  RCL 0        tan α'
40  =            y'
41  STO 3
42  LBL 1
43  RCL 3        y'(y*)
44  INV PRD 0
45  RCL 0
46  1/x          −p'
47  +/−          p'
48  R/S
49  RST
```

Next surface:
Put new n', d and r into registers
0, 1 and 2 as above

Register usage:

STO 0 n';
STO 1 d ; tan α'; $\dfrac{y* + \tan\alpha; \sin i; \sin i'}{r}$;
STO 2 r ;
STO 3 y ; $y + d\tan\alpha = y*$; y';
STO 4 α ; $\alpha - i; \alpha - i + i' (= \alpha')$;
STO 5 n ; n';

APPENDIX 7 LIST OF SYMBOLS USED

This list, which is also given on the inside rear dust cover, contains only those symbols which are carried over a number of chapters.

Symbol	Definition	Section
A, a	Amplitude	12.3, 12.4
B	Luminance	11.4
C	Contrast (C_m Contrast modulation)	11.14
c	decentration of ray (also y)	6.13
d	distance between lenses	6.14
$e(e')$	Principal point to vertex distance	9.4
efl	equivalent focal length	9.2
F	Focal Power of surface, lens, mirror, system	5.11, 6.4 7.2, 9.4
$F_V(F_V')$	Front (back) vertex powers	9.4
$f(f')$	First (second) focal lengths of surface, lens, mirror, system	5.13, 6.6 7.4, 9.2
f/no	f/number	10.15
$f_V(f_V')$	Front (back) vertex focal length	9.4
g	Inter-foci distance, optical tube length	9.4, 10.7
H	Lagrange invariant	5.15, 6.10
$h(h')$	Object (image) height	5.14
I	Luminous intensity	11.2, 12.5
$i(i')$	Angle of incidence (refraction/reflection)	2.6
i_c	Critical angle	4.4
i_b	Brewster angle	16.4
$L(L')$	Reduced vergence of object (image)	5.6
$l(l')$	Object (image distance)	4.6
M	Magnifying power	10.6
m	magnification	5.14
N.A.	Numerical Aperture	15.11
$n(n')$	Refractive index before (after) surface	2.5, 2.6
R	Curvature	5.1
r	radius of curvature	5.1
s	sag	5.2
$u(u')$	slope angle of incident (refracted) ray	5.12
u_{pr}	slope of principal ray	18.4
v	deviation	4.7, 13.10
V	Constringence of optical medium	13.10
$w(w')$	Object (image) angle subtended at surface	5.14
$x(x')$	Object (image) extra-focal distance	5.15, 6.10
y	Ray height at surface	5.12
$\alpha(\alpha')$	Gradient angle of incident (refracted) ray	18.7
λ	Wavelength of light	12.3
μ	Frequency	12.9

Answers to exercises

1. 0.000667, 0.000527, 0.000431 mm.
3. 9° 31'; 240 ft.
5. 300 sq. cm.
6. 5.59, 0.349, 0.785 sq. cm.
7. Circular patch of full illumination 13 cm diameter, surrounding ring of partial illumination 8.33 cm wide.
8. 2916 sq. cm; full illumination 648 sq. cm, partial illumination 4536 sq. cm
9. 52.2 ft.
10. 8 ft. $7^1/_2$ in.
11. $8^2/_3$ ft diameter; umbra $3^2/_3$ ft diameter, penumbra $13^2/_3$ ft diameter.
13. 5.45 in from image.
14. 13.5 in.
15. 53.7 in.
17. 27 in diameter; penumbra.
19. Umbra 5766 miles, penumbra 10234 miles.
21. 144 ft; 20 in.

CHAPTER 2
7. (a) 255.5, (b) 206.8, (c) 181.8 million metres per sec.
8. (a) 1.205, (b) 0.8297.
9. (a) 1.128, (b) 1.819, (c) 0.5496, (d) 0.6060, (e) 1.10.
16. $i'_1 = 28°45'$, $i'_2 = 32°7'$.
17. 122°52'.

CHAPTER 3
3. 100 cm × 75 cm.
4. 12 ft.
5. 0.55 m × 0.367 m; 1.332 m above the ground.
7. 2.087 ft × 1.391 ft.
8. 178°13'.
9. 5 ft 10 in × 4 ft 2 in; lower edge 4 ft $4^1/_2$ in above the ground.
10. 1 ft 10 in × 8 in.
11. 12 cm; two portions 4 cm wide, separated by 4 cm.
12. 1°26'.
22. 90°; 15°W. of N.
23. From M_1, 40, 160, 240 cm.
    From M_2, 60, 140, 260 cm.
24. 35°.
28. 340.3 yds.
30. $1^1/_3$°.

CHAPTER 4

1. $i' = 27°32'$.
2. $i_c = 52°48'$.
3. (a) $i'_1 = i'_2 = i'_3 = 0°$;
    (b) $i'_1 = 32°2'$, $i'_2 = 25°53'$, $i'_3 = 45°$.
4. $i'_1 = 25°53'$; displacement 1.82 cm; angle of emergence in oil $28°20'$.
8. 0.29 in.
11. (a) $41°4'$, (b) $61°7'$, (c) $74°54'$.
12. 3.02 in.
13. 4 in.
15. 1 ft 9 in from side nearer eye.
17. Blue light totally reflected at $40°45'$ to normal.
    Red light emerges at $80°42'$ to normal.
20. $i'_1 = 32°7' = i_2 = r_2$, $i_3 = 57°53'$, $i'_3 = 47°2' = i_4$, light totally internally reflected in glass slide.
22. (a) $i'_2 = 24°28'$, $d = 18°56'$ (b) $i'_2 = 0°$, $d = 23°8'$.
    (c) $i'_2 = 53°8'$, $d = 23°8'$.
23. $i'_1 = 24°51'$, $i_2 = 35°9'$, $i'_2 = 61°45'$, $d = 41°45'$.
    Minimum $v - 39°49'$.
24. 1.556.
26. 1.618.
27. $9°12'$.
28. $i'_1 = 33°19'$, $i_2 = 41°41'$, $r_2 = 41°41'$, $i_3 = 26°41'$, $i_3 = 47°49'$, $v = 107°11'$.
29. $i'_1 = 19°5'$, $i_2 = 40°55'$, $r_2 = 40°55'$, $i_3 = 19°5'$, $i'_3 = 30°$, $v = 120°$, i_1 for minimum deviation = $49°55'$.
30. 366 mm
31. $99°49'$.
32. $65°4'$.
33. Images 1 ft behind prism, 0.64 in apart.
34. $15°53'$; using approximate formula $16°45'$.
35. $2'18''$.
36. (a) Not smaller than $6°24'$, (b) Not greater than $24°19'$ on other side of normal.
37. (a) $16°49'$, (b) $1°42'$, (c) $-18°40'$.
38. $v = 2.19$ prism dioptres, $a = 2.5°$; $n = 1.5$.
39. 1.577.
40. 1.5811.
41. 1.568.
42. $22°23'$.

CHAPTER 5

2. $+ 8.726D$:1.18%.
3. 0.05 in.
4. 2.88 mm.
5. 1.528.
6. 1.543.
7. 3.91 mm; 6.21 mm + allowance for working.
8. 32.6 mm.
9. $+ 5D$; $+ 5.26D$; $+ 1000D$; $- 100D$.
10. $- 1^{1}/_3 D$; $l' = + 37.5$ cm.
11. $- 14.3$ cm from lens; $+ 10.33D$.
15. Virtual erect image $- 37.38$ cm from surface, 1.64 cm high.
    $F - 38.46$ cm, $F' + 58.45$ cm from surface.
17. 0.76 mm.
18. 1.504; 75.1 mm.

19. Centre; 3.02 cm from front surface.
20. − 1.24 in from supper surface; $m = 1.26$.
21. + 22.22 mm; − 22.5 cm.
22. $l' = − 3.05$ mm; 4.52 mm diameter.
23. (a) 20 cm, (b) 80 cm from curved surface.
24. + 53.33D; + 1.875 cm from plane surface.
25. 0.29 mm.

CHAPTER 6

1. 74.2 cm; 12.37 cm.
3. 165 mm; 82.5 mm.
4. 12.4 cm.
5. 2 in; + 3.33D; 12 in.
7. 60 cm.
8. − 6.06 cm virtual; − 11.76 cm real.
9. 418.4 mm.
10. (a) + 25 cm + 4D; (b) + 28.57 cm + 3.5D; (c) + 33.33 cm + 3D; (d) + 50 cm + 2D; (e) ∞, 0D; (f) − 16.67 cm − 6D.
11. −176.5 mm; 1.47 mm erect.
13. +200 mm; +5D.
14. 6.98 in or 29.02 in from lamp; $m = −4$ or $−¼$ (approx.).
15. + 15.55 cm.
17. 40 mm.
18. − 14. in.
19. + 7.58 in.
21. + 229.5 in from lens; 4.5 in.
22. 160 mm from object; $f' = + 128$ mm.
23. 8 in from object; + 6.57D.
24. + 12.25 cm.
25. (a) + 50 cm, 2 mm; (b) + 36.4 cm, 0.9 mm.
26. + 7.41 in.
27. Image 16 in long, one end 14 in from lens; image 13.34 in long, 16.67 in from lens.
28. − 6.22 in from lens.
29. − 80 in; $m = + 5$.
31. − 60 cm.
32. + 40 mm; + 240 mm.
33. 100 mm, + 3D.
34. 185.4 cm.
35. 10.7 cm.
36. 2.65 cm.
37. 91.45 mm.
38. 30 cm from 1st lens ($m = + ^1/_2$).
    110 cm from 1st lens ($m = − ½$).
39. 4.37 cm; 15.63 cm from the 1st lens.
40. + 100 cm.
42. − 1.2 in; − 32.8D.
43. 0.34 cm.
44. Yellow 288.2 mm; + 3.47D.
    Red   289.6 mm; + 3.45D.
    Blue 284.9 mm; + 3.51D.
46. 67.08 cm.
47. 2.17 cm.
49. − 60 cm.
50. + 138.6 cm from lens.

51. + 4.75D; 30 mm.
55. + 6.4 in.
57. − 8D.

CHAPTER 7

3. −99.5 mm from mirror.
4. − 10.53 cm; − 22.22 cm.
6. 34.2 sq. cm; 45.8 cm or 12.5 cm.
9. $^5/_8\,r$; $^3/_8\,r$.
10. − 3.15D; − 21.25 in; − 8.85 in.
11. − 8.53 cm from mirror.
12. (*a*) + 3.33 cm; 1 cm; (*b*) + 15 cm; 9 cm.
14. $l = -3.86$ in.
15. $l = +103.4$ mm; 0.86 mm; virtual.
16. − 18 in.
17. + 5.75D; − 2.875D; − 34.78 cm.
18. 200 mm; 400 mm.
19. 0.6 ft from mirror.
20. $m = -0.6$.
21. $l = -3.5$ in.
22. 5.02 in by 4.18 in; 8.37 in from mirror.
24. − 46.17 cm; 3.34 mm.
25. 48 ft 9 in by 29 ft 3 in.
26. 1.18 mm long; 3.92 mm behind cornea.
27. − 53.33 cm.
28. 2 ft from flame; $f = -1^1/_2$ ft.
29. 9.78 cm; 14.22 cm; 25.78 cm.
32. 591 cm.
33. 70 cm.
34. 12.73 cm in front of concave mirror; 0.273 cm high; inverted.
35. − 13.85 cm.
36. + 10D.
37. Real inverted image 60 cm in front of 2nd mirror; 1 cm high.
39. + 3.97 mm; 4.96 mm.
40. Virtual image 60 cm from lens; 15 cm high.
42. − 8 cm; − 3.87 cm.
43. − 62.86 cm; − 19.14 cm.
44. $r_2 = -15$ cm; $r_1 = +7.5$ cm.
45. 1.415.
46. + 18.34 cm.
47. $n = 1.5$; $r = 24$ cm.
49. − 3.09 cm from front surface; − 5.33 cm from front surface.
50. Light focuses 20.58 cm in front of lens.
52. $a/b = 7$.
53. $AB = 182.1$ mm.
55. 36°.

CHAPTER 8

1. 5.23D; 0; 1.31D.
2. 1.97 mm; 3.42 mm.
4. + 4.0D.S./−6.0D.C., ax. 170° or − 2.0D.S./+6.0D.C., ax. 80°.
5. 13.82 cm; 22.62 cm.
6. (*a*) + 33.3 cm from lens; 6.66 cm long, vertical.
   (*b*) Vertical line 33.3 mm long; + 166.7 mm from lens; horizontal line 200 mm long + 1 m from lens.
   Circle of least confusion 28.6 mm diameter; + 285.7 mm from lens.

7. Horizontal line 37.7 mm long; + 214 mm from lens.
   Vertical line 150.1 mm long; + 857 mm from lens.
8. Horizontal line 1.97 cm long, + 21.88 cm from lens; vertical line 3.88 cm long, + 43.08 cm from lens; circle of least confusion 1.31 cm diameter, + 29.1 cm from lens.
9. Circular, 15 mm diameter; + 2.0D.S./−4.0D.C., ax. H.
10. (*a*) 18 mm horiz., 27 mm vert.; (*b*) 9 mm horiz., 9 mm vert.; (*c*) 36 mm horiz; 9 mm vert.
11. (*a*) Line 45 mm long; (*b*) Patch 22.5 mm square.
12. + 3.0D.C. ax. V./+5.0D.C. ax. H.; toric.
13. 52.3 mm; 87.2 mm.
14. 0.3 in.
15. H line − 27.64 cm or − 72.36 cm.
    V line − 50 cm.
16. − 50 cm; infinity.
17. $l = -500$ mm, Circle 62.6 mm diameter.

CHAPTER 9

1. + 4.65D; + 21.5 cm.
2. Virtual image − 60 cm from 2nd lens.
3. + 15.64D.
4. − 4.21D.
8. 2.92 cm; 9.17 cm from 1st lens.
9. + 5.39 cm from 2nd lens, 2.71 cm high; − 3.85 cm from 1st lens.
11. − 14.5 cm from 2nd lens; virtual vertical line.
12. $^2/_3$ cm high; 2 cm from 2nd lens.
13. 5.11 mm; + 8.1D.
15. + 46.73D.C. ax. V./−12.12D.C. ax. H.
16. Inverted image 33.7 mm high, + 115.4 mm from 2nd lens;
    $f' = + 192.3$ mm image of same size, + 153.9 mm from + 4D lens; f' as before.
17. + 6 cm.
18. $f' = + 45.7$ mm; $f_V = -19.1$ mm; $f'_V = +25.7$ mm;
    $e = + 26.6$ mm, $e' = -20$ mm.
19. $f' = + 41.67$ cm; image + 68.73 cm from 2nd lens, $m = - 1.25$.
20. $F = + 6D$.; $e = - 6.67$ cm; $e' = - 8.33$ cm. Telephoto system.
21. 3.513 inches.
22. $f_V = - \frac{1}{2}$ in, $f'_V = + \frac{1}{2}$ in, $e = + 1$ in, $e' = - 1$ in.
23. $F = - 3D$; $f_V = - 33.3$ cm; $f'_V = + 133.3$ cm; $e = - 66.6$ cm.
    $e' = + 166.6$ cm.
24. + 20 cm from 2nd lens; 8.81 cm high.
25. $f' = + 17.2$ cm; $e = + 5.2$ cm; $e' = - 6.9$ cm; image + 18.2 cm from 2nd lens; 1.38 cm high.
26. $F = + 1.84D$; $e = - 34.8$ cm; $e' = - 26.1$ cm; virtual image − 47.2 cm from 2nd lens; 4.17 cm high.
29. $f' = 1.429$ in, $e = + 2.5$ in; $e' = - 1$ in.
32. + 33.33 cm.
34. + 14.55D.
35. + 9.87D, + 11.12D, + 89.9 mm from 2nd surface.
36. + 9.64 in from 2nd surface; 1.33 in long.
37. Afocal.
38. − 2.25 in.
39. 1.55; 9.09D.
40. + 1.925 in; principal points at centre of sphere.
42. + 7.14 in.
45. $F = + 6D$; $h' = 0.834$ cm.
    f'_V changes from + 8.33 cm to + 23.3 cm; h' is unchanged.

47. PP′ = − 5 mm; (*a*) + 20D, (*b*) 20 mm.
48. + 0.268D.
49. *e* = + 8 mm; *e*′ = − 5 mm.

CHAPTER 10

3. 2° 3′.
4. − 7.78 cm from the lens; 1.8 cm.
5. (*a*) 3.75 (*b*) 4.0.
6. 2.86.
8. 72.
9. 10; 0.08 in.
10. − 2.66 in from objective.
11. − 1.66 cm from objective; 75; 3.33 mm.
13. 0.014 mm away from object; 146.
15. + 2¹/₈ in.
16. (*a*) 10²/₃ in and 1¹/₃ in; (*b*) 13⁵/₇ in and − 1⁵/₇ in.
19. − 36 mm, 138 mm.
20. Inverted and magnified 4 times in horizontal meridian.
22. 1.1 in.
24. 10.7; 15.04 in (assuming 40 in from objective).
25. (*a*) Eyepiece moved inwards 7.58 mm, M = 3.8.
      (*b*) Eyepiece moved outwards 2.57 mm, M = 5.6.
26. Principal points at infinity, telescopic system, angular magnification − 4; + 62.5 mm
      from 20D surface, m = − ¹/₄.
27. f' = + 1.43 in, f_V = + 1.07 in, f'_V = + 0.428 in; M = 14.
30. f' = + 53.30 cm, e = − 100.0 cm, e' = − 40.0 cm, f'_V = + 13.3 cm, h' = 1.86 cm.
32. $f/5.5$, $f/8$, $f/16$; 1 sec, 8 sec.
33. 4.86 metres × 3.89 metres.

CHAPTER 11

1. (*a*) 1.44 lm/ft², (*b*) 15.5 lm/m².
2. 28.8 c.p.
3. 3.08 × 10¹⁶.
6. 1 lm/ft² = 0.00108 lm/cm² = 10.8 lm/m²; 0.347 lm/ft²; 3.75 lm/m².
7. 0.18 lm/ft².
10. 59° 48′.
11. 4.67 lm/m².
12. 4.73 ft.
13. 8.73 lumens; 7.66 lm/ft².
14. (*a*) 0.44 lm/ft² (*b*) 0.096 lm/ft².
15. 1.41 lumens; 260 lm/ft².
16. 1/120 lm/cm²; no change.
17. 0.4 lm/ft²; 1.28 lm/ft².
18. (*a*) 0.78 lm/ft²; (*b*) 0.122 lm/ft².
19. 6.44 lm/ft².
20. 9.39 lm/ft²; 4.8 lm/ft².
21. 19.4 lm/ft²; 17.6 lm/ft².
22. 4.47 ft; 0.88 lm/ft².
23. (*a*) 0.28 lm/ft², (*b*) 171 c.p.
24. 0.205 lm/ft².
25. 5.69 lm/ft².
26. 1.307 lm/ft²; 340 c.p.
27. 3.2, 2.4, 0.267; 1:1.78:144.
28. 1.72 lm/ft².

29. 17.32 candles; 80; 45.
32. 22.5 c.p.
34. 6.37 cd. per sq. mm; 0.8 lm/mm².
35. 110 cm and 1090 cm from 30 c.p. lamp.
36. D = 0.699, T = 4%, D = 1.398.
38. 8.2%.
39. 96.6%.
40. 1 metre from either lamp.
41. 108.5 cm from brighter lamp.
42. 6 ft.
43. (*a*) 50 cd (*b*) 25 cd; 0.0628 lumens.
44. 3¹/₅ lm/mm².
45. (*a*) 0.0014 lm (*b*) 0.0053 lm/ft².
46. 11.1%; 0.95.

CHAPTER 12

2. 7595A, 759.5 nm, 0.7595μ, 2.99 × 10⁻⁵ in.
   5894A, 589.4 nm, 0.5894μ, 2.32 × 10⁻⁵ in.
   4862A, 486.2 nm, 0.4862μ, 1.91 × 10⁻⁵ in.
3. A stationary, above.
   B upwards, level.
   C downwards, below.
9. − 8.
13. 186,700 miles per second.
14. 16²/₃ times per second.
15. 3.08 × 10⁸ metres per second.
16. 5 × 10¹³ miles.

CHAPTER 13

4. 0.01646; Hard Crown.
8. 1°41′.
10. $R_1 - R_2 = - 13.0D$.
11. 0.0821D; −2.96D.
12. r_1 = +10.04 cm, r_2 = −10.04 cm, r_3 = −10.04 cm, r_4 = + 294.1 cm.
13. r_1 = +8.67 cm, r_2 = −13.0 cm, r_3 = −12.2 cm, r_4 = +68.7 cm.
14. r_1 = +8.74 cm, $r_2 = r_3 = - 10.2$ cm, $r_4 = \infty$.
15. 0.365 mm; 0.036 mm.
16.

	D	F	C
F	+ 4.253D	+ 4.312D	+ 4.235D
f'	+235.1 mm	+231.9 mm	+236.2 mm
e	− 18.53 mm	− 18.41 mm	− 18.55 mm
e'	− 29.36 mm	− 29.17 mm	− 29.39 mm
h'	41.46 mm	40.89 mm	41.64 mm

19. 4°; 1.7°.
20. 2.16′.
21. Crown 3°56′, flint 1°58′.
22. Crown 2.77°, flint 1.4°.
23. $a = 5°, v = 2°7.5′$.
28. Red to yellow 30.5′, yellow to blue 1°15.5′.
33. Crown 0.0157, 0.375, 0.625, 1.125
    Flint   0.0357, 0.296, 0.704, 1.333.
36. $B = 2A$.
38. 15 cm, 33 cm.
40. r_1 = 83.1 mm, $r_2 = r_3 = -77.7$ mm.

CHAPTER 14

6. 0.514 mm.
7. 0.00064 mm.
8. 1.2 mm; 120 mm toward film side.
9. 0.000658 mm.
10. 0.0053 mm.
11. 9.96 mm.
12. 0.105D.
14. 12.25 metres.
16. 1.57 mm; 4.96 mm.
20. 1.18 mm.
21. 6×10^{-4} mm.
22. 0.145 mm.
24. 0.006 mm.
30. 1′ 2″.
32. 273 nm; 537.7 nm; 472.8 nm.
33. 90.6 nm.
34. 103.4 nm.
35. 89.28 nm; 1%, 1.64%, 1.3%.
36. 0.5D.

CHAPTER 15

1. 1.54 mm radius; intensity approximately zero.
5. 1.55, 2.19, 2.68, 3.1 mm.
8. 0.93 mm.
9. 3.7 sec.
11. 0.008 mm.
12. 54 mm.
14. 0.99 in.
15. W = 1.85 sec; 0.56 in.
16. 0.00325 mm.
18. 9.39 cm.
19. 11.28 mm.
21. Radius of 1st dark ring 0.000298 in.
    W = 0.198 sec; magnification 454.
24. 1′ 15″.
25. 0.0133, 0.1975, 0.6893 in.
28. 9.6 mm.
31. 45 mm.
32. $-61.6°$, $-43.62°$, $-30°$ (zero order), $-18.07°$, $-6.91°$, 3.98°, 15.03°, 26.69°, 39.72°, 55.98°.
34. 81; assuming d is width of opaque strip between slits.

CHAPTER 16

10. 90°.
11. 53.1°.
12. 1°2′.
13. (a) $0.75I_0$, (b) $0.375I_0$.
14. π.
16. (a) 0.0328 mm (b) 0.0164 mm.
17. 1; 0.75; 0.5; 0.25; 0.
18. $0.00446k$ mm where k is an odd integer.
19. 5.88° using small angle approximation (less than 1% error).
22. a_2 = 29° 29′; ν = 5° 17′.

23. 0.03 mm.
25. 57.7°.
27. $i'_0 = 19.07°$; $i'_e = 21.8°$.
28. 19.7°.

CHAPTER 17

1. 3.052 mm behind cornea; $m = 1.13$.
2. 13.3 cm in front of lens; 10 mm diameter.
3. Entrance pupil 4.29 cm behind 1st lens, 2.68 cm diameter;
   Exit pupil 2.31 cm in front of the 2nd lens, 2.31 cm diameter;
   Entrance port is aperture of 1st lens;
   Exit port 7.5 cm in front of 2nd lens; $m = 1.5$.
4. Field of full illumination 1°52'; total field 68°36'.
6. 28.5 mm; 16.5 cm.
8. (a) 14°14', 3°34'; (b) 7°8', 1°47'; (c) 21°15', 5°19'.
10. Objective $f' = 8$ in, diameter 1.33 in.
    Eyepiece, field lens $f' = 1.33$ in, diameter 1.6 in.
            eye lens $f' = 1.33$ in, diameter 0.8 in.
12. 23.15 metres.
13. 11 ft 5 in to 8 ft 10$^1/_2$ in.
14. 1/7.5 sec, 0.032 in.
16. 13.04 cm to 42.86 cm from the lens.
17. (a) 400 mm behind 1st lens, 30 mm diameter.
    (b) 500 mm in front of 1st lens, 80 mm diameter.
    (c) 47.5 mm diameter.
    (d) 64 mm diameter.
19. 1 cm diameter, 12 cm behind eyepiece.
20. Aperture stop is negative lens.
    Entrance pupil is 10 cm behind negative lens, 16 mm diameter.
    Focal Ratio is F/16.6.
21. 3 cm in front of lens, 5 cm diameter.
22. 3.3 cm behind 1st lens, 1.33 cm, diameter.
    2.85 cm in front of 2nd lens, 1.143 cm diameter.
23. 7.5 cm behind lens, 7.5 cm diameter (for given object).
24. 12 cm behind 1st lens, 60 cm behind 1st lens, 4.76.

CHAPTER 18

4. $l = +25$ cm; $l' = +16.67$ cm; $m = 0.44$.
5. $l = +5.2$ in; $l' = +3.25$ in.
6. $m = (n/n')^2$.
11. $t' = +8.57$ cm; $s' = +11.57$ cm.
12.. $l' = +14.4$ in.
21. $l' = +128.54$ cm.
22. $l' = +15.39$ cm (marginal); $+15.87$ cm (paraxial).
24.

Height	1	20	40	50	60	64
Location	292.46	291.96	290.91	290.68	291.36	292.15

Height	66	68	72
Location	292.73	293.45	295.48 (all values in mm).

27. Longitudinal aberration 47.09 mm; Lateral aberration 28.24 mm (by calculation).
30. Longitudinal aberration 33.55 mm; Lateral aberration 15.68 mm (by calculation).

31. By calculator program with subsidiary calculation for initial slope:
    Initial slope = 1.487°
    $$l' = 126.197 \text{ mm}$$
    Final slope = − 5.848°.

33. By calculator program with subsidiary calculation for initial slope:
    Initial slope = 1.708°
    $$l' = 132.819 \text{ mm}$$
    Final slope = − 6.587°.

34. (*a*) 1.0 mm; (*b*) 1.2 mm.

35. 40 cm; 2.9 cm.

CHAPTER 19

1. l_2 = + 178 mm; h' = − 11.8 mm.

3. F = + 4.68D; + 238.5 mm from lens.

4. f = − 241.5 mm; f' = + 321.3 mm;
   e = + 10 mm; e' = − 22.3 mm;
   N + 49.8 mm and N' + 57.5 mm from 2nd surface.

5. 74.275 mm in front of non-silvered surface.

6. F = + 42D; principal points at surface, nodal points at centre of curvature; F = 11.34D.

7. f_V = − 2.365 in; f'_V = + 2.892 in;
   e = + 0.149 in; e' = − 0.46 in;
   N + 0.237 in; N' + 0.378 in from 2nd surface.

8. F′ is 2.182 cm in front of the first surface.

10. F = + 43.05D; f = − 23.22 mm; f' = + 31.03 mm.

11. + 5.28 in from back surface; principal points at centre of curvature.

12. 2.18 in.

13. Image + 18.25 mm from 3rd surface; 3.52 mm high.

14. f' = + 198.8 mm; 1st Principal point + 13.8 mm from 1st surface.
    2nd Principal point +   7.0 mm from final surface.

15. f' = + 101.6 mm; 1st Principal point +   6.3 mm from 1st surface.
    2nd Principal Point −16.8 mm from final surface.

16. (*a*) + 26.39D; f = − 50.63 mm; f' = + 50.63 mm.
    (*b*) e = + 2.126 mm; e' = − 1.276 mm.
    (*c*) + 42.97D; f = − 23.27 mm; f' = + 31.1 mm.
    (*d*) + 64.5D.

17. 2.65 cm in front of 1st surface.

18. + 6D; 1.25 cm behind 1st surface.

19. 3.74 mm behind cornea.

20. e = 18.18 mm from 1st surface.
    e' = − 18.18 mm from 2nd surface.

21. (*a*) 28 cm.
    (*b*) *i* − 2382 cm; m = −1/48
       ii − 1230 cm; m = −1/24
       iii −   846 cm; m = −1/16
       iv −   654 cm; m = −1/12

22. 8.182 mm behind cornea; 0.1394 mm.

CHAPTER 20

4. No. approximately 165 metres.

5. 66.3D, 23.09 mm.

7. Red   1.32923
   Yellow 1.33293

     Blue   1.33663

     Dispersive power = 0.02222.

  9. Between about 90X and 130X.

10. + 9.2D Hypermetropic.

11. 25.85 mm.

Index